Lecture Notes in Mathema

Edited by A. Dold and E

1201

Curvature and Topology of Riemannian Manifolds

Proceedings of the 17th International Taniguchi Symposium held in Katata, Japan, Aug. 26–31, 1985

Edited by K. Shiohama, T. Sakai and T. Sunada

Springer-Verlag
Berlin Heidelberg New York London Paris Tokyo

Editors

Katsuhiro Shiohama
Department of Mathematics, Faculty of Science
Kyushu University, Fukuoka, 812, Japan

Takashi Sakai
Department of Mathematics, Faculty of Science
Okayama University, Okayama, 700, Japan

Toshikazu Sunada
Department of Mathematics, Faculty of Science
Nagoya University, Nagoya, 464, Japan

Mathematics Subject Classification (1980): 53Cxx

ISBN 3-540-16770-6 Springer-Verlag Berlin Heidelberg New York
ISBN 0-387-16770-6 Springer-Verlag New York Berlin Heidelberg

Printing and binding: Beltz Offsetdruck, Hemsbach/Bergstr.
2146/3140-543210

PREFACE

The seventeenth Taniguchi International Symposium was held at Katata in Japan from 26th August till 31st August, 1985 under the title

Curvature and Topology of Riemannian Manifolds.

It was followed by a conference at the Research Institute for Mathematical Science, Kyoto University, from 2nd September till 4th September, 1985 under the title

Problems in Riemannian Geometry in the Large.

Seventeen mathematicians from France, Switzerland, West Germany, U.S.A. and Japan were invited under the support of the Taniguchi Foundation and they all gave talks in the Taniguchi Symposium. Besides the invited mathematicians from abroad there were six Japanese speakers who gave talks in the Kyoto Conference. The organizing committee was very happy to receive contributions from participants in both the Taniguchi Symposium and the Kyoto Conference. A complete list of the participants in both the Taniguchi Symposium and the Kyoto Conference and the submitted papers together with author's address will be given here. The organizing committee would like to express thanks to all the speakers who gave talks and submitted papers and lecture notes in this proceedings.

All the participants would like to take this opportunity to express their hearty thanks to Mr.Toyosaburo Taniguchi for his support. The Editorial Board of the organizing committee also like to express their thanks to Professor Shingo Murakami who, as the coordinator of the Taniguchi International Symposia, guided them to the success of the symposium and conference.

The Editorial Board

Katsuhiro Shiohama
Takashi Sakai
Toshikazu Sunada

Participants in the Taniguchi International Symposium

Ballmann, Werner	University of Maryland, Department of Mathematics, College Park, Maryland 20742 U.S.A.
Brooks, Robert	Department of Mathematics, University of Southern California, DRB 306, University Park, Los Angeles, California 90089 - 1113 U.S.A.
Cheeger, Jeff	State University of New York at Stony Brook, Department of Mathematics, Stony Brook, New York 11794 U.S.A.
Eberlein, Patrick	Department of Mathematics, The University of North Carolina at Chapel Hill, Chapel Hill, North Carolina 27514 U.S.A.
Fukaya, Kenji	Department of Mathematics, Faculty of Science, Tokyo University, Hongo, Tokyo, 113-Japan.
Gromov, Mikhael	Institut des Hautes Etudes Scientifiques, 91440 Bures-Sur-Yvette, France.
Kasue, Atsushi	Department of Mathematics, Faculty of Science, Osaka University, Toyonaka, 560-Japan.
Katsuda, Atsushi	Department of Mathematics, Faculty of Science, Nagoya University, Chikusa-ku, Nagoya, 464-Japan.
Murakami, Shingo *)	Department of Mathematics, Faculty of Science, Osaka University, Toyonaka, 560-Japan.
Nishikawa, Seiki	Department of Mathematics, Faculty of Science, Kyushu University, Fukuoka, 810-Japan.
Pansu, Pièrre	Ecole Polytechnique, Centre de Mathematiques, 91128 Palaiseau Cedec, France.
Ruh, Ernst	Ohio State University, Department of Mathematics, Columbus, Ohio 43210, U.S.A.
Sakai, Takashi *)	Department of Mathematics, Faculty of Science, Okayama University, Okayama, 700-Japan.
Sato, Hajime	Department of Mathematics, Faculty of Science, Tohoku University, Sendai, 980-Japan.
Shiohama, Katsuhiro *)	Department of Mathematics, Faculty of Science, Kyushu University, Fukuoka, 810-Japan.
Sunada, Toshikazu *)	Department of Mathematics, Faculty of Science, Nagoya University, Chikusa-ku, Nagoya, 464-Japan.
Ziller, Wolfgang	University of Pennsylvania, Faculty of Arts and Sciences, Department of Mathematics, Philadelphia 19104 - 3859, U.S.A.

*) member of organizing committee

Japanese speakers in the Kyoto Conference

Kanai, Masahiko	Department of Mathematics, Keio University, Yokohama, 223-Japan.
Koiso, Norihito	Department of Mathematics, College of General Education, Osaka University, Toyonaka, 560-Japan.
Muto, Hideo	Department of Mathematics, Faculty of Science, Tokyo Institute of Technology, Ohokayama, 152-Japan.
Sakamoto, Kunio	Department of Mathematics, Faculty of Science, Tokyo Institute of Technology, Ohokayama, 152-Japan.
Urakawa, Hajime	Department of Mathematics, College of General Education, Tohoku University, Sendai, 980-Japan.
Yamaguchi, Takao	Department of Mathematics, Faculty of Science and Engeneering, Saga University, Saga, 840-Japan.

Contents

STRUCTURE OF MANIFOLDS OF NONPOSITIVE SECTIONAL CURVATURE

Werner Ballmann
Department of Mathematics
University of Maryland
College Park, Md. 20742

Let M be a complete Riemannian manifold with nonpositive sectional curvature K. Recall that the universal covering space $\tilde{M}$ of M is diffeomorphic to $\mathbb{R}^n$, where n denotes the dimension of M. Therefore the homotopy type of M is determined by the fundamental group Γ of M.

For a tangent vector v of M, denote by γ_v the geodesic with initial velocity v. The <u>geodesic flow</u> g^t of M is defined by $g^t(v) = \dot{\gamma}_v(t)$. The geodesic flow acts on SM, the unit tangent bundle of M, and it leaves the natural measure of SM invariant. This measure is called the <u>Liouville measure</u>, and we denote by μ the normalized Liouville measure.

Given unit speed geodesics γ_1 and γ_2 in M, the function $d(\gamma_1(t),\gamma_2(t))$ is convex in t, and it is bounded if and only if γ_1 and γ_2 bound a common flat strip, see [EO]. This elementary fact is the main reason for many of the strong relations between the dynamics of g^t, the geometry of M and the structure of Γ. Below I discuss various such relations.

Section 1. Entropies of the geodesic flow, curvature and growth of Γ

Throughout this section we assume that M is compact. Then the fundamental group Γ of M is finitely generated. Given a system $G = (\gamma_1,\ldots,\gamma_k)$ of generators of Γ, we denote by $N_G(t)$ the number of different elements of Γ which can be expressed as a word in G of length at most t. We say that Γ is of <u>exponential growth</u> if

$$\lim_{t\to\infty} \frac{1}{t}\ln\left(N_G(t)\right) > 0.$$

This is independent of the choice of G.

Choose a point $x\in\tilde{M}$ and let B(r) be the geodesic ball of radius r about x. Set

$$h_{vol} = h_{vol}(M) = \lim_{t\to\infty} \frac{1}{t} \ln\left(\mathrm{vol}(B(r))\right).$$

This limit always exists and is independent of the choice of x.

We can think of Γ as a properly discontinuous group of isometries of $\widetilde{M}$. Then $\widetilde{M}$ is covered by the Γ - translates of the geodesic ball $B(D)$, where D is the diameter of M. There exists a constant A such that $\gamma\in\Gamma$ can be expressed as a word of length $\leq At$ in G if $d(x,\gamma(x)) \leq t$ and, vice versa, such that $d(x,\gamma(x)) \leq At$ if γ can be expressed as a word of length $\leq t$ in G. Thus

$$\mathrm{vol}\left(B\left(\frac{t}{A}-D\right)\right) \leq N_G(t)\cdot \mathrm{vol}(B(D)) \leq \mathrm{vol}(B(At+D)),$$

see [Sv] and [Mi]. In particular, $N_G(t)$ grows exponentially if and only if $h_{vol} > 0$. For example, if the curvature of M is strictly negative, then $h_{vol} > 0$ and hence Γ is of exponential growth. As for the weaker assumption $K \leq 0$, there is the following beautiful result of Avez [Av].

<u>1.1 Theorem</u>. Either M is flat or Γ is of exponential growth.

Avez proof consists in showing that $h_{vol} > 0$ if M is not flat.

The number h_{vol} is also connected to the geodesic flow g^t of M. Namely, h_{top} is equal to the <u>topological entropy</u> h_{top} of g^t, cf. [Di] and [Ma]. The topological entropy of g^t can be defined in the following way. Choose any metric d^* for SM, and let $n(t,\varepsilon)$ be the maximal number of disjoint ε-balls in SM with respect to the d_t^*-metric,

$$d_t^*(v,w) = \max_{0\leq s\leq t} d^*(g^s v, g^s w).$$

Then the topological entropy of g^t is given by

$$h_{top} = \lim_{\varepsilon\to 0}\lim_{t\to\infty} \frac{1}{t}\ln(n(t,\varepsilon)).$$

The above mentioned equality $h_{vol} = h_{top}$ is an easy consequence of

$$d(\gamma_1(s),\gamma_2(s)) \leq \max\{d(\gamma_1(0),\gamma_2(0)),\ d(\gamma_1(t),\gamma_2(t))\}$$

for $0\leq s\leq t$, where γ_1 and γ_2 are geodesics in $\widetilde{M}$. (The latter follows from $K\leq 0$.)

We now describe another invariant of the geodesic flow, namely its <u>measure theoretic entropy</u>. For any vector $v\in SM$, let $W =$

$W(v) \subset T_v SM$ be the tangent space to the C^1-submanifold of SM consisting of vectors asymptotic to v, with footpoint on the horosphere determined by v. Then the limit

$$(1.2) \qquad \lim_{t\to\infty} \frac{1}{t} \ln|\det(dg^t|_W)| =: \chi(v)$$

exists for almost every v, see [Os], and the measure theoretic entropy of g^t (with respect to the measure μ) is given by

$$h_\mu = - \int_{SM} \chi(v)\,d\mu(v),$$

see [P1] and [P2]. Topological and measure theoretic entropy are related by

$$h_\mu \le h_{top},$$

see [Di]. In particular, h_{top} is positive if h_μ is positive.

Now h_μ can also be expressed in terms of the second fundamental form of the horospheres of M. Denote by $U(v)$ the second fundamental form, in the footpoint of v, of the horosphere determined by the vector $v \in SM$. In our normalization, $U(v)$ is negative semidefinite, and it acts on the orthogonal complement E_v of v in T_pM, $p = \text{foot}(v)$. Pesin proved

$$(1.3) \qquad h_\mu = - \int_{SM} \operatorname{tr}(U(v))\,d\mu(v),$$

see [P2]. Indeed, the space W in (1.2) consists of the vectors $(Y,Z) \in E_v \oplus E_v \subset T_v SM$ such that the Jacobi field $J(t)$ along $\gamma_v(t)$ determined by $J(0) = Y$ and $J'(0) = Z$ is monotonically not increasing. Then

$$Z = U(v)Y.$$

If $-a^2$ is a lower bound for the sectional curvature of M, then $\|Z\| \le a\|Y\|$, see [Eb], and hence

$$\chi(v) = \lim_{t\to\infty} \frac{1}{t} \ln|J_2(t) \wedge \ldots \wedge J_n(t)|,$$

where $J_2(t), \ldots, J_n(t)$ are Jacobi fields as above such that $J_2(0), \ldots, J_n(0)$ are a basis of E_v. Differentiation yields

$$\chi(v) = \lim_{t\to\infty} \frac{1}{t} \int_0^t tr(U(g^t v))dt.$$

Applying the Birkhoff ergodic theorem, cf. for example [AA], we obtain the formula (1.3).

The horosphere determined by $g^t v$ is parallel to the horosphere determined by v, and hence the family $U(v)$, $v \in SM$, satisfies the Ricatti equation

$$(1.4) \qquad U' + U^2 + S = 0,$$

where $U'(v)$ denotes the covariant derivative of U along γ_v and $S(v)\cdot X := R(X,v)v$. The Ricatti equation relates U to the curvature of M.

If the curvature of M is negative, then U is negative definite and hence invertible, and we get

$$U'U^{-1} + U + SU^{-1} = 0$$

Therefore

$$(\ln(\det(U)))' + tr(U) + tr(SU^{-1}) = 0.$$

Since g^t preserves μ we obtain

$$-\int_{SM} tr(U(v))d\mu(v) = \int_{SM} tr(S(v)U^{-1}(v))d\mu(v).$$

Now recall that $tr(AA^t) \geq 0$ for every matrix A, with equality if and only if $A = 0$. Applying this we get

$$0 \leq tr((\sqrt{-U} - \sqrt{-S}\,\sqrt{-U}^{-1})\,(\sqrt{-U} - \sqrt{-S}\,\sqrt{-U}^{-1})^t)$$

$$= tr(-U - 2\sqrt{-S} + \sqrt{-S}(-U)^{-1}\sqrt{-S})$$

$$= -tr(U) + tr(SU^{-1}) - 2tr\sqrt{(-S)}$$

since $\sqrt{-S}$ and $\sqrt{-U}$ are symmetric. Hence

$$(1.5) \qquad h_\mu \geq -\int_{SM} tr(\sqrt{-S(v)})d\mu(v).$$

Equality implies $S = U^2$, and hence $U' = 0$ by the Ricatti equation. So U^2 and therefore S are parallel along geodesics, and it follows easily that M is a locally symmetric space (of noncompact type and rank one). The inequality (1.5) (with the equality discussion) is due to Osserman and Sarnak [OS], and it contains all previously known lower estimates for h_μ for negatively curved manifolds. The above argument is actually a simplification of the argument in [OS] and is due to Wojtkowski. It can be shown that (1.5) is also true under the weaker assumption of nonpositive sectional curvature, see [BW]. The main difficulty in extending the above argument lies in the fact that U is not necessarily invertible.

As for upper estimates of h_μ, these are connected with the question under which circumstances the equality $h_\mu = h_{top}$ can occur.

1.6 Problem (Katok). Suppose M is a compact manifold of negative curvature. Show that $h_\mu = h_{top}$ implies that M is a locally symmetric space of rank one!

Katok proved that this is the case if M is twodimensional of genus ≥ 2 (without assuming negative curvature), see [Ka]. For locally symmetric spaces of noncompact type of rank one, the equality $h_{top} = h_\mu$ is well known, see [Sp]. As for Katok's result about surfaces, he also showed that

$$h_\mu^2 \leq \frac{1}{2\pi \operatorname{vol}(M)} \int_M K(p)\, d\operatorname{Vol}(p) \leq h_{top}^2$$

where both inequalities are strict unless the Gaussian curvature $K(p)$ is constant. In light of this, it is natural to look for upper estimates of h_μ in terms of the sectional curvature.

The Ricatti equation 1.4 implies

$$\int_{SM} \operatorname{tr}(U^2(v))\, d\mu(v) = -\int_{SM} \operatorname{tr}(S(v))\, d\mu(v) = -\int_{SM} \operatorname{Ric}(v)\, d\mu(v)$$

where $\operatorname{Ric}(v) = \operatorname{tr}(S(v))$ is the Ricci curvature of v. Now

$$(n-1)\operatorname{tr}(U^2) = \operatorname{tr}(U^2)\cdot\operatorname{tr}(Id) \geq (\operatorname{tr}(U))^2$$

and hence

$$(1.7) \qquad -\int_{SM} \operatorname{tr}(U(v))\, d\mu(v) \leq \sqrt{\int_{SM} (\operatorname{tr}(U(v))^2\, d\mu(v)} \leq \sqrt{n-1}\sqrt{\int_{SM} -\operatorname{Ric}(v)\, d\mu(v)}$$

This estimate is due to Freire-Mañé [FM]. Since equality in (1.7) implies that the curvature of M is constant, this inequality is not fine enough with respect to problem 1.6. Now let

$$0 \geq \lambda_2(v) \geq \ldots \geq \lambda_n(v)$$

be the eigenvalues of the operator S(v). Then we can restate (1.5) as

$$\sum_{i=2}^{n} \int_{SM} \sqrt{-\lambda_i(v)}\, d\mu(v) \leq h_\mu,$$

and the question is whether h_μ is within the bounds given by the Schwartz inequality.

1.8 Problem (Osserman). Suppose M is a compact manifold of negative curvature.

Show that

$$h_\mu < \sum_{i=2}^{n} \sqrt{\int_{SM} -\lambda_i(v)\, d_\mu(v)}$$

with equality if and only if M is a locally symmetric space of rank one!

One may also consider both problems, (1.6) and (1.8), under the weaker assumption of nonpositive curvature. Note, however, that

$$h_\mu < \sum_{i=2}^{n} \sqrt{\int_{SM} -\lambda_i(v)\, d\mu(v)} < h_{top}$$

if M is a locally symmetric space of higher rank, see [Sp].

Section 2. Manifolds of higher rank

For $v \in SM$, define $J^P(v)$ to be the space of parallel Jacobi fields along γ_v and set

$$\operatorname{rank}(v) = \dim J^P(v) \text{ and } \operatorname{rank}(M) = \min\{\operatorname{rank}(v) \mid v \in SM\}.$$

If the sectional curvature of M is negative, then the rank of M is one. A surface of nonpositive curvature has rank one if and only if it is not flat. It is also easy to see that the above notion of rank coincides with the usual one in the case that M is locally symmetric. Note that rank(M)≥ 2 if $\tilde{M}$ is a Riemannian product.

A manifold of rank one resembles in many ways a manifold of negative curvature.

<u>2.1 Theorem</u>. Suppose M has rank one.

a) If $\mathrm{vol}(M) < \infty$, then the geodesic flow of M is topologically transitive.

b) If M is compact, then the geodesic flow of M is ergodic.

The second part of this theorem, proved in [BB] and independently in [Bu], generalizes the result of Anosov that the geodesic flow on a compact manifold of negative curvature is ergodic [An]. It also generalizes a result of Pesin [P3], namely that the geodesic flow on a compact surface of nonpositive curvature is ergodic if and only if the surface is not flat.

The first part of the above theorem is proved in [Bl] under the at first sight weaker assumption that M has a geodesic which does not bound a totally geodesic immersed flat half plane (and $\mathrm{vol}(M)<\infty$). However, it was shown in [Bu] that this implies that M has rank one. Burns' result was a first step towards the following theorem.

<u>2.2 Theorem [BBE]</u>. Suppose $\mathrm{vol}(M)<\infty$ and $\mathrm{rank}(M) = k \geq 2$. Then every geodesic in M is contained in a k-flat, that is, a totally geodesic and isometrically immersed Euclidean space of dimension k.

This is one of the basic results for the investigations in [BBE] and [BBS]. Another basic result is the Angle Lemma [BBE], in which the further hypothesis $K \geq -a^2$ is needed. To formulate the Angle Lemma, call a vector $v \in SM$ <u>regular</u> if $\mathrm{rank}(v) = \mathrm{rank}(M) = k$, and denote by $F(v)$ the (unique) k-flat containing v. The Angle Lemma asserts that for a dense set of regular vectors v in $S\tilde{M}$ (namely the <u>uniformly recurrent</u> regular vectors) there is an open neighborhood of v in $S_pF(v)$, $p=\mathrm{foot}(v)$, such that any w' asymptotic to some $w \in U$ is regular and $F(w') = F(v')$, where v' denotes the unique vector asymptotic to v such that $\mathrm{foot}(v') = \mathrm{foot}(w')$. This leads to the definition of <u>Weyl chambers</u> in $S\tilde{M}$ (and SM) [BBS]. We say that two unit vectors v and w tangent to a flat F at a point $p \in \tilde{M}$ belong to the same Weyl chamber if $F(v(g)) = F(w(g))$ whenever v(g) and w(g) are regular. Here v(g) and w(g) denote the vectors at g asymptotic to v and w respectively. The Weyl chamber of v is denoted by C(v). In the case that M is a symmetric space, this definition of Weyl chamber corresponds to the usual definition of Weyl chamber; the provision "whenever v(g) and w(g) are regular" is not necessary in this case. Weyl chambers define an equivalence relation on the set of regular vectors.

The Angle Lemma shows that there are v which are contained in the interior of $C(v)$ if $\operatorname{rank}(M) = k \geq 2$. On an open, dense and g^t-invariant subset $\mathcal{O}$ of $S\tilde{M}$, Weyl chambers depend continuously on v, and $C(w)$ is locally isometric to $C(v)$ if w is sufficiently close to v according to the <u>Rigidity Lemma</u> [BBS]. If v is regular, then the map $w \to w(q)$ gives rize to an isometry of $C(v)$ and $C(v(q))$ for all q such that $v(q)$ is regular. In particular, the function

$$\Phi: \mathcal{O} \to \mathbb{R}; \qquad \Phi(v) = \min \{\measuredangle(v,w) \mid w \in \partial C(v)\}$$

is a continuous integral for the geodesic flow on $\mathcal{O}$. Clearly Φ is not constant if $\operatorname{rank}(M) \geq 2$ and thus we get the following result.

<u>2.3 Theorem</u>. Suppose $\operatorname{vol}(M) < \infty$, $K \geq -a^2$, and $\operatorname{rank}(M) = k \geq 2$. Then the geodesic flow of M is not topologically transitive.

One can actually construct k-1 independent differentiable first integrals for g^t [BBS]. Using a generalized version of Anosov's <u>Closing Lemma</u> [An], one can also show that a dense set of vectors $v \in SM$ is contained in an immersed totally geodesic flat k-torus [BBS]. We now come to the main result about manifolds of higher rank.

<u>2.4 Theorem</u>. Suppose $\operatorname{vol}(M) < \infty$ and $K \geq -a^2$. Then $\tilde{M}$ is a space of rank one or a symmetric space or a Riemannian product of such spaces.

This theorem is proved in [B2] and [BS]. Both proofs rely on the previous results in [BBE] and [BBS], but apart from that they are completely different.

The proof in [B2] uses Berger's theorem that M is a symmetric space of higher rank if it is irreducible and if the holonomy group at some point $p \in M$ does not act transitively on $S_p\tilde{M}$ see [Be], [Si]. Consider the function Φ as above. It turns out that Φ is a Lipschitz function on $\mathcal{O}$ and thus has a continuous extension to a <u>Lipschitz</u> function $\Phi: S\tilde{M} \to \mathbb{R}$. (Recall that $\mathcal{O}$ is dense in $S\tilde{M}$). Moreover, Φ is constant along the C^1- submanifolds $A^+(v) = \{v(q) \mid q \in \tilde{M}\}$ of $S\tilde{M}$. One can also show that Φ is constant along the C^1- submanifolds $A^-(v) = -A^+(-v)$, $v \in S\tilde{M}$. This implies that the directional derivative of Φ in the direction of any $X \in T_vA^+(v)$ or $X \in T_vA^-(v)$, $v \in S\tilde{M}$, exists and vanishes. Since the families $A^+(v)$ and $A^-(v)$, $v \in S\tilde{M}$, foliate $S\tilde{M}$ and depend continuously on v in the C^1 - topology, and since $T_vA^+(v) + T_vA^-(v)$ contains the horizontal subspace of $T_vS\tilde{M}$ for every $v \in S\tilde{M}$, it follows that the directional derivative of Φ in any horizontal direction exists and vanishes. Thus $\Phi(w(t)) = \Phi(w(o))$ if w(t) is a parallel vector field along a curve in M. If $\mathrm{rank}(M) \geq 2$, then Φ is not constant on $S_p\tilde{M}$, where $p \in M$ is arbitrary. Since the orbits of the holonomy group at p are contained in the level surfaces of Φ, they do not fill $S_p\tilde{M}$. Hence $\tilde{M}$ is either reducible or a symmetric space of higher rank according to Berger's theorem. The argument we outlined here gives rise to several simplifications in the published proof of Theorem 2.4, and it allows the following extension to foliations.

2.5 Theorem. Suppose F is a smooth foliation, with transverse measure, of a manifold N such that every leaf of F is a complete Riemannian manifold with sectional curvature in $[-a^2, 0]$ for some $a > 0$. Assume that N has finite volume with respect to the product measure of the transverse measure with the Riemannian measure of the leaves. If no leaf $F(x)$ is flat, if

$$\min\{\mathrm{rank}(F(x)) \mid x \in N\} \geq 2,$$

and if one leaf of F is irreducible, then all leaves of F are locally isometric and locally symmetric spaces of noncompact type.

The proof of (2.4) by Burns and Spatzier [BS] relies on an extension of the theory of Tits buildings to a theory of Tits buildings with topology. Assume for simplicity that $\tilde{M}$ is irreducible and $\mathrm{rank}(\tilde{M}) \geq 2$. The Tits building T associated to $\tilde{M}$ is given by the points at infinity, $\tilde{M}(\infty)$, together with the subsets defined by the Weyl chambers. Each element of the fundamental group Γ of M induces a continuous automorphism of T. The group of all continuous automorphisms of T is a centerless Lie group G of noncompact type, and Γ is a discrete lattice in G. Vice versa, the Tits building is also determined by the group G, that is, T is also the Tits building associated to the symmetric space G/K, where K is a maximal compact subgroup of G. Given a point $p \in \tilde{M}$, the geodesic flip about p defines an involutive automorphism of T (not necessarily an isometry of $\tilde{M}$ a priori). Such involutive automorphisms correspond to geodesic symmetries in G/K. Thus we can associate to p the fixed point $\psi(p)$ of that geodesic symmetry. This idea is due to Gromov, and he proved that ψ is an isometry up to a scaling factor, see [GS]. Note that ψ is Γ-invariant, and hence (2.4) follows.

I now want to mention some applications of Theorem 2.4 and conclude this section by stating several problems.

2.6 Theorem. Suppose $\mathrm{vol}(M)<\infty$ and $K \geq -a^2$. If M is not flat, then $\pi_1(M)$ contains a nonabelian free subgroup.

This follows from results in [Bl] and [Ti], see [BE]. Theorem 2.6 generalizes Avez' result, see (1.2).

Using ideas of Prasad and Raghunathan [PR] one can introduce the notion of rank for the fundamental group Γ of M. The main result in [PR] can then be stated as $\mathrm{rank}(\Gamma) = \mathrm{rank}(M)$ if M is locally symmetric of noncompact type. It can be shown that this equality is also true without the latter restriction, see [BE], and as an application one obtains:

2.7 Theorem [BE]. Suppose $\mathrm{vol}(M)<\infty$ and $K \geq -a^2$. Then M is an irreducible locally symmetric space of noncompact type of rank $k \geq 2$ if and only if the following three conditions are satisfied:

(a) Γ is finitely generated

(b) Γ does not contain a product subgroup of finite index

(c) $\mathrm{rank}\,(\Gamma) = k$.

Here a Riemannian manifold is called irreducible if none of its finite covering spaces is a Riemannian product.

2.8 Problem. Show that the assumption in (2.3) - (2.7) that the curvature has a lower bound $-a^2$ is not necessary.

2.9 Problem. In part b) of Theorem 2.1, is it necessary to assume that M is compact?

REFERENCES

[An] D. ANOSOV, "Geodesic flows on closed Riemannian manifolds with negative curvature", Proc. Steklov Instit. Math. 90, Amer. Math. Soc., Providence, R.I. 1969

[AA] V. ARNOLD and A. AVEZ, "Problèmes ergodiques de la méchanique classique", Gauthier-Villars, Paris 1967.

[Av] A. AVEZ, "Varietes Riemanniennes sans points focaux", C.R. Acad. Sc. Paris 270 (1970), 188-191.

[B1] W. BALLMANN, "Axial isometries of manifolds of non-positive curvature", Math. Ann. 259 (1982), 131-144.

[B2] W. BALLMANN, "Nonpositively curved manifolds of higher rank", Annals of Math. 122 (1985).

[BB] W. BALLMANN and M. BRIN, "On the ergodicity of geodesic flows", Erg. Th. Dyn. Syst. 2 (1982), 311-315.

[BBE] W. BALLMANN, M. BRIN and P. EBERLEIN, "Structure of manifolds of nonpositive curvature. I", Annals of Math. 122 (1985), 171-203.

[BBS] W. BALLMANN, M. BRIN and R. SPATZIER, "Structure of manifolds of nonpositive curvature. II", Annals of Math. 122 (1985), 205-235.

[BE] W. BALLMANN and P. EBERLEIN, "Fundamental group of manifolds of nonpositive curvature", Preprint, College Park-Chapel Hill 1985.

[BW] W. BALLMANN and M. WOJTKOWSKI, " Estimates for the measure entropy of geodesic flows", in preparation.

[Be] M. BERGER, "Sur les groupes d'holonomie homogène des variétés a connexion affine et des variétés Riemanniennes", Bull. Soc. Math. France 83 (1955), 279-330.

[Bu] K. BURNS, "Hyperbolic behaviour of geodesic flows on manifolds with no focal points", Erg. Th. Dyn. Syst. 3 (1983), 1-12.

[BS] K. BURNS and R. SPATZIER, "Manifolds of nonpositive curvature and their buildings", in preparation.

[Di] E. DINABURG, "On the relations among various entropy charasteristics of dynamical systems", Math USSR Izv.5(1971)

[Eb] P. EBERLEIN, "When is a geodesic flow of Anosov type?, I", J. Diff. Geom. 8 (1973), 437-463.

[EO] P. EBERLEIN and B. O'NEILL, "Visibility manifolds", Pac. J. Math. 46 (1973), 45-109.

[FM] A. FREIRE and R. MAÑÉ, "On the entropy of the geodesic flow in manifolds without conjugate points", Invent. math. 69 (1982), 375-392.

[GS] M. GROMOV and V. SCHROEDER, "Lectures on manifolds of nonpositive curvature", Preprint, Berkeley 1985.

[Ka] A. KATOK, "Entropy and closed geodesics", Erg. Th. Dyn. Syst. 2 (1982), 339-367.

[Ma] A. MANNING, "Topological entropy for geodesic flows", Annals of Math. 110 (1979), 567-573.

[Mi] J. MILNOR, "A note on curvature and fundamental group", J. Diff. Geom. 2 (1968), 1-7.

[Os] V. OSELEDEČ "A multiplicative ergodic theorem" Trans. Moscow Math. Soc. 19 (1968), 197-231.

[OS] R. OSSERMAN and P. SARNAK, "A new curvature invariant and entropy of geodesic flows", Invent. math. 77 (1984), 455-462.

[P1] JA. PESIN, "Characteristic Lyapunov exponents and smooth ergodic theory", Russian Math. Surveys 32:4 (1977), 55-114.

[P2] JA. PESIN, "Equations for the entropy of a geodesic flow on a compact Riemannian manifold without conjugate points", Math. Notes 24 (1978), 796-805.

[P3] JA. PESIN, "Geodesic flows on closed Riemannian manifolds without focal points", Math. USSR Izv. 11 (1977), 1195-1228

[PR] G. PRASAD and M. RAGHUNATHAN, " Cartan subgroups and lattices in semisimple Lie groups", Annals of Math. 96 (1972) 296-317.

[Si] J. SIMONS, "On the transitivity of holonomy systems", Annals of Math. 76 (1962), 213-234.

[Sp] R. SPATZIER, "Dynamical properties of algebraic systems", Thesis, Warwick University 1983.

[Sv] A. ŠVARC, "The volume invariant of a covering", Doklady Akad. Nauk SSSR 105 (1955), 32-34.

[Ti] J. TITS, "Free subgroups of linear groups", J. of Algebra 20 (1972), 250-270.

COMBINATORIAL PROBLEMS IN SPECTRAL GEOMETRY

Robert Brooks*
Department of Mathematics
University of Southern California
Los Angeles, CA 90089-1113

Let M be a compact manifold. Then, as is well-known, there is a close relationship between the covering space theory of M and the fundamental group of M.

It was first noticed by Milnor [14], and later and more extensively by Gromov [11], that this relationship picks up more power as one brings geometry into the picture. The idea is that various asymptotic combinatorial properties of the fundamental group should be reflected in the "geometry at infinity" of covering spaces of M. One thus has Milnor's theorem [14] that the growth type of $\pi_1(M)$ is the same as the growth type of the universal cover $\tilde{M}$ of M.

Another result of this type is given in [2] (see also [11]), and relates the bottom λ_0 of the L^2-spectrum of the Laplacian on $\tilde{M}$ with the amenability of $\pi_1(M)$.

In this paper, we would like to consider this point of view in light of the following problem: given M, how does one understand the behavior of the first eigenvalue of the Laplacian $\lambda_1(M')$ when M' ranges over finite covering spaces of M?

In the special case when M = S is a Riemann surface of genus greater than 1, this problem has a rather fascinating history.

In [13], Henry McKean claimed to have proved the following elegant result, which he called the "Riemann hypothesis for Riemann surfaces": if S is endowed with a metric of constant curvature -1, then $\lambda_1(S) > \frac{1}{4}$.

The number $\frac{1}{4}$ occurs because it is the bottom of the L^2-spectrum $\lambda_0(H^2)$ of hyperbolic space. The content of McKean's result was thus to compare the first eigenvalue of S with λ_0 of its universal covering.

Shortly after his paper appeared, B. Randol [15] observed this could not be right, and in fact proved the following strong counter example if M is any manifold with a non-trivial homomorphism $\pi_1(M) \to Z$, then M has coverings $M^{(i)}$ with $\lambda_1(M^{(i)})$ tending to 0 as i tends to infinity.

It turned out that in the case of surfaces, Selberg [16] had already considered these questions in the non-compact case, with the following elegant twist: Let Γ be the modular group PSL(2,Z), and let Γ_n be the congruence subgroup

$$\{\Gamma_n = \begin{pmatrix} a & b \\ c & d \end{pmatrix} \in PSL(2,Z) : \begin{pmatrix} a & b \\ c & d \end{pmatrix} \equiv \pm\begin{pmatrix} 1 & 0 \\ 0 & 1 \end{pmatrix} \pmod{n}\}.$$

Then H^2/Γ_n is for each n a finite volume Riemann surface, and furthermore

*Partially Supported by NSF grant DMS-83-15552; Alfred P. Sloan fellow.

$$\lambda_1(H^2/\Gamma_n) \geq \frac{3}{16}$$

Note that the spectrum of H^2/Γ_n is discrete below $\frac{1}{4}$, so it makes sense to talk of λ_1.

He conjectured that the bound $\frac{3}{16}$ could be improved to $\frac{1}{4}$, and also mentioned that, for arbitrary subgroups Γ' of finite index in Γ, one may have $\lambda_1(\Gamma')$ arbitrarily small.

It turned out that in coming to grips with Selberg's theorem, one must make a number of enlargements in the philosophy expressed above. Indeed, it is not too difficult, given [2], to find necessary and sufficient conditions of a combinatorial nature on the fundamental group to decide whether or not the first eigenvalue is bounded away from 0 in a given family of finite coverings of M. The delicate part is to give a workable method to solve this combinatorial problem. The solution that we propose here is to enlarge the picture so that it includes also the representation theory of discrete groups. When one does this, one has at his disposal a powerful new idea: Kazhdan's Property T [12].

The outline of this paper is as follows: in §1 we sketch the relationship between combinatorial problems and eigenvalue problems via Cheeger's inequality. In §2 we then take this combinatorial discussion and place it in the setting of representation theory. §3 then presents a discussion of Kazhadan's Property T, and in §4 we show how the theorem of Selberg fits into this picture.

It is a pleasure to thank John Millson and Peter Sarnak for helpful discussion.

§1: Cheeger's inequality

Let M be a compact Riemannian manifold. The Laplacian $\Delta = -{*}d{*}d$ acting on L^2-functions on M is then a second order differential operator which is self-adjoint in the usual L^2-inner product.

If M' is a Riemannian covering space of M, then it makes sense to compare the Laplacian of M' to that of M. There are two cases to consider:

Case (i): M' is a finite covering of M. In this case, M' is again compact, so that, for instance, one may lift an eigenfunction on M to an eigenfunction on M', to see that $\lambda_i(M') \leq \lambda_i(M)$. In general, this estimate should be very weak, so a natural question to ask is: how can one estimate $\lambda_i(M')$ from below in terms of M?

Case (ii): M' is an infinite covering of M.

In this case, the spectrum of M' will no longer be discrete, so that it no longer makes sense to talk about λ_1, λ_2, etc. At the same time in this case we may no longer lift L^2-eigenfunctions on M to L^2-eigenfunctions on M', so that, for instance, the constant function is no longer an L^2-function on M'. In general, we expect $\lambda_0(M') \geq \lambda_0(M) = 0$, but the question is when we get equality.

In [2], we considered case (ii). To state our results, let us assume for simplicity that M' is a normal covering of M, so that $\pi_1(M)/\pi_1(M')$ is a well-defined group. Then we showed ([2], see also [11]):

Theorem ([2]): $\lambda_0(M') = 0$ if and only if $\pi_1(M)/\pi_1(M')$ is an amenable group.

Let us recall the argument briefly, with an eye to adapting it to case (i). The argument involved the following three steps:

Step 1: Given a choice of generators $g_1,\ldots,g_k$ for $\pi_1(M)/\pi_1(M')$, we consider the following combinatorial graph Γ: the vertices of Γ are the elements of the group $\pi_1(M)/\pi_1(M')$, and an edge joins two vertices if they differ by left-multiplication by a generator.

We now observe that, according to a theorem of Folner [9] (see also [2]), there is a nice interpretation of the amenability of $\pi_1(M)/\pi_1(M')$ in terms of an isoperimetric constant of this graph: Let $h(\Gamma) = \inf_E \frac{\#(E)}{\#(A)}$ where E ranges over finite sets of edges of Γ such that Γ-E has a bounded component A, and a possibly unbounded component:

Theorem (Folner): $\pi_1(M)/\pi_1(M')$ is amenable if and only if $h(\Gamma) = 0$

Step 2: One has an analogous, purely geometric estimate for the bottom of the spectrum of a Riemannian manifold, due to Jeff Cheeger [6]. When N is a complete manifold, let h(N) denote the following:

$$h(N) = \inf_S \frac{\text{area}(S)}{\text{vol}(\text{int}(S))}$$

where S ranges over compact hypersurfaces whic divide N into a bounded piece int(S) and a possibly unbounded piece. Note that $h(N) = 0$ if N is compact.

Then one has Cheeger's inequality [6]:

Theorem (Cheeger): $\lambda_0(N) \geq \frac{1}{4}(h^2(N))$.

Furthermore, one has a partial converse to Cheeger's inequality, due to P. Buser, [5], that says that when the geometry of N is bounded, then

$$\lambda_0(N) \leq (\text{const})\cdot h(N)$$

where (const) depends on the bounds on the geometry of N. Thus, in the presence of bounded geometry, $\lambda_0(N) = 0$ if and only if $h(N) = 0$.

Step 3: The theorem is now equivalent to the assertion that the combinatorial constant $h(\Gamma)$ is zero if and only if the geometric constant $h(M')$ is zero. We will prove slightly more; that there are constants c_1, c_2, depending on the choice of generators and the choice of a fundamental domain F, such that

$$c_1 h(\Gamma) \leq h(M') \leq c_2\, h(\Gamma)$$

Given F, if we choose generators for $\pi_1(M)/\pi_1(M')$ to be $\{g : g(F) \text{ adjoins } F\}$, we can see the second inequality easily, because dividing Γ is the same as dividing M' into unions of fundamental domains. The difficult part is the first inequality, where the problem is to see how an arbitrary hypersurface dividing M' into a bounded and unbounded piece implies the same kind of division of the graph Γ.

To see this, one takes a hypersurface S, and considers the problem of minimizing h(T) for hypersurfaces T whose support lies in the union of fundamental domains which meet int(S). From geometric measure thery, one can show that h(T) is minimized by an integral current of bounded mean curvature, where the bound depends only on h(S) and the geometry of F. It then follows from boundedness of the mean curvature that $h(T) \geq c_1\, h(\Gamma)$, taking the division of Γ into fundamental domain which meet int(T).

See [2] for details.

We now consider the modifications in this approach necessary to handle case (i). Here the main point is not to look at a fixed covering M' of M, but rather at a family of covering $\{M_i\}$, to decide whether $\lambda_1(M_i) \to 0$ as $i \to \infty$.

One has an analogue of Cheeger's inequality, valid for studying λ_1 of compact manifolds-indeed, it was in this context that Cheeger first defined his constant. Namely for N compact, let $h_1(N)$ be defined by

$$h_1(N) = \inf_S \frac{\text{area}(S)}{\min(\text{vol}(N_1),\ \text{vol}(N_2))}$$

where $N - S = N_1 \quad N_2$.

Then one has

Theorem (Cheeger [6]): $\lambda_1(N) \geq \frac{1}{4} h_1^2(N)$.

This immediately suggests the following changes in Step 1: for each covering M_i of M, let Γ_i be the following finite graph: the vertices of Γ_i are the cosets $\pi_1(M)/\pi_1(M_i)$, and an edge joins two vertices whenever the corresponding group elements differ by a generator.

Then let $h_1(\Gamma_i)$ be defined by:

$$h_1(\Gamma_i) = \inf_E \frac{\#(E)}{\min(\#(A),\ \#(B))} \cdot$$

where E ranges over sets of edges such that Γ-E has two components, A and B.

We now run into a problem that did not occur in our discussion of case (ii), but which is easily solved. In order to make much sense out of $h_1(\Gamma_i)$ as $i \to \infty$, we must make a uniform choice of generators of $\pi_1(M)/\pi_1(M_i)$. This is done easily enough - it suffices to take fixed generators $g_1,\ldots,g_k$ for $\pi_1(M)$, as these clearly generate the quotient group.

It remains to show the analogue of Step 3. One would like to show the existence of constants c_1 and c_2 such that $c_1 h_1(\Gamma_i) \leq h_1(M_i) \leq c_2 h_1(\Gamma_i)$.

Once again, the right-hand inequality is straightfoward, but the left-hand side runs into some technical problems. For instance, it may happen that the hypersurface T_i minimizing the isoperimetic constant for M_i may wander through all the fundamental domains of M_i, so that one does not see directly how to divide Γ_i into two pieces from T_i. However, it is not hard to see that this implies that T_i itself must have large isoperimetric constant and one can then show:

Lemma: There exist constants c_1, c_2, and d such that

$$c_1 h_1(\Gamma_i) \leq h_1(M_i) \leq c_2 h_1(\Gamma_i)$$

whenever $h_1(M_i) \leq d$ (see [4] for details).

In particular, we find:

<u>Theorem 1</u>: For any family of coverings $\{M_i\}$ of M,

$$\lambda_1(M_i) \to 0 \quad \text{as } i \to \infty$$

if and only if

$$h_1(\Gamma_i) \to 0 \text{ as } i \to \infty.$$

§2: The Representational Laplacian

In §1, we showed how the problem of studying the behavior of λ_1 under coverings reduces to a combinatorial problem on the fundamental group. This combinatorial problem, however, appears to be quite delicate, and in this section and the next we consider some approaches to solving it.

The main idea can be understood from the point of view of Dodziuk's combinatorial Laplacian [8], and proceeds as follows: Considering the graphs Γ_i corresponding to the coverings M_i of M, one may define a combinatorial Laplace difference operator Δ_i on the space of functions on the vertices of Γ_i, with its obvious L^2-inner product. Then one may once again invoke the inequalities of Cheeger and Buser to say that $h_1(\Gamma_i)$ tends to zero as i tends to ∞ if and only if the first eigenvalue $\lambda_1(\Delta_i)$ tends to zero as i tends to ∞.

Actually, as we will see below (see also [8]) the analogues of the Cheeger and Buser estimates for the combinatorial Laplacian are quite straightforward.

At this point, the argument looks almost completely circular, because it appears that we have merely unwound the Laplacian to get an isoperimetric constant, only to rewind it again to get another Laplacian. But in actuality, we have gained two things by this sequence of moves: first of all, when we rewind our discussion, it is still on a combinatorial level. Secondly, and this is the main point, we are now in a position to make use of representation theory, which was unavailable until now.

To that end, let us be given a finitely generated group π, and let us fix generators $g_1,\ldots,g_k$ for π. We will assume that the set of generators is symmetric, that is, if $g \in \{g_1,\ldots,g_k\}$, then $g^{-1} \in \{g_1,\ldots,g_k\}$.

Suppose that H is a Hilbert space on which π acts unitarily. Let Δ_π be the operator

$$\Delta_\pi(X) = \sum_i (X - g_i(X))$$

We claim:

__Lemma:__ Δ_π is a positive definite self-adjoint operator, satisfying

$$\langle \Delta_\pi(X), Y\rangle = \frac{1}{2}\sum_i \langle X - g_i(X), Y - g_i(Y)\rangle$$

__Proof:__ Expanding the right-hand side gives

$$\sum_i \langle X - g_i(X), Y - g_i(Y)\rangle$$

$$= \sum_i \langle X,Y\rangle + \langle g_i(X), g_i(Y)\rangle$$

$$- \langle g_i(X), Y\rangle - \langle X, g_i(Y)\rangle$$

$$= \sum_i (2 \langle X,Y\rangle - \langle g_i(X) + g_i^{-1}(X), Y\rangle)$$

$$= 2 \sum_i \langle X - g_i(X), Y\rangle = 2\langle \Delta_\pi(X), Y\rangle$$

as desired, where in the second equality we have used the fact that π acts unitarily on H, and in the third that the set $\{g_i\}$ is symmetric.

From this we immediately have the following Rayleigh formula for the bottom of the spectrum $\lambda_0(H)$ of Δ_π on H:

$$\lambda_0(H) = \inf_X \frac{\frac{1}{2} \sum_i \|X - g_i(X)\|^2}{\|X\|^2}$$

Using the fact that all norms on a finite-dimensional space are equivalent, we see readily that, for a fixed number of generators, $\lambda_0(H)$ stays in bounded proportion to similar expression such as

$$h(H) = \inf_X \frac{\sum \|X - g_i(X)\|}{\|X\|}$$

and

$$k(H) = \inf_X \Big(\sup_i \frac{\|X - g_i(X)\|}{\|X\|}\Big) .$$

We now observe that when $H = L^2(\pi')$, where π' is equal to π or a quotient of π, the representational Laplacian is identical to Dodziuk's combinatorial Laplacian. One thus has at one's disposal the machinery of Cheeger's inequality. Let us quickly review this machinery in the combinatorial context, keeping in mind the reference [8].

Cheeger's inequality consists of two steps. The first step is using a Cauchy-Schwartz argument, to estimate the Rayleigh quotient

$$\lambda_0 = \inf \frac{\int \|df\|^2}{\int |f|^2} \geq \frac{1}{4} \Big(\inf \frac{\int \|df\|}{\int \|f\|}\Big)^2 .$$

In the combinatorial case, one may replace df by the difference operator $\delta f(\text{edge})$ = the difference of f along an edge, in a manner similar to the operator $\sum (X - g_i(X))$ of the Lemma. What does not have an analogue in the representational setting is the right-hand side, because it involves an L^1-estimate rather than an L^2-estimate. In the combinatorial setting, there is no difficulty here, since one may use the L^1-norm on functions on π'.

The second step is the "coarea formula," which one may think of as saying that it suffices, in estimating $\frac{\int \|df\|}{\int \|f\|}$, to consider only characteristic functions (or at least smooth approximants to them). There seems to be no representational analogue of this.

In the combinatorial analogue of this, we simply list all the values (finitely many) b_i of the function f, setting

$$L_i = \{x : f(x) > b_i\}$$

and setting

$$f_i(x) = 0 \quad \text{if } x \notin L_i$$

$$f_i(x) = b_i - b_{i-1} \quad \text{if } x \in L_i .$$

Then, for a non-negative function f, with $b_0 = 0$, we have

$$|\delta f|_{L^1} = \sum |\delta f_i|_{L^1}$$

$$|f|_{L^1} = \sum |f_i|_{L^1}$$

so that

$$\frac{|\delta f|_{L^1}}{|f|_{L^1}} = \frac{\sum |\delta f_i|_{L^1}}{\sum |f_i|_{L^1}} > \inf_i \left(\frac{|\delta f_i|}{|f_i|}\right) .$$

On the other hand, the right-hand side is clearly bounded below by the combinatorial isoperimetric constant of the graph, establishing the combinatorial analogue of Cheeger's inequality.

§3: Kazhdan's Property T

We begin by discussing the following construction, due to Kazhdan [12], which puts a topology on the space of all unitary representations of π.

Given a Hilbert space H_1, on which π acts unitarily, and given a unit-lenth vector $X \in H_1$, we will define an (X,ε)-neighborhood $N(X,\varepsilon)$ of H_1 in the space of all representations by the formula: $H_2 \in N(X,\varepsilon)$ iff there exists $Y \in H_2$ $\|Y\| = 1$, such that $|\langle g_i(Y),Y\rangle - \langle g_i(X), X\rangle| < \varepsilon$ for all i.

Note that when $H_2 = H_1 \oplus H_3$, then any $N(X,\varepsilon)$ of H_1 contains H_2.

When we specialize to the case where H_1 is the trivial 1-dimensioanl representation, then there is an essentially unique choice for X, and one may speak of the Kazhdan distance $k(H_2)$ from H_2 to the trivial representation. In this case, the formula (*) simplifies as follows: from the formulas

$$\|g_i(Y) - Y\|^2 = \langle g_i(Y), g_i(Y)\rangle - 2\langle g_i(Y), Y\rangle + \langle Y,Y\rangle = 2\langle Y,Y\rangle - 2\langle g_i(Y), Y\rangle$$

we see that the expression

$$|\langle g_i(Y), Y\rangle - \langle g_i(X), X\rangle| < \varepsilon , \quad |Y| = 1$$

is equivalent, for X the unit vector in the trivial representation, to

$$\|g_i(Y) - Y\| < \sqrt{2\varepsilon} .$$

We may therefore define the Kazhdan distance by

$$(**) \quad k(H_2) = \inf_{\|Y\| = 1} \sup_i \|g_i(Y) - Y\|.$$

Note that from our remarks in §2, we see that there are constants c_1, c_2, depending on the number of generators for π, such that

$$c_1\lambda_0(H_1) \leq k(H_1) \leq c_2\,\lambda_0(H_1).$$

As an exercise in working with these ideas, we will show the following well-known

Lemma: A group π is amenable if and only if the trivial representation is in the closure of the regular representation of π on $L^2(\pi)$ (or equivalently, if and only if $k(L^2(\pi)) = 0$).

Proof: We distinguish two cases.

First, when π is a finite (and hence amenable) group, the constant function occurs as a trivial subrepresentation of the regular representation. It follows that $k(L^2(\pi)) = 0$.

Now suppose that π is infinite.

If π is amenable, then we may choose Y_E to be the characteristic function for the sets E given by Folner's characterization of amenability, to see that for all ε,

$$\frac{\|g_i(Y_E) - Y_E\|}{\|Y_E\|} < \varepsilon \quad \text{for all } i$$

can be satisfied, showing that $k(L^2(\pi)) < \varepsilon$.

To show the converse statement, one concludes from $k(L^2(\pi)) = 0$ that $\lambda_0(L^2(\pi)) = 0$, and hence from the first part of Cheeger's inequality that there are functions f with

$$\frac{\|\delta f\|_{L^1}}{\|f\|_{L^1}} < \varepsilon, \text{ for all } \varepsilon.$$

From the second part of Cheeger's inequality, one may conclude that there are characteristic functions f_E with these properties. One then uses the sets E to establish Folner's criterion.

We then have the following ([12], [17])

Definition: π has Kazhdan's Property T if and only if the trivial representation is isolated among all irreducible representations of π.

Equivalently, π has property T if and only if there exists $\varepsilon > 0$ with the following property: for any unitary representation $_2$, if, for some X with $\|X\| = 1$,

$$\|g_i(X) - X\| < \varepsilon \quad \text{for all } i,$$

then for some $Y \in H_2$, $g_i(Y) = Y$ for all i, and consequently $g(Y) = Y$ for all $g \in \pi$.

The equivalence of these two conditions is not quite obvious, as was observed in [17].

In [12], Kazhdan gives a version of Property T for Lie groups, and proves the following Theorem ([12]): If π is discrete of cofinite volume in G, then G has Property T if and only if π does.

This is the main ingredient which allows Property T to be a useful tool in settling the combinatorial problem of §1, since the question of whether a Lie group has Property T can be settled from the representation theory of G. For the record, we note the following result of Kazhdan [12], as expanded on by Wang [17]:

Theorem: ([12], [17]) Let G be a semi-simple Lie group all of whose factors have R-rank ≥ 2. Then G has Property T, and hence also its cofinite-volume discrete subgroups.

As an example, SL(3,Z) has Property T, while SL(2,Z) does not. The groups of isometries of quaternionic hyperbolic spaces are rank 1 groups which also have

Property T. As an immediate consequence of the definitions, together with §1 and §2, we have:

Theorem: Suppose $\pi = \pi_1(M)$ has Property T, with M compact.

Then there exists $C > 0$ such that for all finite coverings M' of M, $\lambda_1(M') > C$.

Proof: From §1 and §2, it suffices to show that there is an $\varepsilon > 0$ with the property:

for any normal subgroup π' of finite index in π, $\lambda_1(L^2(\pi/\pi')) > \varepsilon$.

Setting $H_{\pi'}$ to be the subspace of $L^2(\pi/\pi')$ orthogonal to the constant function, we see that

$$\lambda_1(L^2(\pi/\pi')) = \lambda_0(H_{\pi'}) \sim k(H_{\pi'}).$$

From the definition of Property T, it follows that if $k(H_{\pi'})$ is sufficiently small, then $H_{\pi'}$ must contain a π-invariant vector, that is a constant function. But $H_{\pi'}$ is orthogonal to the constant functions, a contradiction.

The converse to this theorem is not quite true. It is clear from the proof that not all representations of π enter into λ_1 ~ only those representations which factor through a finite quotient enter in. Thus we have the following variation of the theorem:

Theorem': Let M be a compact manifold. Then there exists a constant $C > 0$ such that $\lambda_1(M') > C$ for all finite coverings C if and only if there is a neighborhood of the trivial representation in the space of all irreducible unitary representations of $\pi_1(M)$ containing no representations which factor through a finite quotient of $\pi_1(M)$.

We remark that even in reasonably well-behaved examples of groups π, there may well exist representations which are not well-approximated (in the Kazhdan topology) by representations which factors through finite quotients. Thus, the Theorem is far from giving a characterization of Property T in terms of λ_1. On the other hand, the condition given in Theorem' is spiritually quite close to Property T, and we know of no way to establish that a group π has this property without relating π to a group with Property T.

§4: Selberg's Theorem

In this section, we will bring the discussion of the previous sections to bear on the following theorem of Selberg [16]: Let Γ_n denote the congruence subgroup of PSL(2,Z) defined by:

$$\Gamma_n = \{\begin{pmatrix} a & b \\ c & d \end{pmatrix} \in PSL(2,Z) : \begin{pmatrix} a & b \\ c & d \end{pmatrix} \equiv \begin{pmatrix} 1 & 0 \\ 0 & 1 \end{pmatrix} \pmod{n}\}.$$

Then

<u>Theorem (Selberg)</u>: $\lambda_1(H^2/\Gamma_n) \geq \frac{3}{16}$ for all n.

In what follows, we will show that Selberg's theorem is equivalent to a purely number-theoretic statement, which we will present below, at least after replacing "$\frac{3}{16}$" by "some positive constant." At present, we do not have a proof of this statement independent of Selberg's theorem, although we initially beileved that such would be the case. We have no idea whether or not the number-theoretic statements we present are known or accessible by other means, or whether they are in some sense new results.

We begin by collecting some information from the previous section. We may recast the discussion of §1 - §3 by saying:

<u>Theorem</u>: Given a compact manifold M, let $\{M_i\}$ be a family of finite normal coverings of M. Let $\pi^i = \pi_1(M)/\pi_1(M_i)$.

Then there is a constant $C > 0$ such that $\lambda_1(M_i) > C$ for all i if and only if there is a positive constant ε such that, for all i and for all non-trivial irreducible representations H of π^i, the Kazhdan distance satisfies $k(H) > \varepsilon$.

<u>Proof</u>: By §1, we have that there is a constant $C > 0$ such that $\lambda_1(M_i) > C$ if and only if there exists $C > 0$ such that $h(\Gamma_i) > D$, this last condition being equivalent by §2 to the condition that, for some $D' > 0$, $\lambda_1(L^2(\pi^i)) > D'$. We now decompose $L^2(\pi^i)$ into the orthogonal direct sum $L^2(\pi^i) = \quad H_{i,j}$. It is standard from the representation theory of finite groups that each irreducible representation of π^i occurs in the direct sum with a multiplicity equal to the dimension of the representation. Furthermore, the representational Laplacian also decomposes under this direct sum, so that the condition that $\lambda_1(L^2(\pi^i)) > D'$ is equivalent to $\lambda_0(H_{i,j}) > D'$ for all non-trivial irreducible representation $H_{i,j}$ of π^i.

Finally, we observe that $\lambda_0(H_{i,j}) \sim k(H_{i,j})$, completing the proof of the theorem.

There is one obstacle to overcome before applying this to Selberg's theorem. Namely, we must remove, or at least weaken, the restriction of compactness of M. When M has a standard cusp, or more generally when M satisfies an "isoperimetric condition at infinity," the techniques of [1] and [3] apply directly to show that the theorem remains valid.

The main idea behind our argument below is as follows: Suppose that π_1, and hence π^i, is generated by two elements U and V, and suppose we want to find an ε such that $k(\mathcal{H}_{i,j}) > \varepsilon$ for all i,j.

Let us fix $\delta > 0$, and for an arbitrary irreducible representation of π^i, let us split

$$\mathcal{H} = \mathcal{H}_U^{\text{small}} \oplus \mathcal{H}_U^{\text{large}}$$

where $\mathcal{H}_U^{\text{small}}$ is spanned by the eigenvectors of U with eigenvalue within δ of 1, and $\mathcal{H}_U^{\text{large}}$ is spanned by the eigenvectors whose eigenvalues lie farther than δ from 1, and similarly for V.

Let us denote by α the cosine of the angle between $\mathcal{H}_U^{\text{small}}$ and $\mathcal{H}_V^{\text{small}}$:

$$\alpha = \sup \frac{\langle u,v\rangle}{\|u\|\|v\|}, \quad u \in \mathcal{H}_U^{\text{small}} \quad v \in \mathcal{H}_V^{\text{small}}.$$

Then we have:

<u>Lemma</u>: The Kazhdan distance $k(\mathcal{H})$ from $\mathcal{H}$ to the trivial representation satisfies

$$\max\left(\delta,\ \delta\cdot\alpha + 2\sqrt{1-\alpha^2}\right) \geq k(\mathcal{H}) \geq \delta\sqrt{\frac{1-\alpha^2}{2}}$$

<u>Proof</u>: Given a vector $v \in \mathcal{H}$, we decompose

$$v = v_U + v_V + v^{\perp}$$

where $v_U \in \mathcal{H}_U^{\text{small}}$, $v_V \in \mathcal{H}_V^{\text{small}}$, and $v^{\perp}$ is perpendicular to $\mathcal{H}_U^{\text{small}}$ and $\mathcal{H}_V^{\text{small}}$. We wish to estimate $\|U(v) - v\|$ and $\|V(v) - v\|$.

If we write $v = v'_U + v'^{\perp}_U$, where $v'_U \in \mathcal{H}_U^{\text{small}}$, $v'^{\perp}_U$ perpendicular to $\mathcal{H}_U^{\text{small}}$, then

$$\|U(v) - v\| \geq \delta\|v'^{\perp}_U\|$$

and similarly

$$\|V(v) - v\| \geq \delta \|v'^{\perp}_V\|$$

and the problem is to estimate $\|v'^{\perp}_U\|$, $\|v'^{\perp}_V\|$ in terms of $\|v\|$.

But $\|v'^{\perp}_U\|^2 \geq \|v^{\perp}\|^2 + (1 - \alpha^2) \|v'_V\|^2$ and similarly for v'_V, so that

$$\|v'_U\|^2 + \|v'_V\|^2 \geq 2\|v^{\perp}\|^2 + (1 - \alpha^2)(\|v_U\|^2 + \|v_V\|^2 \geq (1 - \alpha^2) \|v\|^2$$

from which it follows that one of the two summands on the left must be at least half the right-hand term, establishing the right-hand inequality of the lemma.

To show the left-hand inequality, we first observe that, for any v,

$$\|V(v) - v\| \leq 2\|v\|.$$

Now let us choose $v \in H^{small}_U$ such that its orthogonal projection v^{small} onto H^{small}_V has length $\alpha \cdot \|v\|$. Then

$$\begin{aligned}\|U(v) - v\| &\leq \|V(v^{small}) - v^{small}\| + \|V(v^{\perp}) - v^{\perp}\| \\ &\leq \delta\|v^{small}\| + 2\|v^{\perp}\| = (\delta \cdot \alpha + 2 \sqrt{1 - \alpha^2}) \|v\|\end{aligned}$$

establishing the left-hand inequality.

We may paraphrase the lemma in the follow way:

Corollary: Suppose $\{H_i\}$ are a family of representations of π. Then the following are equivalent:

(a) $k(H_i)$ is bounded away from 0 as $i \to \infty$.

(b) For δ sufficiently small, the angle between $H^{small}_{i,U}$ and $H^{small}_{i,V}$ is bounded away from 0 as $i \to \infty$.

Let us now restrict attention to the case when $n = p$, a prime number. Then, if $\Gamma = PSL(2,Z)$, we have $\Pi^p = \Gamma/\Gamma_p \cong PSL(2, Z/p)$. Let us fix as generators for $PSL(2,Z)$ the two elements

$$U = \begin{pmatrix} 1 & 0 \\ -1 & 1 \end{pmatrix} \qquad V = \begin{pmatrix} 1 & 1 \\ 0 & 1 \end{pmatrix}$$

noting that $V = E\, U\, E^{-1}$, where

$$E = \begin{pmatrix} 0 & 1 \\ -1 & 0 \end{pmatrix}, \qquad E^2 = id.$$

For a and b relatively prime to p, and χ a multiplicative character (mod p (that is, $\chi(ab) = \chi(a)\,\chi(b)$), the Kloosterman sum $S_\chi(a,b,p)$ is defined by:

$$S_\chi(a,b,p) = \sum_{y \not\equiv 0 (\mathrm{mod}\ p)} \overline{\chi(y)}\ \zeta^{(ay + by^{-1})}$$

where $\zeta = e^{2\pi i/p}$, and y^{-1} is the multiplicative inverse of y (mod p).

One has the deep estimate of Weil [18]:

$$(***) \qquad |S_\chi(a,b,p)| \leq 2\sqrt{p}$$

sharpening the estimate of Davenport [7] that $|S_\chi(a,b,p)| \leq O(p^{2/3})$.

For our purpose, we will also introduce the "Kloosterman - like "sums $T_\chi(a,b,p)$, defined as follows:

There is a unique quadratic extension F_{p^2} of the field Z/p. For each multiplicative character χ on F_{p^2}, let

$$T_\chi(a,b,p) = \sum_{t\bar{t} = ba^{-1}} \overline{\chi(t)}\ \zeta^{(at + bt^{-1})}$$

noting that from $t\bar{t} = ba^{-1}$, we get $at + bt^{-1} = a(t + \bar{t}) \in A/p$. ($\bar{t}$ denotes the conjugate of t in the unique field automorphism of F_{p^2} over F_p).

We have no idea whether there is an analogue of Weil's estimate for the T_χ's.

Our main result is now:

Theorem: The following two statements are equivalent:

(a) There exists $C > 0$ such that $\lambda_1(H^2/\Gamma_p) \geq C$ for all p

(b) For θ sufficiently small, there exist $\beta < 1$ such that, for all p, χ, and p-tuple

$$(\xi) = (\xi_1 \xi_2, \ldots)$$
$$(\eta) = (\eta_1, \ldots)$$

we have:

$$(\S) \qquad \frac{1}{p} \left| \sum_{|a| < \theta\cdot p} \ \sum_{|b| < \theta\cdot p} \xi_a \cdot \eta_b\ S_\chi(a,b,p) \right| \leq \beta \cdot \|\xi\| \|\eta\|$$

and

$$(\S\S)\qquad \frac{1}{p}\Big|\sum_{|a|<\theta\cdot p}\ \sum_{|b|<\theta\cdot p} \xi_a\cdot\eta_b\, T_\chi(a,b,p)\Big| \leq \beta\|\xi\|\|\eta\|.$$

The idea of the proof is to begin from the fact that the irreducible representations of PSL(2,Z/p) are known - we dug this, and the presentation of it below, out of [10]. The irreducible representations of PSL(2,Z/n) for n arbritrary are known, but the theory is substantially more delicate, and it is for this reason that we restrict the discussion to the case where p is a prime.

The irreducible representations of PSL(2,Z/p) fall into two classes: the continuous series and the discrete series (so named from analogy with the real case).

To describe the continuous series, we consider a multiplicative character χ on Z/p, and the space H_χ of functions f on Z/p × Z/p - {0}, which transform according to the rule

$$(\dagger)\quad f(t\cdot x,\ t\cdot y) = \chi(t)\ f(x,y),\quad t \in (Z/p)^*$$

SL(2,Z/p) acts on H_χ on the right by the rule

$$\left(\begin{pmatrix} a & b\\ c & d\end{pmatrix} f\right)(x,y = f(ax + cy,\ bx + dy)$$

so to insure that this action descends to PSL(2,Z/p), we must have that

$$f(x,y) = \left(\begin{pmatrix} -1 & 0\\ 0 & -1\end{pmatrix} f\right)(x,y) = f(-x_1 - y) = \chi(-1)\ f(x,y),$$

i.e. that χ is an even character.

To write this representation out in coordinates, let us fix the orthonormal basis

$$\begin{aligned} f_a(x,1) &= 1 \quad \text{if } x = a \qquad a = 0,\ldots,p-1\\ &\ 0 \quad \text{otherwise}\\ f_a(1,0) &= 0\\ f_\infty(x,1) &= 0 \quad \text{for } x = 0,\ldots,p-1\\ f_\infty(1,0) &= 1\end{aligned}$$

where we extend the f_a's to all Z/p × Z/p - {0} by (†). We now can compute the actions of U and V on H_χ by the formula

$$(\dagger\dagger)\qquad \begin{aligned} U(f_a)\ (x,1) &= f_a(x + 1,1)\ U(f_a)(1;0) = f_a(1,0)\\ &= f_{a-1}(x,1) \qquad\qquad a = 0,\ldots,p-1,\ \infty.\end{aligned}$$

$$(Vf_a)(\tfrac{1}{x}, a) = f_a(\tfrac{1}{x}, 1 - \tfrac{1}{x})$$
$$= f_a(\tfrac{1}{x}, \tfrac{x-1}{x})$$
$$= \chi(1 - \tfrac{1}{x})\, f_a(\tfrac{1}{x-1}, 1)$$

$$(E)(f_a)(x,1) = f_a(1, -x) = \chi(-x)(f_a(-\tfrac{1}{x}, 1) = \chi(x) f_a(-\tfrac{1}{x}, 1)$$

or $E(f_a) = \overline{\chi}(a)\ f(-\tfrac{1}{a})$ $\qquad = \chi(x)\ f_{-\frac{1}{a}}(x,1).$

From (††) we may write down an eigenbasis for U. Let $\{v_b\}$ be the basis

$$v_b = \frac{1}{\sqrt{p}}\left(\sum_{a=0}^{p-1} \zeta^{ba}\cdot f_a\right)$$
$$v_\infty = f_\infty.$$

Then $U(v_b) = \frac{1}{\sqrt{p}}\left(\sum_{a=0}^{p-1} \zeta^{b\cdot a}\, U(f_a)\right) = \frac{1}{\sqrt{p}}\left(\sum_{a=0} \zeta^{b\cdot a} f_{a-1}\right)$

$$= \zeta^b v_b$$
$$(v_\infty) = v_\infty$$

From V = EUE, we seethat an eigenbasis for V is given by $\{w_a\}$, here

$$w_b = E(v_b) = \frac{1}{\sqrt{p}} \sum \zeta^{ab} E(f_a) = \frac{1}{\sqrt{p}} \sum \zeta^{ab}\, \overline{\chi}(a)\ f(-\tfrac{1}{a}) = \frac{1}{\sqrt{p}} \sum \zeta^{-ba^{-1}} \chi(a) f_a$$

with $V(w_b) = \rho^b \cdot w_b$.

The basic formula we need is:

<u>Lemma</u>: $\langle v_a, w_b\rangle = \frac{1}{p} S_\chi(a,b,p)$

<u>Proof</u>: From $v_a = \frac{}{\sqrt{p}} \sum_c \zeta^{ac}\cdot f_c$, $w_b = \frac{1}{\sqrt{p}} \sum_c \zeta^{-bc^{-1}} \chi(c) f_c$, we see

$$\langle v_a, w_b\rangle = \frac{1}{p}\left(\sum_c \zeta^{ac} \cdot \overline{\chi(c)\zeta^{-bc^{-1}}}\right) = \frac{1}{p}\sum_c \overline{\chi}(c)\zeta^{ac+bc^{-1}} = \frac{1}{p} S_\chi(a,b,p)$$

as desired.

When χ is the trivial character, H_χ has a one-dimensional trivial subrepresentation, but for all other χ, H_χ is irreducible. Also, H_χ is isomorphic to $H_{\overline{\chi}}$.

Leaving the case of χ the trivial character aside as a special case, we see that, for $\delta = 2\sin(\theta)$, the space $H^{small}_{\chi,U}$ is spanned by $\{v_a, |a| < \theta\cdot p\} \cup U\{v_\infty\}$, and $H^{small}_{\chi,V}$ is spanned by $\{w_b, |b| < \theta\cdot p\} \cup \{w_\infty\}$.

Notice that the Weil estimate may be rephrased as saying that $|\langle v_a, w_b\rangle| < \frac{2}{\sqrt{p}}$, or in other words that the orthonormal bases $\{v_a\}$, $\{w_b\}$ are about as perpendicular from each other as possible, given that they span the same p+1-dimensional space.

It is now evident that formula (b)(§) is equivalent to the assertion: the angle between $H^{small}_{\chi,U}$ and $H^{small}_{\chi,V}$ is bounded away from 0, independent of χ and p.

The analogous estimates (5) §§ for the exponential sums T_χ follow in an analogous manner, using the discrete series representations of PSL(2,Z/p). As the discrete series representations of finite fields are somewhat more cumbersome to describe, we leave the details to the reader, referring to [10] as a guide.

The equivalence of (a) and (b) now follows from the fact that these are all the irreducible representations of PSL(2,Z/p), as can be seen from a standard counting argument.

References

1. R. Brooks, "The Bottom of the Spectrum of a Riemannian Covering" Crelles J. 357(1985) pp. 101-114.

2. R. Brooks, "The Fundamental Group and the Spectrum of the Laplacian," Comm. Math. Helv. 56(1981), pp. 581-598.

3. R. Brooks, "The Spectral Geometry of the Apollonian Packing," Comm. P & Appl. Math. XXXVIII (1985), pp. 357-366.

4. R. Brooks, "The Spectral Geometry of a Tower of Coverings," to appear.

5. P. Buser, "A Note on the Isoperimetric Constant," Ann. Sci. Ec. Norm. Sup. 15(1982) pp. 213-230.

6. J. Cheeger, "A Lower Bound for the Smallest Eigenvalue of the Laplacian," in Gunning, Problems in Analysis Princeton U. Press 1970, pp. 195-199.

7. H. Davenport, "On Certain Exponential Sums," Jour. fur Math. 169(1933) pp. 158-176.

8. J. Dodziuk, "Difference Equations, Isoperimetric Inequality, and Transience of Certain Random Walks," Trans. AMS 284(1984), pp. 787-794.

9. E. Folner, "On Groups with Full Banach Mean Values," Math. Scand. 3(1955), pp. 243-254.

10. Gelfand, Graev, and Pyatetskii-Shapiro, Representation Theory and Automorphic Functions, W.B. Saunders Co. (1969).

11. M. Gromov, Structures Metriques pour les Varietes Riemanniennes Ferdnand Nathan, 1981.

12. D.A. Kazhdan, "Connections of the Dual Space of a Group with the Structure of its closed Subgroups," Funct. and Appl. 1(1968) pp. 63-65.

13. H.P. McKean, "Selberg's Trace Formula as Applied to a Compact Riemann Surface," Comm. P & Appl. Math. 25(1972), pp. 225-246.

14. J. Milnor, "A Note on Curvature and Fundamental Group," J. Diff. Geom. 2(1968), pp. 1-7.

15 B. Randol, "Small Eigenvalues of the Laplace Operator on Compact Riemann Surfaces," Bull. AMS. (1974), pp. 990-1000.

16. A. Selberg, "On the Estimation of Fourier Coefficients of Modular Forms," Proc. Symp. Pure Math. VIII (1965), pp. 1-15.

17. S.P. Wang, "The Dual Space of Semi-Simple Lie Groups, "Amer. J. Math 91(1969), pp. 921-937.

18. A. Weil, "On Some Exponential Sums," Proc. Nat. Acad. Sci. USA 34(1948), pp. 204-207.

A Vanishing Theorem for Piecewise Constant Curvature Spaces

by

Jeff Cheeger

Let X^n be a triangulated closed normal pseudomanifold (see [GM] for definitions) equipped with a metric of piecewise constant curvature. Recall that X^n can be described as follows. Start with a collection of simplices, $\{\sigma^n\}$, whose interiors have a metric of (fixed) constant curvature, K, and whose faces are all totally geodesic. Then identify various faces by isometries in such a way that underlying topological space so obtained is a normal pseudomanifold (see [CMS], $[C_1]$-$[C_4]$, which provide the general background for this paper).

Associated to X^n is a natural stratification, $X^n = \bigcup_j \mathcal{S}^i$, where $\mathcal{S}^j$ is a smooth manifold of codimension i and constant curvature K ; $\mathcal{S}^1$ is empty. If $\mathfrak{C}$ is any piecewise constant curvature triangulation as above, then the j-skeleton, Σ^j of $\mathfrak{C}$, contains $\mathcal{S}^{n-j}$. The <u>link</u>, $L(\mathcal{S}^j,p)$, of $\mathcal{S}^j$ at p $\mathcal{S}^j$ is, by definition, the <u>cross section</u> (or <u>base</u>) of the <u>normal cone</u>, $C^{\perp}(\mathcal{S}^j,p)$, to $\mathcal{S}^j$ at p. The link, $L(\Sigma^{n-j},p)$, at $p \in \Sigma^j$ is defined similarly.

We say that X^n has <u>positive curvature at its singularities</u> if for each $p \in \mathcal{S}^2$, the link, $L(\mathcal{S}^2,p)$ is a circle of length $< 2\pi$. If, in addition $K > 0$ (respectively $K = 0$) we say that X^n has <u>positive curvature</u> (respectively <u>nonnegative curvature</u>).

<u>1. Observation.</u> If $K = 0$ and the triangulation $\mathfrak{C}$ can be chosen such that $\mathcal{S}^2 = \Sigma^{n-2}$ and the curvature at the singularities is positive, then X^n admits a metric of positive curvature.

Proof : Choose small $K' > 0$ and replace each (totally geodesic flat), simplex of $\mathfrak{C}$ by the (totally geodesic) simplex of constant curvature K', which has the same edge lengths. For K' sufficiently small, the curvature at the singularities will still be positive.

<u>2. Example</u>. If X^n is the surface of a tetrahedron in R^{n+1}, then X^n admits a metric of positive curvature. Moreover, by "rounding the corners", this metric can be approximated (in an obvious sense) by a smooth metric of positive curvature on the n-sphere, S^n.

For some time, it was assumed that the condition that X^n has positive or negative curvature in the above sense, was the analog of the corresponding condition for the <u>sectional</u> curvature in the smooth case. However (as M. Gromov pointed out) in view of the following result, it may be more accurate to think of these conditions as replacing conditions on the curvature <u>operator</u> (compare [Gal, Mey]).

<u>3. Theorem</u>. Let X^n be a closed normal pseudomanifold with piecewise constant curvature metric.

i) If X^n has nonnegative curvature then X^n is a real homology manifold. Moreover,

$$b^i(X^n) \leq \binom{n}{i}. \tag{4}$$

ii) If X^n has positive curvature then it is a real homology sphere.

Theorem 3 was discovered in 1977 and announced in $[C_2]$, $[C_3]$. Here we will indicate the proof, but some of the more technical analytical details will be omitted.

<u>5. Remark</u>. It is conceivable that an even stronger version of Theorem 3 could be proved by other means, perhaps even by a direct geometric argument.

<u>6. Remark</u>. M. Gromov has suggested that the method of [Ham] might eventually be brought to bear on our situation.

In proving Theorem 3, essentially, one attempts to repeat the argument of the Bochner Vanishing Theorem in the smooth case. If the curvature at the singularities is positive this goes through.

To fix ideas, first consider the case in which X^n is actually a piecewise flat real homology <u>manifold</u> of nonnegative curvature. Let $\bar{\mathfrak{S}}^2 = \bigcup_{j\geq 2} \mathfrak{S}^j$ and let $\mathcal{H}^i$ denote the space of L_2-harmonic forms on $X^n \setminus \bar{\mathfrak{S}}^2$ which are closed and coclosed. According to the Hodge Theorem proved in $[C_3]$, we have

(7) $$\dim \mathcal{H}^i = b^i(X^n).$$

Let $h \in \mathcal{H}^i$ and let $\{e_i\}$ be a local orthonormal frame field near $x \in X^n \setminus \bar{\mathcal{S}}^2$ satisfying $\nabla e_i = 0$ at x. The standard local computation at x gives

(8)
$$\begin{aligned} 0 &= \langle (d\delta+\delta d)h, h\rangle, \\ &= \langle -\Sigma \nabla_{e_i} \nabla_{e_i} h, h\rangle, \\ &= -\tfrac{1}{2} \operatorname{div}(\operatorname{grad} \|h\|^2) + \langle \nabla h, \nabla h\rangle, \end{aligned}$$

(where we have used $K = 0$ in going from the first line to the second). Assuming for the moment that the integrals exist, we have

(9) $$0 = -\frac{1}{2} \int_{X^n \setminus \bar{\mathcal{S}}^2} \operatorname{div}(\operatorname{grad} \|h\|^2) + \int_{X^n \setminus \bar{\mathcal{S}}^2} \|\nabla h\|^2.$$

If X^n were actually smooth we could replace the domain of integration in the first integral by X^n and conclude by the divergence theorem that this integral vanishes. Then (11) would imply that $\nabla h \equiv 0$ giving (4). Since X^n is not smooth, we take a suitable tubular neighborhood $T_\varepsilon(\bar{\mathcal{S}}^{n-2})$ (as in [C_3], [C_4]) and by Stokes' Theorem, write

(10)
$$\begin{aligned} &\pm\frac{1}{2} \int_{\partial T_\varepsilon(\bar{\mathcal{S}}^2)} *d(\|h\|^2) \\ &= -\int_{X^n \setminus T_\varepsilon(\bar{\mathcal{S}}^2)} \|\nabla h\|^2. \end{aligned}$$

We claim that the condition that X^n has positive curvature implies that in the limit as $\varepsilon \to 0$, the left hand side of (10) vanishes. This yields (4).

We begin by deriving an analytic condition on the links which implies the above vanishing and then show how positive curvature guarantees that this condition holds. Observe that near $p \in \mathcal{S}^j$, the geometry of X^n is locally a product, $U^{n-j} \times C^\perp_{0,\varepsilon}(\mathcal{S}^j, p)$. Here $U^{n-j} \subset \mathcal{S}^{n-j}$ is flat and $C^\perp_{0,\varepsilon}(\mathcal{S}^j, p) \subset C^\perp(\mathcal{S}^j, p)$ denotes the set of points whose radial polar coordinate, r, satisfies $r < \varepsilon$. One can show that on $U^{n-j} \times C^\perp_{0,\varepsilon}(\mathcal{S}^j, p)$, a closed and coclosed L_2-harmonic i-form, h, can be written as a convergent series of products

(11) $$h = \sum_k h_{1,k} \wedge h_{2,k},$$

where $h_{1,k}$ is a closed and coclosed harmonic form on U^{n-j} and $h_{2,k}$ is a closed and coclosed L_2-harmonic form on $C^{\perp}(S^j,p)$ ($\deg h_{1,k}+\deg h_{2,k}=i$). Since the forms $h_{1,k}$ are smooth and the dimension of the fibre of $\partial T_\varepsilon(S^j)$ is $j-1$, in order for the left hand side of (10) to vanish in the limit, we must have*

(12) $$\|*d(\|h\|^2)\| = o(\varepsilon^{-(j-1)}),$$

or equivalently, for all k,

(13) $$\|*d(\|h_{2,k}\|^2)\| = o(\varepsilon^{-(j-1)}).$$

To see the meaning of (13), we recall the representation of the forms $h_{2,k}$ in polar coordinates (r,y) on $C^{\perp}(S^j,p)$ (see [C_1], [C_4] for details). Put $m = j-1 = \dim L(S^j,p)$. It is a consequence of the method of separation of variables that the closed and coclosed harmonic $(i+1)$-forms on $C^{\perp}(S^j,p)$ can be written as convergent sums of forms with the following description. Let ϕ be a coexact eigen i-form of the Laplacian, $\tilde{\Delta}$, on $L(S^{m+1},p)$, with eigenvalue $\mu > 0$ (see [C_4] for a discussion of anlaysis on $L(S^{m+1},p)$). Put

(14) $$\alpha = \frac{1+2i-m}{2}.$$

(15) $$\nu = \sqrt{\alpha^2+\mu}$$

(16) $$a^+ = \alpha+\nu.$$

Then corresponding to ϕ, we have the closed and coclosed L_2-harmonic $(i+1)$-form on $C^{\perp}(S^{m+1},p)$,

(17) $$r^{a^+} d\phi + a^+ r^{a^+-1} dr\wedge\phi.$$

For the case in which X^n is a manifold, we must also include the constant function $h_{2,0} \equiv 1$ and its dual, $*1$. Although these are not of the above type, they can be ignored since they satisfy

(18) $$*d(\|1\|^2) = *d(\|*1\|^2) \equiv 0.$$

---

* It is no essential loss of generality to assume X^n is oriented.

If X^n is not a real homology manifold, in general, there are analogous exceptional closed and coclosed L_2-harmonic forms corresponding to the L_2-cohomology of $L(\tilde{\mathcal{S}}^{n+1},p)$ $(d\phi=\delta\phi=0,\mu=0)$ whose pointwise norms are not constant. But the inductive argument below shows that for the case in which X^n has nonnegative curvature, the possible existence of these forms can be ruled out before they need be considered (i.e. positive curvature implies X^n is a real homology manifold).

We now examine (13) for the form in (17). The pointwise norms of the forms $\phi(y)$, $d\phi(y)$ in (17) satisfy

(19) $$\|\phi\| = O(r^{-2i}),$$

(20) $$\|d\phi\| = O(r^{-2(i+1)}) \ ;$$

see $[C_2]$. Thus,

(21) $$\|r^{a^+}+a^+r^{a^+-1}dr\wedge\phi\| = O(r^{2a^+-2(i+1)}),$$

(22) $$\|*d(\|r^{a^+}d\phi+a^+r^{a^+-1}dr\wedge\phi\|)\| = O(r^{2a^+-2i-3})$$
$$= O(r^{2(\nu-1)-m})$$

Thus, we must verify that $\nu > 1$. In view of (14) and the condition $\mu > 0$, this is automatic unless

(23) $$\alpha = 0,$$

(24) $$\alpha = 1/2,$$

in which case,

(25) $$i = (m-1)/2,$$

respectively,

(26) $$i = m/2.$$

In these cases, we still have $\nu > 1$, provided,

(27) $$\mu > 1,$$

respectively,

(28) $$\mu > 3/4.$$

Here, the hypothesis that X^n has positive curvature at its singularities will intervene.

Suppose for the moment that $L(\mathfrak{S}^{m+1},p)$ is actually <u>smooth</u> (of curvature $\equiv 1$). Then for i-forms on $L(\mathfrak{S}^{m+1},p)$, the Weitzenbock formula is

$$\tilde{\Delta}\phi = -\nabla^2\phi + i(m-i)\phi. \tag{29}$$

The same integration by parts argument whose validity we are investigating for X^n, shows that $-\nabla^2$ is a positive semidefinite operator and we immediately obtain

$$\mu \geq i(m-i). \tag{30}$$

For the case, (23), this gives

$$\mu \geq \frac{(m-1)^2}{2}, \tag{31}$$

m odd, while for (24), it gives

$$\mu \geq (\tfrac{m}{2})^2. \tag{32}$$

$m \geq 2$, even.

By (30) we get $\mu \geq 1 \geq 3/4$. The remaining cases for (31) are $m = 3$, $i = 1$ and $m = 1$, $i = 0$. In the former case, (29) still yields $\mu > 1$ unless $\nabla\phi \equiv 0$. Since, by de Rham's decomposition theorem, a space of curvature $K \equiv 1$ admits no parallel vector field (and hence, no parallel 1-form) even locally, we obtain, $\mu > 1$ in this case.

Finally, suppose $m = 1$, $i = 0$. Here, $L(\mathfrak{S}^2,p)$ is a circle and the hypothesis of positive curvature at the singularities of X^n says precisely that the length of this circle is $> 2\pi$. Thus, the smallest nonzero eigenvalue, μ, of $\tilde{\Delta} = \frac{-d^2}{dy^2}$, satisfies $\mu > 1$.

It remains to remove the hypotheses that X^n is a rational homology manifold and that $L(\mathfrak{S}^m,p)$ is smooth. For this, we point out the obvious fact that for the natural stratification of $L(\mathfrak{S}^m,p)$, every link is isometric to the link of some stratum of X^n, which contains p in its closure. Similarly, links of links on $L(\mathfrak{S}^m,p)$ and in fact, all such iterated links, are isometric to links of strata of X^n. Thus, such iterated links are actually spaces of positive curvature (in our sense) and have dimension strictly smaller than that of X^n.

An analysis like that just performed for the space X^n (and which we omit) shows that $\tilde{\nu} > 1$ is the condition justifies the integration by parts argument needed to show $\langle -\nabla^2\phi,\phi\rangle \geq 0$ in (29). Here $\tilde{\nu}$ is defined as in (15) but μ in (15) is replaced by an eigenvalue of the Laplacian $-\tilde{\tilde{\nabla}}^2$ on a link of a stratum of $L(\mathcal{S}^{m+1},p)$.

Now an obvious inductive argument shows that for all spaces of positive curvature, Y^ℓ (and in particular all iterated links above) the smallest nonzero eigenvalue of the Laplacian on forms is nonzero (except for the zero eigenvalues in dimensions $0,\ell$) and that the integration by parts argument $(\langle -\nabla^2\phi,\phi\rangle \geq 0)$ is valid. By the Hodge-de Rham Theory, of [C_3], the spaces, Y^ℓ, are real homology spheres. Thus, X^n is a real homology manifold and in the same way, the inequality (4) follows. q.e.d.

<u>33 Remark</u>. For general piecewise constant curvature pseudomanifolds the Laplacian on $C_0^\infty(\Lambda^+)$ need not be essentially self adjoint and one must choose "ideal boundary conditions". Even if the Laplacian is essentially self adjoint the closed and coclosed harmonic forms represent the L_2-cohomology of X^n (or equivalently, the middle intersection cohomology) which in general is different from the simplicial cohomology of X^n (see [C_3], [GM]). The hypothesis that X^n is a normal pseudomanifold rules out these possibilities via the inductive argument.

<u>34 Remark</u>. If we allow 1-dimensional links consisting of several circles, each of length $<2\pi$, then our conclusions apply to the L_2-cohomology of the (non-normal) space X^n, which coincides with the simplicial cohomology of an associated normal space $\hat{X}^n$ called the normalization. As an example, let X^2 be a tetrahedron $\hat{X}^2$ with all vertices identified to a point.

<u>35 Remark</u>. One can start with a smooth manifold M^n of nonnegative curvature and form a piecewise flat approximation as in [CMS]. By Rauch comparison one obtains a piecewise flat space for which all links, L, with their natural induced triangulations, have the following property. There is a combinatorially isomorphic totally geodesic triangulation of the unit sphere such that all corresponding edge lengths are at most equal to the corresponding edge lengths on L. It would be interesting to know if any useful information for the smooth case can be derived from this condition (which coincides with our definition nonnegative curvature only in dimension 2).

References

[C_1] J. Cheeger, Spectral geometry of spaces with cone-like singularities, preprint 1978.

[C_2] J. Cheeger, On the spectral geometry of spaces with cone-like singularities, Proc. Nat. Acad. Sci., Vol. 76, 1979, 2103-2106.

[C_3] J. Cheeger, On the Hodge Theory of Riemannian pseudomanifolds, A.M.S. Proc. Sym. Pure. Math., Vol. XXXVI, 1980, 91-146.

[C_4] J. Cheeger, Spectral Geometry of singular Riemannian spaces, J. Dif. Geo. 18, 1983, 575-657.

[CMS] J. Cheeger, W. Muller and R. Schrader, On the curvature of Piecewise flat spaces, Commun. Math. Phys. 92, 1984, 405-545.

[Gal, Mey] S. Gallot and D. Meyer, Opérateur de courbure et Laplacien des formes differentielles d'une varieté riemannienne, J. Math. Pures et Appl. 54, (1975), 285-304.

[GM] M. Goresky and R. MacPherson, Intersection Homology Theory, Topolog 19, 1980, 135-162.

[Ham] R. Hamilton, Four-manifolds with positive curvature operator (preprint 1985).

L-subgroups in spaces of nonpositive curvature

by

Patrick Eberlein*
Department of Mathematics
University of North Carolina
Chapel Hill, North Carolina 27514
U.S.A.

Table of Contents

*Supported in part by NSF Grant MCS-8219609, the Taniguchi Foundation and the Japanese Society for the Promotion of Science

Introduction

Let M denote a smooth Riemannian manifold with finite Riemannian volume and bounded nonpositive sectional curvature, and let $\tilde{M}$ denote the universal Riemannian cover of M. In recent years it has become evident that many geometric properties of M and $\tilde{M}$ can be expressed entirely in terms of algebraic properties of the fundamental group of M; one calls such geometric properties rigid. This phenomenon is not completely unexpected when one recalls that $\pi_k(M) = 0$ for all $k \geq 2$. One of the first striking rigidity results ([GW],[LY],[Sch 3]) says that if $\pi_1(M)$ has no center and is a direct product $A_1 \times A_2$, then M splits as a Riemannian product $M_1 \times M_2$, where $\pi_1(M_j) = A_j$ for $j = 1,2$. In later years other rigidity results have been obtained, several of them using the result just stated as an essential ingredient of the proof. In particular, there exist fairly simple algebraic conditions on $\pi_1(M)$ that characterize manifolds with ergodic geodesic flow or, in another direction, manifolds M whose universal cover $\tilde{M}$ is a symmetric space of noncompact type and rank at least two. For a survey of many of these results see section 7 of [E5]. Recent rigidity results include those in [A], [AS], [BE], [Sch 1] and [Sch 2].

The manifold M may be regarded as a quotient space $\tilde{M}/\Gamma$, where $\Gamma \subseteq I(\tilde{M})$ is a discrete (lattice) subgroup isomorphic to $\pi_1(M)$ and having no elements of finite order. In this paper we consider the algebraic properties of L-subgroups of $I(\tilde{M})$, a class of discrete subgroups of $I(\tilde{M})$ that includes the lattice subgroups, and we relate these algebraic properties to the geometric properties of $\tilde{M}$. The name L-subgroup is taken from [R]. Our main result is the existence and uniqueness of an irreducible decomposition of an L-subgroup $\Gamma \subseteq I(\tilde{M})$ (or more precisely, of some finite index subgroup of Γ) into a direct product of algebraically irreducible pieces (Theorem 5.3). This irreducible decomposition is an analogue of the de Rham decomposition of $\tilde{M}$, and in fact, the irreducible decomposition of an L-subgroup $\Gamma \subseteq I(\tilde{M})$ is related to the de Rham decomposition of $\tilde{M}$ in a way made precise in section 5. The proof of the main result is rather simple when $\tilde{M}$ has no Euclidean de Rham factor, but annoying technical complications arise when $\tilde{M}$ has a nontrivial Euclidean de Rham factor (see for example the proofs of Propositions 2.4 and 2.6).

The second focal point of the paper concerns the finer algebraic structure of a uniform (cocompact) lattice $\Gamma \subseteq G = I_0(\tilde{M})$, where $\tilde{M}$ is a symmetric space of noncompact type and rank $k \geq 2$. The group G is a connected, semisimple Lie group with trivial center and no compact factors. It is known that

a) Γ contains infinitely many nonconjugate subgroups isomorphic to $\mathbb{Z}^k$ (cf. section 8 of [M]).

b) The rank k of M can be expressed in terms of the algebraic structure of the centralizer $\Gamma_\phi = \{\psi \in \Gamma : \psi\phi = \phi\psi\}$ of a "generic" element ϕ of Γ (cf. Theorem 3.9 of [PR], Theorem 6.2 of [W1] and section 6 of this paper).

Following b) and a general principle that a lattice Γ imitates its Lie group G, we believe that a more refined knowledge of the structure of the centralizers Γ_ϕ, $\phi \in \Gamma$, particularly the centralizers of "singular" or "nongeneric" elements ϕ, should be sufficient to identify the Lie group G. Such an intrinsic characterization of G by the internal algebraic structure of its lattices Γ would be an interesting contrast to known "extrinsic" characterizations of G, for example by the representations of Γ and G [BMS,§16]. In section 7 we sketch an approach to this intrinsic characterization that involves what we call the centralizer complex of a uniform lattice $\Gamma \subseteq G = I_0(\tilde{M})$.

The table of contents gives an adequate description of the organization of this paper.

Finally, I would like to express my great appreciation for the financial support and superb hospitality provided by the Taniguchi Foundation, the Japanese Society for the Promotion of Science and by Professors T. Ochiai, S. Murakami, T. Sakai, T. Sunada and K. Shiohama.

Section 1 Preliminaries

(1.1) Notation

In general we shall use the notation and basic results of [EO]. In this paper $\tilde{M}$ will always denote a complete, simply connected Riemannian manifold of nonpositive sectional curvature. The unit tangent bundle of $\tilde{M}$ is denoted by $S\tilde{M}$, and $\pi : S\tilde{M} \to \tilde{M}$ denotes the projection map. All geodesics of $\tilde{M}$ will be assumed to have unit speed. If v is a unit vector in $S\tilde{M}$, then γ_v denotes the geodesic with initial velocity v. The distance function on $\tilde{M}$ arising from the Riemannian metric is denoted by d. The isometry group of $\tilde{M}$ is denoted by $I(\tilde{M})$, and the connected component of $I(\tilde{M})$ that contains the identity is denoted by $I_0(\tilde{M})$.

(1.2) Asymptotes and Points at Infinity

Two unit speed geodesics γ, σ of $\tilde{M}$ are said to be <u>asymptotes</u> if $d(\gamma t, \sigma t) \leq c$ for some positive constant c and for all $t \geq 0$. An equivalence class of asymptotes is a <u>point at infinity</u> for $\tilde{M}$, and $\tilde{M}(\infty)$ denotes the set of all points at infinity for $\tilde{M}$. Given a geodesic γ of $\tilde{M}$ we let $\gamma(\infty)$ and $\gamma(-\infty)$ denote the points at infinity determined by γ and $\gamma^{-1}: t \to \gamma(-t)$. Given points $p \in \tilde{M}$ and $x \in \tilde{M}(\infty)$ there is a unique unit speed geodesic γ_{px} such that $\gamma_{px}(o) = p$ and γ_{px} belongs to x. The space $\tilde{M}^* = \tilde{M} \cup \tilde{M}(\infty)$ with a natural topology is homeomorphic to the closed unit n-ball, and for each point p in $\tilde{M}$ the map $x \to \gamma_{px}'(o)$ is a homeomorphism of $\tilde{M}(\infty)$ onto the unit (n-1)-sphere in $T_p\tilde{M}$. Isometries of $\tilde{M}$ extend in an obvious way to homeomorphisms of $\tilde{M}(\infty)$.

(1.3) de Rham decomposition (KN)

A complete, simply connected Riemannian manifold X is called <u>reducible</u> if X can be written as the Riemannian product $X_1 \times X_2$ of two Riemannian manifolds of positive dimension. X is called <u>irreducible</u> if it is not reducible. A reducible space X has a Riemannian product decomposition

$$X = X_0 \times X_1 \times \ldots \times X_k$$

where X_0 is a Euclidean space and each X_i is irreducible for $1 \leq i \leq k$. This decomposition of de Rham is unique up to isometric equivalence and ordering of the factors $X_1, \ldots, X_k$.

Let $\tilde{M} = \tilde{M}_1 \times \ldots \times \tilde{M}_r$ be any Riemannian product decomposition of a complete, simply connected manifold $\tilde{M}$ of nonpositive sectional curvature, and let $\Gamma \subseteq I(\tilde{M})$ be any group of isometries. The splitting of $\tilde{M}$ above is said to be <u>Γ-invariant</u> if the foliations of $\tilde{M}$ induced by the factors $\{\tilde{M}_i\}$ are left invariant by Γ. In this case every element ϕ of Γ can be written uniquely as $\phi = \phi_1 \times \ldots \times \phi_r$, where $\phi_i \in I(\tilde{M}_i)$, and one obtains projection homomorphisms $p_i : \Gamma \to I(\tilde{M}_i)$ given by $p_i(\phi) = \phi_i$, for all $1 \leq i \leq r$. If $\tilde{M}$ has no Euclidean de Rham factor, then each factor $\tilde{M}_i$ above is a Riemannian product of some subset of the de Rham factors of $\tilde{M}$. Every subgroup Γ of $I(\tilde{M})$ has a finite index subgroup Γ^* such that the de Rham decomposition of $\tilde{M}$ is Γ^*-invariant. It follows that every Riemannian product splitting of $\tilde{M}$ is Γ^*-invariant if $\tilde{M}$ has no Euclidean de Rham factor.

(1.4) <u>Rank of $\tilde{M}$</u> [BBE]

For each unit vector v in $S\tilde{M}$ we define r(v) to be the dimension of the space of parallel Jacobi vector fields defined on the maximal geodesic $\gamma_v : \mathbb{R} \to \tilde{M}$. We define $r(\tilde{M})$, the rank of $\tilde{M}$, to be the minimum of the integers r(v) over all vectors v in $S\tilde{M}$. It is easy to see that

$r(\tilde{M}) = r(\tilde{M}_1) + r(\tilde{M}_2)$ if $\tilde{M}$ is a Riemannian product $\tilde{M}_1 \times \tilde{M}_2$. A space $\tilde{M}$ is flat if and only if $r(\tilde{M}) = \dim \tilde{M}$, and in general $1 \leq r(\tilde{M}) \leq \dim \tilde{M}$. If $\tilde{M}$ has negative sectional curvature than $r(\tilde{M}) = 1$. A vector v in $S\tilde{M}$ is called regular if $r(v) = r(\tilde{M})$. If $\tilde{M}$ is a Riemannian covering of a smooth manifold of finite volume, then the regular vectors of $S\tilde{M}$ form a dense open subset $\mathcal{R}$ of $S\tilde{M}$ by Theorem 2.6 of [BBE]. The set $\mathcal{R}$ is invariant under the geodesic flow in $S\tilde{M}$.

If $\tilde{M}$ is a symmetric space, then one may define the rank of $\tilde{M}$ equivalently in terms of maximal r-flats. A space $\tilde{M}$ is said to be symmetric if for each point p in $\tilde{M}$ the geodesic symmetry $S_p : \tilde{M} \to \tilde{M}$ given by $S_p(\gamma(t)) = \gamma(-t)$, where t lies in $\mathbb{R}$ and γ is any geodesic with $\gamma(o) = p$, is an isometry of $\tilde{M}$. A symmetric space $\tilde{M}$ is said to be of noncompact type if $\tilde{M}$ has no Euclidean de Rham factor. An r-flat in a symmetric space $\tilde{M}$ is defined to be a submanifold of $\tilde{M}$ equipped with the induced topology and metric that is complete, totally geodesic and isometric to an r-dimensional Euclidean space. In particular, a 1-flat is a maximal geodesic of $\tilde{M}$. One can show that for a symmetric space $\tilde{M}$ the rank of $\tilde{M}$ equals the largest integer r for which $\tilde{M}$ contains an r-flat. Moreover, it is known that if $\tilde{M}$ is a symmetric space with rank $k \geq 1$, then for every vector v in $S\tilde{M}$ the corresponding geodesic γ_v is contained in at least one k-flat and in exactly one k-flat if and only if v lies in $\mathcal{R}$, the set of regular vectors in $S\tilde{M}$ defined above.

(1.5) Discrete subgroups of $I(\tilde{M})$

The full group of isometries of $\tilde{M}$, $I(\tilde{M})$, is a Lie group with respect to the compact-open topology. The following sequential compactness criterion which is Theorem 2.2 of [H , p. 167], is useful.

Proposition 1.5.1. Let $\{\phi_n\} \subseteq I(\tilde{M})$ be a sequence such that for some point $p \in \tilde{M}$ the sequence $\{\phi_n(p)\}$ is bounded in $\tilde{M}$. Then $\{\phi_n\}$ has a subsequence converging to some element ϕ of $I(\tilde{M})$.

This result is true if $\tilde{M}$ is replaced by any complete, Riemannian manifold N.

Definition 1.5.2. A subgroup $\Gamma \subseteq I(\tilde{M})$ is discrete if for any compact subset $C \subseteq \tilde{M}$ there are only finitely many isometries $\gamma \in \Gamma$ such that $\gamma(C) \cap C$ is nonempty.

It follows from the definition that $\Gamma \subseteq I(\tilde{M})$ is discrete if and only if the orbit $\Gamma(p) = \{\phi(p): \phi \in \Gamma\}$ has no accumulations points in $\tilde{M}$ for any point p of $\tilde{M}$. Moreover, a discrete subgroup of $I(\tilde{M})$ is closed in $I(\tilde{M})$. If ϕ is an element of a discrete group $\Gamma \subseteq I(\tilde{M})$ such that ϕ fixes some point p of $\tilde{M}$, then $\phi^n = 1$ for some positive integer n. The set of points of $\tilde{M}$ fixed by any nonidentity element of $I(\tilde{M})$ is a closed, proper, totally geodesic submanifold of $\tilde{M}$.

<u>Definition 1.5.3</u> Let $\Gamma \subseteq I(\tilde{M})$ be a discrete group. Let $d_\Gamma : \tilde{M} \to [o,\infty)$ be the <u>displacement function</u> defined by

$$d_\Gamma(p) = \min\{d(p,\phi p) : \phi \in \Gamma,\ \phi \neq 1\}$$

for any point p of $\tilde{M}$.

From the definition of d_Γ, the remarks above, the fact that Γ is a countable group and the Baire category theorem it is easy to show that

a) d_Γ is positive on a dense open subset of $\tilde{M}$ for any discrete group $\Gamma \subseteq I(\tilde{M})$.

b) $d_\Gamma \circ \phi = d_\Gamma$ for any element $\phi \in I(\tilde{M})$ such that $\phi\Gamma\phi^{-1} = \Gamma$.

The displacement function d_Γ was first used systematically by Gromov in [G]. The next result relates the displacement function of a discrete group Γ to that of a finite index subgroup Γ^*.

<u>Proposition 1.5.4.</u> Let $\Gamma \subseteq I(\tilde{M})$ be a discrete group, and let Γ^* be a subgroup of Γ with finite index n_o. Given positive numbers a and ε there exists a positive number $\delta = \delta(a,\varepsilon,n_o)$ such that if $d_{\Gamma^*}(p) \geq a$ for some point $p \in \tilde{M}$, then $d_\Gamma(q) \geq \delta$ for some point $q \in B_\varepsilon(p) = \{p^* \in \tilde{M} : d(p,p^*) < \varepsilon\}$.

<u>Proof</u> See the appendix. □

We remark that $d_{\Gamma^*} \geq d_\Gamma$ in $\tilde{M}$ by definition of the displacement functions d_Γ, d_{Γ^*}. In the statement of the result above we cannot avoid the complication of replacing the original point p by a nearby point q when comparing d_Γ and d_{Γ^*}. It may happen that Γ contains an element of finite order that fixes p but has no power in Γ^* except the identity. In such a case we could have $d_{\Gamma^*}(p) > o$ and $d_\Gamma(p) = o$.

<u>Definition 1.5.5</u> (Dirichlet fundamental domain). Let $\Gamma \subseteq I(\tilde{M})$ be a discrete group, and let $p \in \tilde{M}$ be a point such that $d_\Gamma(p) > o$. Define

$$R_p = \{q \in \tilde{M} : d(p,q) < d(p,\phi q) \quad \text{for all } \phi \in \Gamma,\ \phi \neq 1\}$$

We call R_p the <u>Dirichlet fundamental domain</u> for Γ with center p.

It is routine to show that

a) R_p is an open subset of $\tilde{M}$

b) $\bigcup_{\phi \in \Gamma} \phi(\overline{R_p}) = \tilde{M}$ for any $p \in \tilde{M}$ with $d_\Gamma(p) > o$

c) $\phi(R_p) \cap \psi(R_p)$ is empty if $\phi, \psi \in \Gamma$, $\phi \neq \psi$

<u>Definition 1.5.6</u> A discrete subgroup $\Gamma \subseteq I(\tilde{M})$ is called a <u>lattice</u> if some Dirichlet fundamental domain R_p has finite Riemannian volume in $\tilde{M}$. A lattice Γ is called <u>uniform</u> (<u>nonuniform</u>) if R_p is compact (noncompact) for some $p \in \tilde{M}$ with $d_\Gamma(p) > o$.

One can show that if any of the conditions above hold for one Dirichlet fundamental domain R_p then they hold for all Dirichlet fundamental domains R_p.

(1.6) Duality condition [CE], [Bl]

Two points x,y in $\tilde{M}(\infty)$ are said to be __dual__ relative to a group $\Gamma \subseteq I(\tilde{M})$ if there exists a sequence $\{\phi_n\} \subseteq \Gamma$ such that $\phi_n(p) \to x$ and $\phi_n^{-1}(p) \to y$ as $n \to \infty$ for any point p of $\tilde{M}$. Given a point $x \in \tilde{M}(\infty)$ the set of points y in $\tilde{M}(\infty)$ that are dual to x relative to Γ is closed in $\tilde{M}(\infty)$ and invariant under Γ.

__Definition 1.6.1__ A group $\Gamma \subseteq I(\tilde{M})$ (not necessarily discrete) is said to satisfy the __duality condition__ if for every geodesic γ of $\tilde{M}$ the points $\gamma(\infty)$ and $\gamma(-\infty)$ are dual relative to Γ.

The duality condition may be restated in the following useful form (cf. [Bl, p. 137]). For any group $\Gamma \subseteq I(\tilde{M})$ one defines a nonwandering set $\Omega(\Gamma) \subseteq S\tilde{M}$ as follows: a vector $v \in S\tilde{M}$ lies in $\Omega(\Gamma)$ if and only if for every neighborhood $0 \subseteq S\tilde{M}$ of v and every positive number A there exists $t \geq A$ and $\phi \in \Gamma$ such that $[(d\phi \circ g^t)(0)] \cap 0$ is nonempty, where $\{g^s\}$ denotes the geodesic flow in $S\tilde{M}$. One can show that Γ satisfies the duality condition if and only if $\Omega(\Gamma) = S\tilde{M}$.

If $\Gamma \subseteq I(\tilde{M})$ is a lattice then one can show that Γ satisfies the duality condition. The following result is useful.

__Proposition 1.6.2__ Let $\Gamma \subseteq I(\tilde{M})$ satisfy the duality condition. Then the following properties hold:

1) If $\Gamma^* \subseteq I(\tilde{M})$ is a subgroup that contains Γ, then Γ^* also satisfies the duality condition.

2) If Γ^* is a subgroup of Γ with finite index in Γ, then Γ^* also satisfies the duality condition.

3) If $\tilde{M} = \tilde{M}_1 \times \tilde{M}_2$ is a Γ-invariant splitting, and if $p_i : \Gamma \to I(\tilde{M}_i)$ denotes the corresponding projection homomorphism for $i = 1,2$, then $\Gamma_i = p_i(\Gamma)$ satisfies the duality condition for $i = 1,2$.

The proof of 1) is immediate from the definition of the duality condition, and the proof of 3) is routine. Property 2) is proved in Appendix I of [E3].

Note that the projection $p_i(\Gamma)$ need not be a discrete subgroup of $I(\tilde{M}_i)$ in 3) above, even if Γ is a discrete subgroup of $I(\tilde{M})$.

__Definition 1.6.3__ Let $\Gamma \subseteq I(\tilde{M})$ be an arbitrary group. A subset $A \subseteq \tilde{M}$ is said to be __compact modulo Γ__ if $A \subseteq \Gamma \cdot C = \bigcup_{\phi \in \Gamma} \phi(C)$ for some compact subset C of $\tilde{M}$.

The next result will be useful later, but the proof, because of its length, is omitted here and may be found in the appendix.

__Proposition 1.6.4__ Let $\tilde{M}$ be a Riemannian product $\tilde{M}_1 \times \tilde{M}_2$, where $\tilde{M}_1$ is a Euclidean space. Let $\Gamma \subseteq I(\tilde{M})$ be a subgroup that preserves this splitting, and let $p_i : \Gamma \to I(\tilde{M}_i)$ be the corresponding projection homomorphisms for $i = 1,2$. Suppose that Γ satisfies the following properties:

1) $p_2(\Gamma)$ satisfies the duality condition in $\tilde{M}_2$.
2) $p_1(\Gamma)$ consists of translations of $\tilde{M}_1$.
3) $\tilde{M}_1$ is compact modulo N = kernel (p_2)

Then under these conditions Γ satisfies the duality condition in $\tilde{M}$.

(1.7) Clifford translations

Definition 1.7.1 An isometry $\phi \in I(\tilde{M})$ is called a Clifford translation if the displacement function $d_\phi : p \to d(p,\phi p)$ is a constant function on $\tilde{M}$. If $\Gamma \subseteq I(\tilde{M})$ is any group, then $C(\Gamma)$ will denote the set of Clifford translations in Γ.

The next result, due to J. Wolf [W2], shows that for any group $\Gamma \subseteq I(\tilde{M})$ the set $C(\Gamma)$ is a normal abelian subgroup of Γ, called the Clifford subgroup of Γ.

Proposition 1.7.2 Let $\tilde{M}$ be expressed as the Riemannian product $\tilde{M}_0 \times \tilde{M}_1$, where $\tilde{M}_0$ is the Euclidean de Rham factor of $\tilde{M}$ and $\tilde{M}_1$ is the Riemannian product of all nonEuclidean de Rham factors of $\tilde{M}$. Then the following are equivalent for an arbitrary element $\phi \in I(\tilde{M})$:

1) ϕ is a Clifford translation.
2) $\phi = T \times \{1\} \in I(\tilde{M}_0) \times \{1\}$, where T is a translation of $\tilde{M}_0$.

The next result, which is Theorem 2.4 of [CE], gives a useful existence criterion for Clifford translations.

Proposition 1.7.3 Let $A \subseteq I(\tilde{M})$ be an abelian subgroup whose normalizer in $I(\tilde{M})$ satisfies the duality condition. Then A consists of Clifford translations.

From this result and remarks above we obtain immediately the following characterization of $C(\Gamma)$ for groups $\Gamma \subseteq I(\tilde{M})$ that satisfy the duality condition. In particular, this result or the one above implies that the center of such a group Γ lies in $C(\Gamma)$.

Proposition 1.7.4 Let $\Gamma \subseteq I(\tilde{M})$ satisfy the duality condition. Then $C(\Gamma)$ is the unique maximal normal abelian subgroup of Γ.

The next result is a useful observation proved in Lemma 3.3 of [Y].

Proposition 1.7.5 Let $\Gamma \subseteq I(\tilde{M})$ be any discrete group, and let $\Gamma^* = \{\phi \in \Gamma : \phi\psi = \psi\phi \text{ for all } \psi \in C(\Gamma)\}$. Then Γ^* has finite index in Γ.

(1.8) Splitting theorems.

If $\Gamma \subseteq I(\tilde{M})$ satisfies the duality condition, then an algebraic splitting of Γ corresponds very closely to a geometric splitting of $\tilde{M}$.

Theorem 1.8.1 Let $\Gamma \subseteq I(\tilde{M})$ satisfy the duality condition and suppose that Γ is a direct product $\Gamma_1 \times \Gamma_2$ of subgroups. Then $\tilde{M}$ splits Γ-invariantly as a Riemannian product $\tilde{M}_0 \times \tilde{M}_1 \times \tilde{M}_2$, where $\tilde{M}_0$ is a Euclidean space of dimension $s \geq 0$. Every element $\alpha \in \Gamma_1$ may be written as

$\alpha = \alpha_0 \times \alpha_1 \times \{1\} \in I(\tilde{M}_0) \times I(\tilde{M}_1) \times \{1\}$ and every element $\beta \in \Gamma_2$ may be written as $\beta = \beta_0 \times \{1\} \times \beta_2 \in I(\tilde{M}_0) \times \{1\} \times I(\tilde{M}_2)$, where α_0 and β_0 are translations on $\tilde{M}_0$ At least two of the three factors $\tilde{M}_0, \tilde{M}_1$ and $\tilde{M}_2$ must be nontrivial.

Corollary 1.8.2 Let $\Gamma \subseteq I(\tilde{M})$ be a lattice whose center is trivial. If Γ is a direct product $\Gamma_1 \times \Gamma_2$, then $\tilde{M}$ splits Γ-invariantly as a Riemannian product $\tilde{M}_1 \times \tilde{M}_2$ such that $\Gamma_1 \subseteq I(\tilde{M}_1) \times \{1\}$ and $\Gamma_2 \subseteq \{1\} \times I(\tilde{M}_2)$ are lattices.

Theorem 1.8.3 Let $\tilde{M}$ be a Riemannian product $\tilde{M}_1 \times \tilde{M}_2$ such that $\tilde{M}$ has no Euclidean de Rham factor and $I(\tilde{M}_2)$ is discrete. Let $\Gamma \subseteq I(\tilde{M})$ be a lattice. Then Γ has a finite index subgroup Γ^* that is a direct product $\Gamma_1^* \times \Gamma_2^*$ where $\Gamma_1^* \subseteq I(\tilde{M}_1) \times \{1\}$ and $\Gamma_2^* \subseteq \{1\} \times I(\tilde{M}_2)$.

The first two results above are due to V. Schroeder [Sch 3]. The second of these results was first proved for uniform lattices by Lawson and Yau [LY] when $\tilde{M}$ is real analytic and by Gromoll and Wolf [GW] when $\tilde{M}$ is C^∞. The third of these results follows from Theorem 4.1 and Proposition 2.2 of [E4]. The second and third results are generalized in the next section to L-subgroups, and are Propositions 2.5 and 2.4 respectively.

If $I(\tilde{M})$ satisfies the duality condition, then $\tilde{M}$ admits the following decomposition, which is Proposition 4.1 of [E3].

Theorem 1.8.4 If $I(\tilde{M})$ satisfies the duality condition, then there exist spaces $\tilde{M}_0$, $\tilde{M}_1$, $\tilde{M}_2$, two of which may have dimension zero, such that

1) $\tilde{M}$ is isometric to the Riemannian product $\tilde{M}_0 \times \tilde{M}_1 \times \tilde{M}_2$.
2) $\tilde{M}_0$ is a Euclidean space.
3) $\tilde{M}_1$ is a symmetric space of noncompact type.
4) $I(\tilde{M}_2)$ is discrete but satisfies the duality condition.

If $\tilde{M}$ has bounded nonpositive sectional curvature and if $I(\tilde{M})$ admits a lattice Γ, then the results of section 1 of [B2] and [BuS2] strengthen this result as follows.

Theorem 1.8.5 Let $\tilde{M}$ have sectional curvature satisfying $-a^2 \leq K \leq 0$ for some positive constant a, and let $\Gamma \subseteq I(\tilde{M})$ be a lattice. Then $\tilde{M}$ is a Riemannian product

$$\tilde{M} = \tilde{M}_0 \times \tilde{M}_s \times \tilde{M}_1 \times \ldots \times \tilde{M}_k$$

where $\tilde{M}_0$ is a Euclidean space, $\tilde{M}_s$ is a symmetric space of noncompact type and for $1 \leq i \leq k$, $\tilde{M}_i$ is a manifold of rank one such that $I(\tilde{M}_i)$ is discrete and satisfies the duality condition.

Section 2 L-subgroups and their basic properties

In this section we discuss certain discrete subgroups $\Gamma \subseteq I(\tilde{M})$, called L-subgroups, that have properties generalizing those of lattices. L-subgroups may in fact always be lattices under appropriate restrictions on $\tilde{M}$.

Definition 2.1 A discrete subgroup $\Gamma \subseteq I(\tilde{M})$ is called a lattice if the volume in $\tilde{M}$ of some Dirichlet fundamental domain of Γ is finite (see section 1). A lattice $\Gamma \subseteq I(\tilde{M})$ is called uniform (nonuniform) if the quotient space $\tilde{M}/\Gamma$ with the quotient topology is compact (noncompact).

Definition 2.2 A subgroup $\Gamma \subseteq I(\tilde{M})$ is called an L-subgroup if Γ satisfies the following three conditions

1) Γ is discrete
2) Γ satisfies the duality condition
3) $\tilde{M}^a_\Gamma = \{p \in \tilde{M} : d_\Gamma(p) \geq a\}$ is compact modulo Γ for every positive number a.

Remarks: 1) It is easy to see that any lattice $\Gamma \subseteq I(\tilde{M})$ is an L-subgroup. 2) The terminology L-subgroup is taken from [R, p. 28] where it was defined and studied for discrete subgroups Γ of a semi-simple Lie group G with finite center and no compact factors. See also chapter 11 of [R]. Using Proposition 1.5.1 in section 1 it is not difficult to see that in this context our definition of L-subgroup is equivalent to the definition of [R], regarding G as a group of isometries of the associated symmetric space G/K, K a maximal compact subgroup of G. In this context it is also possible, perhaps likely, that if G has real rank $k \geq 2$ then any L-subgroup of G must be a lattice in G.

In the remainder of this section we derive some elementary properties of L-subgroups. Many of these properties are known in the special case that Γ is a lattice.

Proposition 2.3 Let $\Gamma \subseteq I(\tilde{M})$ be an L-subgroup of $I(\tilde{M})$. Then the following properties hold:

1) If $\Gamma^* \subseteq I(\tilde{M})$ is a discrete subgroup that contains Γ, then Γ^* is also an L-subgroup of $I(\tilde{M})$ and Γ has finite index in Γ^*.

2) If Γ^* is a subgroup of Γ with finite index in Γ, then Γ^* is also an L-subgroup of $I(\tilde{M})$.

3) If $\tilde{M} = \tilde{M}_1 \times \tilde{M}_2$ is a Γ-invariant splitting and if $p_i : \Gamma \to I(\tilde{M}_i)$ denotes the projection homomorphism for $i = 1,2$, then $\Gamma_i = p_i(\Gamma)$ is an L-subgroup of $I(\tilde{M}_i)$ for $i = 1,2$ whenever Γ_i is discrete.

Proof 1) Under the conditions on Γ^* it follows routinely from the definitions and properties of subgroups of $I(\tilde{M})$ that satisfy the duality condition that Γ^* is an L-subgroup. See Proposition 1.6.2. We show that Γ has finite index in Γ^*. Choose a point $p \in \tilde{M}$ such that $d_{\Gamma^*}(p) = a > 0$. It follows that $d_{\Gamma^*}(\phi p) = d_{\Gamma^*}(p) = a > 0$ for all $\phi \in \Gamma_*$. Now $d_\Gamma(q) \geq d_{\Gamma^*}(q)$ for all $q \in \tilde{M}$ since Γ is a subgroup of Γ^* and hence $\Gamma^*(p) \subseteq \tilde{M}^a_\Gamma =$

$\{q \in \tilde{M} : d_\Gamma(q) \geq a\}$. Let $S = \{\phi_\alpha^*\} \subseteq \Gamma^*$ be any set such that the left cosets $\{\phi_\alpha^*\Gamma\}$ are all distinct. Since $\Gamma^*(p) \subseteq \tilde{M}_\Gamma^a$ and Γ is an L-subgroup we may choose a positive number R and elements $\{\phi_\alpha\} \subseteq \Gamma$ such that $d(\phi_\alpha(p), (\phi_\alpha^*)^{-1}(p)) \leq R$ for every α. If $\psi_\alpha^* = \phi_\alpha^*\phi_\alpha$ then $d(p, \psi_\alpha^* p) \leq R$ for every α, and hence there are only finitely many distinct elements $\{\psi_\alpha^*\}$ by the discreteness of Γ^*. However, $\psi_\alpha^*\Gamma = \phi_\alpha^*\Gamma$ for every α, and it follows that Γ has finite index in Γ^*.

We prove 2). Let Γ^* be a finite index subgroup of an L-subgroup Γ. Clearly Γ^* is discrete and it follows from Proposition 1.6.2 that Γ^* satisfies the duality condition. Let $a > 0$ be given. By Proposition 1.5.4 there exists $\delta > 0$ such that if $d_{\Gamma^*}(p) \geq a$ for some point $p \in \tilde{M}$, then $d_\Gamma(q) \geq \delta$ for some point $q \in B_1(p) = \{p' \in \tilde{M} : d(p,p') < 1\}$. The set $\tilde{M}_\Gamma^\delta = \{q \in \tilde{M} : d_\Gamma(q) \geq \delta\}$ is compact modulo Γ since Γ is an L-subgroup, and it follows that $\tilde{M}_{\Gamma^*}^a$ is compact modulo Γ^* since Γ^* has finite index in Γ and $\tilde{M}_{\Gamma^*}^a$ is contained in the tubular neighborhood of radius 1 around $\tilde{M}_\Gamma^\delta$.

We prove 3). By Proposition 1.6.2 each group $p_i(\Gamma)$ satisfies the duality condition in $\tilde{M}_i$ for $i = 1,2$. Suppose now that one of these groups, say $\Gamma_1 = p_1(\Gamma)$, is discrete. To complete the proof we shall need the following.

<u>Lemma 2.3</u> There exists a point $p_2 \in \tilde{M}_2$ with the following property: for every $a > 0$ there exists $\delta > 0$ such that if $d_{\Gamma_1}(q_1) \geq a$ for some point $q_1 \in \tilde{M}_1$, then $d_\Gamma(q) \geq \delta$ where $q = (q_1, p_2)$.

Deferring the proof of the lemma for the moment we complete the proof. Choose $p_2 \in \tilde{M}_2$ as in the lemma. Let $a > 0$ be given and choose $\delta > 0$ as in the lemma. Since Γ is an L-subgroup there exists a compact subset $B \subseteq \tilde{M}$ such that $\tilde{M}_\Gamma^\delta = \{p \in \tilde{M} : d_\Gamma(p) \geq \delta\} \subseteq \bigcup_{\phi \in \Gamma} \phi(B)$. Let $p_1 \in \tilde{M}_1$ be any point such that $d_{\Gamma_1}(p_1) \geq a$, and let $p = (p_1, p_2)$. Since $d_\Gamma(p) \geq \delta$ there exists $\phi \in \Gamma$ and $b \in B$ such that $\phi(b) = p$. Hence $\phi_1(b_1) = p_1$, where $\phi = \phi_1 \times \phi_2 \in \Gamma_1 \times \Gamma_2$ and $b = (b_1, b_2) \in \tilde{M}_1 \times \tilde{M}_2$. We have shown that $\{p_1 \in \tilde{M}_1 : d_{\Gamma_1}(p_1) \geq a\} \subseteq \bigcup_{\phi_1 \in \Gamma_1} \phi_1(B_1)$, where $B_1 = \pi_1(B)$ and $\pi_1 : \tilde{M} \to \tilde{M}_1$ is the projection. Therefore Γ_1 is an L-subgroup.

We now prove Lemma 2.3. Let $p = (p_1, p_2) \in \tilde{M}$ be a point that is not fixed by any nonidentity element of Γ. We show that $p_2 = \pi_2(p) \in \tilde{M}_2$ satisfies the assertions of the lemma. Suppose that the lemma is false for some $a > 0$. Then we can find a sequence $\{p_n\} \subseteq \tilde{M}_1$ such that $d_{\Gamma_1}(p_n) \geq a$ for every n but $d_\Gamma(q_n) \to o$ as $n \to +\infty$, where $q_n = (p_n, p_2)$. Choose $\{\phi_n\} \subseteq \Gamma$, $\phi_n \neq 1$, such that $d(q_n, \phi_n q_n) = d_\Gamma(q_n) \to o$ as $n \to +\infty$. Write $\phi_n = \alpha_n \times \beta_n$, where $\alpha_n = p_1(\phi_n) \in \Gamma_1$ and $\beta_n = p_2(\phi_n) \in \Gamma_2$. If $\alpha_{n_k} \neq 1$ for some subsequence $\{n_k\}$, then $d(q_{n_k}, \phi_{n_k}(q_{n_k})) \geq d(p_{n_k}, \alpha_{n_k}(p_{n_k})) \geq d_{\Gamma_1}(p_{n_k}) \geq a$ for every k, contradicting the choice of $\{q_n\}$. Hence for any sufficiently large n $\alpha_n = 1$ and $d(p,\phi_n p) = d(p_2,\beta_n(p_2)) = d(q_n,\phi_n(q_n)) \to 0$ as $n \to +\infty$. By the discreteness of Γ only finitely many of the elements $\{\phi_n\}$ are distinct and hence $\phi_n \equiv \phi \neq 1$, passing to a subsequence. Since $d(p,\phi_n p) \to 0$ as $n \to +\infty$ it follows that ϕ fixes the point p, contradicting the choice of p. This completes the proof of the lemma. $\square$

Proposition 2.4 Let $\Gamma \subseteq I(\tilde{M})$ be an L-subgroup, and let $\tilde{M} = \tilde{M}_1 \times \tilde{M}_2$ be a Γ-invariant splitting with corresponding projection homomorphisms $p_1 : \Gamma \to I(\tilde{M}_1)$ and $p_2 : \Gamma \to I(\tilde{M}_2)$. Let $\Gamma_2 = p_2(\Gamma)$ be a discrete subgroup of $I(\tilde{M}_2)$. Then

1) N = kernel (p_2) is an L-subgroup of $I(\tilde{M}_1)$. If $\tilde{M}_1$ is a Euclidean space, then N is a uniform lattice in $\tilde{M}_1$ and $C(\Gamma) \cap N = C(N)$, the Clifford subgroup of N, is a normal abelian subgroup of N with finite index in N.

2) One of the following occurs:

   a) $\Gamma_1 = p_1(\Gamma)$ is discrete. If $\Gamma_1^* = N$ = kernel $(p_2) \subseteq I(\tilde{M}_1) \times \{1\}$ and Γ_2^* = kernel $(p_1) \subseteq \{1\} \times I(\tilde{M}_2)$, then Γ_i^* has finite index in Γ_i for $i = 1,2$, and $\Gamma^* = \Gamma_1^* \times \Gamma_2^*$ is a subgroup of Γ of finite index.

   b) Γ_1 is not discrete, $\tilde{M}_1$ has a nontrivial Euclidean de Rham factor and N contains Clifford translations of $I(\tilde{M})$.

Remark: This result is a sharpened version of Theorem 4.1 of [E4]. Because of its length we omit the proof here and place it in the appendix. If Γ is a uniform lattice, then the proof of assertion 1) becomes much shorter; in fact, Lemma 4.1a of [E4] shows that N is a uniform lattice in $\tilde{M}_1$ in this case.

The next two results show that if Γ is an L-subgroup that is a direct product $A \times B$ of subgroups, then each of the factors A,B is either itself an L-subgroup in a suitable nonpositively curved space of lower dimension or isomorphic to such an L-subgroup.

Proposition 2.5 Let $\Gamma \subseteq I(\tilde{M})$ be an L-subgroup with trivial center that is a direct product $A \times B$ of subgroups A,B. Then there exists a Γ-invariant splitting

$$\tilde{M} = \tilde{M}_A \times \tilde{M}_B$$

such that $A \subseteq I(\tilde{M}_A) \times \{1\}$ and $B \subseteq \{1\} \times I(\tilde{M}_B)$ are L-subgroups in $I(\tilde{M}_A)$, $I(\tilde{M}_B)$ respectively.

Proposition 2.6 Let $\Gamma \subseteq I(\tilde{M})$ be an L-subgroup with nontrivial center such that Γ is a direct product $A \times B$ of subgroups A,B. Then there exist complete, simply connected manifolds $\tilde{M}_A$, $\tilde{M}_B$ of nonpositive sectional curvature such that A,B are isomorphic to L-subgroups A^*, B^* in $I(\tilde{M}_A)$, $I(\tilde{M}_B)$ respectively.

The proof of Proposition 2.6 is lengthy and may be found in the appendix. We prove Proposition 2.5. Since Γ has trivial center by hypothesis it follows from Proposition 3.1 below that the centralizer $Z(\Gamma)$ of Γ in $I(\tilde{M})$ is also trivial. In particular $\tilde{M}$ admits no Γ-invariant splitting $\tilde{M}_1 \times \tilde{M}_2$ such that $\tilde{M}_1$ is a Euclidean space and $p_1(\Gamma)$ consists of translations in $\tilde{M}_1$, where $p_1 : \Gamma \to I(\tilde{M}_1)$ is the projection homomorphism; under these conditions the translations in $\tilde{M}_1$ would lie in $Z(\Gamma)$. The result now follows from Theorem 1.8.1. $\square$

We conclude this section by showing that a reducible space $\tilde{M}$ with no Euclidean factor admits an irreducible L-subgroup $\Gamma \subseteq I(\tilde{M})$ only if $\tilde{M}$ is a symmetric space of noncompact type. (A group Γ is <u>irreducible</u> if it contains no direct product subgroup of finite index). Reducible symmetric spaces $\tilde{M}$ of noncompact type do in fact admit irreducible lattices. Such examples have been known to exist for many years in the case that $\tilde{M}$ is a Riemannian product of hyperbolic planes - see for example [Sh] for a discussion. In the case of an arbitrary reducible symmetric space $\tilde{M}$ the existence of an irreducible lattice $\Gamma \subseteq I(\tilde{M})$ follows from [Bo]. Our result is a slight strengthening and generalization of Proposition 4.7 of [E3].

<u>Proposition 2.7</u> Let $\tilde{M}$ be a reducible space with no Euclidean de Rham factor $\tilde{M}_0$. Let $\Gamma \subseteq I(\tilde{M})$ be an L-subgroup that preserves the de Rham splitting of $\tilde{M}$. Then the following conditions are equivalent.

1) Γ is irreducible.

2) If $\tilde{M} = \tilde{M}_1{}^* \times \tilde{M}_2{}^*$ is any Riemannian product decomposition of $\tilde{M}$ and if $q_j : \Gamma \to I(\tilde{M}_j{}^*)$ are the projection homomorphisms for $j = 1,2$, then $q_j(\Gamma)$ is a nondiscrete subgroup of $I(\tilde{M}_j{}^*)$ for $j = 1,2$. (The splitting is Γ-invariant by (1.3)).

3) $\tilde{M}$ is a symmetric space of noncompact type. If $\tilde{M} = \tilde{M}_1{}^* \times \tilde{M}_2{}^*$ is any Riemannian product decomposition of $\tilde{M}$ and if $q_j : \Gamma \to I(\tilde{M}_j{}^*)$ are the projection homomorphisms for $j = 1,2$, then $\overline{q_j(\Gamma)} \supseteq I_0(M_j{}^*)$ for $j = 1,2$.

4) If $\tilde{M} = \tilde{M}_1{}^* \times \tilde{M}_2{}^*$ is any Riemannian product decomposition of $\tilde{M}$ and if $q_j : \Gamma \to I(M_j{}^*)$ are the projection homomorphisms for $j = 1,2$, then the kernel of $q_j = \{1\}$ for $j = 1,2$.

<u>Proof.</u> We shall prove the equivalence of the 4 conditions in the order 1) $\Rightarrow$ 2), 2) $\Rightarrow$ 1), 2) $\Rightarrow$ 3), 3) $\Rightarrow$ 4) and 4) $\Rightarrow$ 1).

The assertion 1) $\Rightarrow$ 2) follows from assertion 2) of Proposition 2.4 since $\tilde{M}$ has no Euclidean de Rham factor. We prove 2) $\Rightarrow$ 1). Assume that 2) holds and that Γ is reducible. Let Γ^* be a finite index subgroup of Γ that is a nontrivial direct product $A \times B$. By Proposition 1.6.2 Γ^*

satisfies the duality condition in $\tilde{M}$ and since $\tilde{M}$ has no Euclidean factor it follows from Theorem 1.8.1 that there exists a splitting $\tilde{M} = \tilde{M}_1{}^* \times \tilde{M}_2{}^*$ such that $A \subseteq I(\tilde{M}_1{}^*) \times \{1\}$ and $B \subseteq \{1\} \times I(\tilde{M}_2{}^*)$. If $q_j : \Gamma \to I(\tilde{M}_j{}^*)$ are the projection homomorphisms for $j = 1,2$, then $A = q_1(\Gamma^*)$ and $B = q_2(\Gamma^*)$ are finite index subgroups of $q_1(\Gamma)$ and $q_2(\Gamma)$ respectively. The groups $q_1(\Gamma)$ and $q_2(\Gamma)$ must therefore be discrete since A and B are discrete, but this contradicts 2).

We prove 2) $\Rightarrow$ 3). If $\tilde{M} = \tilde{M}_1 \times \ldots \times \tilde{M}_r$ is the de Rham decomposition of $\tilde{M}$ with corresponding projection homomorphisms $p_j : \Gamma \to I(\tilde{M}_j)$, $1 \leq j \leq r$, then $p_j(\Gamma)$ is a nondiscrete subgroup of $I(\tilde{M}_j)$ for every j by hypothesis. If G_j denotes the connected component of $\overline{p_j(\Gamma)}$ that contains the identity, then G_j is a closed, connected Lie subgroup of $I_0(\tilde{M}_j)$ of positive dimension and G_j is normalized in $I_0(\tilde{M}_j)$ by $p_j(\Gamma)$. Since $p_j(\Gamma)$ satisfies the duality condition in $\tilde{M}_j$ and since $\tilde{M}_j$ is irreducible it follows from the main theorem in section 3 of [E3] that $\tilde{M}_j$ is a symmetric space of noncompact type for $1 \leq j \leq r$. Hence $\tilde{M}$ is a symmetric space of noncompact type.

Now let $\tilde{M} = \tilde{M}_1{}^* \times \tilde{M}_2{}^*$ be any Riemannian product decomposition of $\tilde{M}$ with projection homomorphisms $q_j : \Gamma \to I(\tilde{M}_j{}^*)$ for $j = 1,2$. By 2) the groups $q_j(\Gamma)$ are nondiscrete subgroups of $I(\tilde{M}_j{}^*)$ for $j = 1,2$, and hence if G_j denotes the connected component of $\overline{q_j(\Gamma)}$ that contains the identity, then G_j is a closed, connected Lie subgroup of $I(\tilde{M}_j{}^*)$ of positive dimension for $j = 1,2$. It suffices to prove that $G_1 = I_0(\tilde{M}_1{}^*)$. If this were not the case then it would follow that Γ is reducible by the proof in Appendix II of [E3] of the assertion 1) $\Rightarrow$ 3) of Proposition 4.7 of [E3]. This would contradict the fact that 2) $\Rightarrow$ 1), which completes the proof of 2) $\Rightarrow$ 3).

The proof of 3) $\Rightarrow$ 4) is the same as the proof of 3) $\Rightarrow$ 2) of Proposition 4.7 of [E3], Appendix II. The proof of 4) $\Rightarrow$ 1) follows by an argument very similar to that used above in the proof of 2) $\Rightarrow$ 1). □

Section 3 Structure of L-subgroups with nontrivial center

Proposition 3.1 Let $\Gamma \subseteq I(\tilde{M})$ be an L-subgroup, and let $Z(\Gamma) = \{\phi \in I(\tilde{M}) : \phi\psi = \psi\phi \text{ for all } \psi \in \Gamma\}$ denote the centralizer in $I(\tilde{M})$ of Γ. If $Z(\Gamma) \neq \{1\}$, then there exists a Γ-invariant splitting

$$\tilde{M} = \tilde{M}_1 \times \tilde{M}_2$$

with corresponding projection homomorphisms $p_i : \Gamma \to I(\tilde{M}_i)$ for $i = 1,2$ such that

1) $\tilde{M}_1$ is a Euclidean space of positive dimension.
2) $p_1(\Gamma)$ consists of translations in $\tilde{M}_1$ and $Z(\Gamma)$ is the subgroup

of $I(\tilde{M}_1) \times \{1\}$ consisting of all translations in $\tilde{M}_1$.

3) $Z_\Gamma = \Gamma \cap Z(\Gamma)$, the center of Γ, is a uniform lattice in $\tilde{M}_1$. Moreover, Z_Γ = kernel (p_2).

4) $\Gamma_2 = p_2(\Gamma)$ is an L-subgroup of $I(\tilde{M}_2)$ whose center is trivial.

Proof. Since Γ satisfies the duality condition the existence of the Γ-invariant splitting $\tilde{M} = \tilde{M}_1 \times \tilde{M}_2$ and the validity of assertions 1) and 2) follow from Theorem 4.2 of [CE]. The fact that $\Gamma_2 = p_2(\Gamma)$ is discrete follows from Lemma 5.1 of [E4]. From 3) of Proposition 2.3 we conclude that Γ_2 is an L-subgroup of $I(\tilde{M}_2)$. If $p_2(\phi)$ lies in the center of $\Gamma_2 = p_2(\Gamma)$ for some $\phi \in \Gamma$, then ϕ lies in $Z(\Gamma)$ since $p_1(\Gamma)$ is abelian by 2). Therefore $p_2(\phi) = 1$ by the second assertion in 2), which shows that Γ_2 has trivial center and completes the proof of 4).

We prove 3). Clearly 2) implies that Z_Γ = kernel (p_2). Fix a point $p \in \tilde{M}$ such that $d_\Gamma(p) = a > 0$ and let $\tilde{M}_1(p)$ denote the integral manifold through p of the Euclidean foliation of $\tilde{M}$ that is induced by $\tilde{M}_1$. By assertion 2) the group $Z(\Gamma)$ leaves $\tilde{M}_1(p)$ invariant and acts transitively on $\tilde{M}_1(p)$. It follows that $d_\Gamma(q) \equiv a > 0$ for all $q \in \tilde{M}_1(p)$ since $d_\Gamma \circ \phi = d_\Gamma$ for all $\phi \in Z(\Gamma)$. Since Γ is an L-subgroup there exists a positive number R such that $\tilde{M}^a_\Gamma = \{q \in \tilde{M} : d_\Gamma(q) \geq a\} \subseteq \Gamma \cdot \overline{B_R(p)}$, where $\overline{B_R(p)} = \{q \in \tilde{M} : d(p,q) \leq R\}$.

To show that $Z(\Gamma) \cap \Gamma = Z_\Gamma$ is a uniform lattice in $\tilde{M}_1$ it suffices to show that for every sequence $\{p_n\} \subseteq \tilde{M}_1(p)$ there exists a subsequence $\{p_{n_k}\}$, a sequence $\{\xi_k\} \subseteq Z_\Gamma$ and a positive number R' such that

$$d(p_{n_k}, \xi_k p) \leq R' \quad \text{for every } k$$

Let $\{p_n\} \subseteq \tilde{M}_1(p)$ be any sequence, and choose a sequence $\{\phi_n\} \subseteq \Gamma$ such that

$$d(p_n, \phi_n p) \leq R \qquad \text{for every } n$$

This can be done by the way in which R was chosen. We write

$$\phi_n = T_n \times \psi_n$$

where $T_n = p_1(\phi_n)$, a translation of $\tilde{M}_1$, and $\psi_n = p_2(\phi_n)$. Next we write

$$p = (a^1, a^2), \quad p_n = (a^1_n, a^2)$$

for suitable points a^1_n in $\tilde{M}_1$, a^i in $\tilde{M}_i$, $i = 1,2$. It follows that

$$\phi_n(p) = (T_n(a^1), \psi_n(a^2))$$

and we conclude

$$(*) \quad \begin{aligned} d(T_n(a^1), a^1_n) &\leq d(\phi_n p, p_n) \leq R \\ d(\psi_n(a^2), a^2) &\leq d(\phi_n p, p_n) \leq R \end{aligned} \qquad \text{for every } n$$

By the discreteness of $\Gamma_2 = p_2(\Gamma)$ there are only finitely many distinct elements in the sequence $\{\psi_n\}$, and hence $\psi_{n_k} \equiv \psi_{n_1}$ for all k by passing to some subsequence $\{n_k\}$. Let

$$\xi_k = \phi_{n_k} \phi_{n_1}^{-1} = (T_{n_k} T_{n_1}^{-1}, 1) \in \Gamma_1 \times \Gamma_2$$

which lies in Z_Γ for every k by 2). Finally, $d(p_{n_k}, \xi_k p) \leq d(p_{n_k}, \phi_{n_k} p) \cdot d(\phi_{n_k} p, \xi_k p) \leq R + d(\xi_k \phi_{n_1} p, \xi_k p) = R + d(\phi_{n_1} p, p) = R'$ for every k, which completes the proof of 3). □

Next we show that every finitely generated L-subgroup with nontrivial center admits a direct product subgroup of finite index. Our proof is a modification of the proof of the first part of the main theorem in [E1].

Proposition 3.2 Let $\Gamma \subseteq I(\tilde{M})$ be a finitely generated L-subgroup whose center Z is nontrivial and has rank $k \geq 1$. Then Γ admits a finite index subgroup Γ* that is isomorphic to $\mathbb{Z}^k \times \Gamma_2{}^*$, where $\Gamma_2{}^*$ is an L-subgroup of $I(\tilde{M}_2)$ and $\tilde{M}_2$ is a Riemannian factor of $\tilde{M}$.

Proof. Let $\tilde{M} = \tilde{M}_1 \times \tilde{M}_2$ be a Γ-invariant splitting that satisfies the properties in the statement of Proposition 3.1. For $i = 1,2$ let Γ_i denote $p_i(\Gamma)$, where $p_i : \Gamma \to I(\tilde{M}_i)$ is the projection homomorphism. We begin by defining a homomorphism $\rho : \Gamma_2 \to T^k$, where T^k is an appropriate flat k-torus. By Proposition 3.1 the center Z of Γ lies in $I(\tilde{M}_1) \times \{1\}$ and equals the kernel of p_2. Moreover, Z is a uniform lattice in $\tilde{M}_1$, and hence Z is free abelian of rank $k = \dim \tilde{M}_1$. Regarding $\tilde{M}_1$ as a Euclidean space $(\mathbb{R}^k, +)$ we let T^k denote the k-torus $\tilde{M}_1/Z$ and let $P : \tilde{M}_1 \to T^k$ denote the Riemannian covering homomorphism. We may also regard the group of translations Γ_1 as a subset of $\tilde{M}_1 = (\mathbb{R}^k, +)$ in the obvious way. Finally, we define $\rho : \Gamma_2 \to T^k$ by $\rho(p_2\phi) = P(p_1(\phi))$ for $\phi \in \Gamma$. It is easy to check that ρ is a well defined homomorphism since $Z = \text{kernel }(p_2) = \text{kernel }(P)$.

If $F = \rho(\Gamma_2)$ then F is a finitely generated abelian group since by hypothesis Γ is finitely generated. Write F as a direct product $A_0 \times T_0$, where A_0 is a free abelian group and T_0 is a finite abelian group. Let $\Gamma_2{}^* = \rho^{-1}(A_0) \subseteq \Gamma_2$ and let $\Gamma^* = p_2^{-1}(\Gamma_2{}^*) \subseteq \Gamma$. Clearly $\Gamma_2{}^*$ has finite index in Γ_2 and Γ* has finite index in Γ since A_0 has finite index in F.

We show that Γ* is isomorphic to $Z \times \Gamma_2{}^*$ and that $\Gamma_2{}^*$ is an L-subgroup of $I(\tilde{M}_2)$, which will complete the proof. Observe first that the group Γ_2 is an L-subgroup of $I(\tilde{M}_2)$ by 4) in the statement of Proposition 3.1. Hence $\Gamma_2{}^*$ is an L-subgroup of $I(\tilde{M}_2)$ by 2) of Proposition 2.3. To construct an isomorphism $T : Z \times \Gamma_2{}^* \to \Gamma^*$ we use the fact that $A_0 = \rho(\Gamma_2{}^*)$ has no elements of finite order to define a homomorphism $\tilde{\rho} : \Gamma_2{}^* \to \tilde{M}_1 = (\mathbb{R}^k, +)$ such that $P \circ \tilde{\rho} = \rho : \Gamma_2{}^* \to T^k = \tilde{M}_1/Z$ (See Lemma 2 of [E1], for example, for details). Then given $(z, p_2(\phi)) \in Z \times \Gamma_2{}^*$, where $\phi \in \Gamma^*$, we define $T(z, p_2(\phi))$ to be the unique element ϕ^* in Γ such that

a) $p_1(\phi^*) = z + \tilde{\rho}(p_2\phi)$

b) $p_2(\phi^*) = p_2(\phi)$

Clearly ϕ^*, if it exists, is uniquely determined by a) and b). To prove that ϕ^* exists we consider the element $\xi = \tilde{\rho}(p_2\phi) - p_1(\phi)$, a translation in $\tilde{M}_1$ or equivalently, an element in $(\mathbb{R}^k, +) = \tilde{M}_1$. By the definitions of ρ and $\tilde{\rho}$ it is clear that $\xi \in$ kernel $(P) = Z$. If $\phi^* = (z + \xi) \cdot \phi \in \Gamma$ then ϕ^* satisfies a) and b) since $Z =$ kernel $(p_2) =$ kernel (P). Hence $T(z, p_2(\phi)) = \phi^*$ is well defined for all $\phi \in \Gamma^*$. It is routine to prove that T is a homomorphism with trivial kernel.

It remains only to prove that Γ^* is the image of T. By property b) of the definition of T it follows that $\Gamma^* = p_2^{-1}(\Gamma_2{}^*)$ contains the image of T. Conversely given $\phi \in \Gamma^*$ we may express ϕ as $T(z, p_2(\phi))$, where $z = p_1(\phi) - \tilde{\rho}(p_2\phi) \in$ kernel $(P) = Z$. □

Section 4 L-subgroups in spaces with nontrivial Euclidean factors

The result of this section is a generalization to L-subgroups of the main theorem of [E2]. The proof given here simplifies the proof in [E2] to some extent.

Theorem 4 Let $\tilde{M}$ be a Riemannian product $\tilde{M}_0 \times \tilde{M}_1$, where $\tilde{M}_0$ is the Euclidean de Rham factor of $\tilde{M}$ and $\tilde{M}_1$ is the Riemannian product of all non Euclidean de Rham factors of $\tilde{M}$. Let $\Gamma \subseteq I(\tilde{M})$ be an L-subgroup. Then

1) The dimension of $\tilde{M}_0$ is the rank of the Clifford subgroup $C(\Gamma)$ of Γ.

2) If Γ is finitely generated and if $\dim \tilde{M}_0 = k \geq 1$, then Γ admits a finite index subgroup Γ^{**} that is isomorphic to a direct product $\mathbb{Z}^k \times \Gamma_1{}^{**}$, where $\Gamma_1{}^{**} \subseteq I(\tilde{M}_1)$ is an L-subgroup with trivial center.

Proof. The splitting $\tilde{M} = \tilde{M}_0 \times \tilde{M}_1$ is Γ-invariant since every isometry of $\tilde{M}$ leaves invariant the Euclidean de Rham foliation of $\tilde{M}$ and also its orthogonal complementary foliation. If $p_0 : \Gamma \to I(\tilde{M}_0)$ and $p_1 : \Gamma \to I(\tilde{M}_1)$ denote the projection homomorphisms, then $p_1(\Gamma)$ is discrete by Lemma A of [E2]. If $N =$ kernel (p_1), then $C(N)$ is a uniform lattice in $\tilde{M}_0$ by Proposition 2.4 above. Hence rank $C(\Gamma) \geq$ rank $C(N) = \dim \tilde{M}_0$. On the other hand rank $C(\Gamma) \leq \dim \tilde{M}_0$ by Proposition 1.7.2. This proves 1). Moreover, it follows that $C(\Gamma)$ is a uniform lattice in $\tilde{M}_0$.

We prove 2). Assume that $\dim \tilde{M}_0 = k \leq 1$ and Γ is finitely generated. If Γ^* is the centralizer in Γ of $C(\Gamma)$, then Γ^* has finite index in Γ by

Proposition 1.7.5, and hence Γ^* itself is an L-subgroup of $I(\tilde{M})$ by Proposition 2.3. Clearly $C(\Gamma^*) = C(\Gamma) = Z_{\Gamma^*}$, the center of Γ^*, since any central element of Γ^* lies in $C(\Gamma^*)$ by Proposition 1.7.3. Now apply Proposition 3.1 to the group Γ^* to obtain a Γ^*-invariant splitting $\tilde{M} = \tilde{M}_\alpha \times \tilde{M}_\beta$ that satisfies the properties listed there. The first of these properties implies that $\tilde{M}_\alpha \subseteq \tilde{M}_0$, and the third of these properties implies that equality holds since $Z_{\Gamma^*} = C(\Gamma)$ is a uniform lattice in $\tilde{M}_0$. Hence $\tilde{M}_\alpha = \tilde{M}_0$ and $\tilde{M}_\beta = \tilde{M}_1$.

We wish to apply Proposition 3.2 to the group Γ^*. We first need the following elementary result.

Lemma. Let G be a finitely generated group, and let G* be a subgroup of G with finite index. Then G* is also finitely generated.

Proof. Let F be a free group on n generators and $p : F \to G$ a surjective homomorphism. If $F^* = p^{-1}(G^*) \subseteq F$, then F* has finite index in F and hence F* is finitely generated by a well known result (see for example [Ku, p. 36] for a proof). It follows that $G^* = p(F^*)$ is finitely generated. □

The lemma above shows that Γ^* is finitely generated since Γ is finitely generated by hypothesis. It follows from the statement and proof of Proposition 3.2 that Γ^* admits a finite index subgroup Γ^{**} that is isomorphic to $\mathbb{Z}^k \times \Gamma_1^{**}$, where $\Gamma_1^{**} = p_1(\Gamma^{**})$ is an L-subgroup of $\tilde{M}_1$. To conclude the proof of 2) we need only show that Γ_1^{**} has trivial center. The second of the properties listed in Proposition 3.1 says that $p_0(\Gamma^{**})$ consists of translations in $\tilde{M}_0$. If $p_1(\phi)$ is a central element of Γ_1^{**} for some $\phi \in \Gamma^{**}$, then $p_1(\phi)$ is a Clifford translation of $\tilde{M}_1$ by Proposition 1.7.3 and hence $\phi = p_0(\phi) \times p_1(\phi)$ is a Clifford translation of $\tilde{M}$. Therefore $\phi \in C(\Gamma) \subseteq I(\tilde{M}_0) \times \{1\}$ by Proposition 1.7.2, which shows that $p_1(\phi) = \{1\}$ and Γ_1^{**} has trivial center. □

Section 5 Irreducible decomposition of an L-subgroup of nonpositive curvature

An L-subgroup $\Gamma \subseteq I(\tilde{M})$ admits an irreducible decomposition analogous to the de Rham decomposition of $\tilde{M}$. In this section we describe this irreducible decomposition and make precise the analogy.

Definition 5.1 An abstract group G is said to be reducible if there exists a finite index subgroup G* that is a nontrivial direct product $A \times B$ of subgroups A and B. A group G is said to be irreducible if it is not reducible.

Definition 5.2 An irreducible decomposition of an abstract group G is a direct product decomposition

$$G = G_0 \times G_1 \times \ldots \times G_k$$

of subgroups G_i, $0 \leq i \leq k$, such that G_0 is abelian and G_i is irreducible for $1 \leq i \leq k$.

For L-subgroups $\Gamma \subseteq I(\tilde{M})$ we have the following

Theorem 5.3 Let $\tilde{M}$ be a complete, simply connected manifold of nonpositive sectional curvature, and let $k \geq o$ denote the dimension of the Euclidean de Rham factor $\tilde{M}_0$ of $\tilde{M}$. Let $\Gamma \subseteq I(\tilde{M})$ be an L-subgroup and assume furthermore that Γ is finitely generated if $k \geq 1$. Then 1) there exists a finite index subgroup Γ^* of Γ that admits an irreducible decomposition. Conversely, 2) let $\Gamma^* = \Gamma_0^* \times \Gamma_1^* \times \ldots \times \Gamma_r^*$ be any irreducible decomposition of a finite index subgroup Γ^* of Γ. Then

a) For $1 \leq i \leq r$ the irreducible group Γ_i^* contains no nonidentity abelian subgroup whose normalizer in Γ_i^* has finite index in Γ_i^*.

b) Γ_0^* is the center of Γ^* and is isomorphic to $\mathbb{Z}^k$. Moreover, Γ_0^* is the Clifford subgroup of Γ^*.

c) The irreducible decomposition of Γ^* is unique. Moreover, the integer r depends on Γ and not on Γ^*. In fact, if Γ^{**} is another finite index subgroup of Γ that admits an irreducible decomposition $\Gamma^{**} = \Gamma_0^{**} \times \Gamma_1^{**} \times \ldots \times \Gamma_s^{**}$, then $r = s$ and for every $o \leq i \leq r$ the group $\Gamma_i^* \cap \Gamma_i^{**}$ has finite index in both Γ_i^* and Γ_i^{**} after a suitable reordering of the factors $\Gamma_1^{**}, \ldots, \Gamma_r^{**}$.

Before beginning the proof we make a few remarks:

(5.4) a) We shall see later in (5.5) that the integer r in the statement above is at most the number of non Euclidean de Rham factors of $\tilde{M}$.

b) An L-subgroup $\Gamma \subseteq I(\tilde{M})$ may not itself admit an irreducible decomposition. For example, if $\tilde{M} = \mathbb{R}^2$ with its usual flat Euclidean metric and if $\Gamma \subseteq I(\mathbb{R}^2)$ is the discrete group generated by $\phi_1 : (x,y) \to (x+1,y)$ and $\phi_2 : (x,y) \to (-x,y+1)$, then the quotient space $\mathbb{R}^2/\Gamma$ is a flat Klein bottle. The group Γ is reducible since it admits an index 2 subgroup isomorphic to $\mathbb{Z} \times \mathbb{Z} = \mathbb{Z}^2$. However, the center of Γ is an infinite cyclic group generated by ϕ_2^2 so Γ itself does not admit an irreducible decomposition. In general, if $\Gamma \subseteq I(\mathbb{R}^n)$ is a uniform lattice, then by a theorem of Bieberbach [W3, p. 100] Γ admits a finite index subgroup Γ^* that is isomorphic to $\mathbb{Z}^n$. Hence Γ^* has a trivial irreducible decomposition $\Gamma^* = \Gamma_0^*$ in which the factors Γ_i^*, $i \geq 1$, are absent.

c) In view of b) and the result above we make the following

<u>Definition</u> Let $\Gamma \subseteq I(\tilde{M})$ be an L-subgroup that is in addition finitely generated if $\tilde{M}$ has a nontrivial Euclidean de Rham factor. If Γ itself admits an irreducible decomposition $\Gamma = \Gamma_0 \times \Gamma_1 \times \ldots \times \Gamma_r$, then we refer to Γ_0 as the <u>Euclidean factor</u> of Γ and to the groups $\{\Gamma_i\}$, $1 \leq i \leq r$, as the <u>non Euclidean factors</u> of Γ. If Γ does not admit an irreducible decomposition, then we define the Euclidean factor and non Euclidean factors of Γ to be those of a finite index subgroup Γ^* that does admit an irreducible decomposition. In this case the factors of Γ are well defined up to order and subgroups of finite index by 2c) of the result above.

d) A theorem of A. Kurosh [Ku, pp. 81, 114] states the existence and uniqueness of an irreducible decomposition of an abstract group G that either has a trivial center or equals its own commutator subgroup [G,G]. However, this decomposition result is not as refined as the one we prove here since an irreducible (or indecomposable) group in the sense of Kurosh is one that is not a direct product of proper subgroups, a definition more restrictive than our own. For example, the fundamental group of the Klein bottle is irreducible in the sense of Kurosh but reducible in our sense.

<u>Proof of the theorem</u> 1) We first show the existence of a finite index subgroup Γ^* of Γ that admits an irreducible decomposition. Let k denote the dimension of the Euclidean de Rham factor $\tilde{M}_0$ of $\tilde{M}$. We consider first the case $k = o$; that is, $\tilde{M}$ has no Euclidean de Rham factor. It follows from Theorem 1.8.1 that if Γ^* is a finite index subgroup of Γ of the form $\Gamma^* = \Gamma_1{}^* \times \ldots \times \Gamma_r{}^*$, direct product, then there exists a Γ^*-invariant splitting $\tilde{M} = \tilde{M}_1 \times \tilde{M}_2 \times \ldots \times \tilde{M}_r$ such that $\Gamma_i{}^* \subseteq I(\tilde{M}_i) \times \{1\}$ for every i. In particular the number r of direct factors is at most $\frac{1}{2}\dim \tilde{M}$, and it now follows immediately that we can find a finite index subgroup Γ^* of Γ that admits an irreducible decomposition.

We consider next the case that $k \geq 1$. The group Γ is by hypothesis finitely generated in this case, and hence by Theorem 4 Γ contains a finite index subgroup Γ^* that is a direct product $\Gamma^* = Z^* \times \tilde{\Gamma}^*$, where Z^* is the center of Γ^*, isomorphic to $\mathbb{Z}^k$, and $\tilde{\Gamma}^*$ is a subgroup of Γ^* that is isomorphic to an L-subgroup with trivial center in $I(\tilde{M}_1)$, $\tilde{M}_1$ the product of all non Euclidean de Rham factors of $\tilde{M}$. Applying the case considered above to $\tilde{\Gamma}^*$ we obtain a finite index subgroup of $\tilde{\Gamma}^*$ of the form $\Gamma_1{}^* \times \Gamma_2{}^* \times \ldots \times \Gamma_s{}^*$, where each $\Gamma_i{}^*$ is irreducible for $1 \leq i \leq s$. The group $\Gamma^{**} = \Gamma_0{}^* \times \Gamma_1{}^* \times \ldots \times \Gamma_s{}^*$, where $\Gamma_0{}^* = Z^*$, has finite index in Γ and is already in irreducible decomposition form. This completes the proof of 1), the existence of an irreducible decomposition of some finite index subgroup of Γ.

Now let Γ^* be an arbitrary finite index subgroup of Γ, and suppose that Γ^* admits an irreducible decomposition $\Gamma^* = \Gamma_0^* \times \Gamma_1^* \times \dots \times \Gamma_r^*$. We prove the three assertions of statement 2) of the theorem, beginning with 2a). Let an integer i with $1 \leq i \leq r$ be given, and suppose that A is a nonidentity abelian subgroup of Γ_i^* whose normalizer Γ_i^{**} in Γ_i^* has finite index in Γ_i^*. Using Theorem 4 we shall show that Γ_i^* is reducible, a contradiction to the hypothesis.

We show first that Γ_i^{**} is finitely generated. If $p_i : \Gamma^* \to \Gamma_i^*$ denotes the projection homomorphism, then $\Gamma^{**} = p_i^{-1}(\Gamma_i^{**})$ has finite index in Γ^* and is the normalizer of A in Γ^*. By Proposition 1.6.2 it follows that Γ^{**} satisfies the duality condition, and by Proposition 1.7.3 we conclude that A consists of Clifford translations in $\tilde{M}$. Therefore the Euclidean factor of $\tilde{M}$ is nontrivial by Proposition 1.7.2, and it follows that Γ is finitely generated by the hypothesis of Theorem 5.3. Hence Γ^{**} and Γ_i^{**} are finitely generated by the lemma that occurs in the proof of Theorem 4.

By Propositions 2.5 and 2.6 the group Γ_i^* is isomorphic to an L-subgroup in $I(\tilde{M}_i)$, where $\tilde{M}_i$ is some complete, simply connected space of nonpositive sectional curvature. By Proposition 2.3 the group Γ_i^{**} is isomorphic to an L-subgroup $\tilde{\Gamma}_i$ in $I(\tilde{M}_i)$. If $\tilde{A} \subseteq \tilde{\Gamma}_i$ is a normal abelian subgroup isomorphic to A, then $\tilde{A}$ consists of Clifford translations in $\tilde{M}_i$ since $\tilde{\Gamma}_i$ satisfies the duality condition in $\tilde{M}_i$. Therefore the Euclidean de Rham factor of $\tilde{M}_i$ is nontrivial by Proposition 1.7.2, and since $\tilde{\Gamma}_i$ is finitely generated by the previous paragraph it follows from Theorem 4 that $\tilde{\Gamma}_i$ is reducible. Therefore Γ_i^{**} and Γ_i^* are reducible, contradicting the hypothesis that $\Gamma^* = \Gamma_0^* \times \Gamma_1^* \times \dots \times \Gamma_r^*$ is an irreducible decomposition. Hence Γ_i^* has no nonidentity abelian subgroup whose normalizer in Γ_i^* has finite index in Γ_i^*, which proves 2a).

We prove 2b). Let Z^* and $C(\Gamma^*)$ denote respectively the center and Clifford subgroup of Γ^*. Clearly $\Gamma_0^* \subseteq Z^* \subseteq C(\Gamma^*)$ since Γ_0^* is an abelian direct factor of Γ^*. On the other hand $C_i^* = p_i(C(\Gamma^*))$ is a normal abelian subgroup of Γ_i^* for $1 \leq i \leq r$, and by 2a) it follows that $C_i^* = \{1\}$ for $1 \leq i \leq r$. Therefore $C(\Gamma^*) \subseteq \Gamma_0^*$, which proves that $\Gamma_0^* = Z^* = C(\Gamma^*)$. The fact that Γ_0^* is isomorphic to $\mathbb{Z}^k$ follows from Theorem 4. The proof of 2b) is complete.

Before beginning the proof of 2c) we need three preliminary results.

Lemma 5.3a If $\Gamma \subseteq I(\tilde{M})$ satisfies the duality condition, then Γ has no finite normal subgroups except the identity.

Proof Let $N \subseteq \Gamma$ be a finite normal subgroup. By the Cartan fixed point theorem N fixes some point p of $\tilde{M}$. If Fix $(N) \subseteq \tilde{M}$ denotes the points of $\tilde{M}$ fixed by N, then Fix (N) is closed, convex, nonempty and invariant under Γ. Since the limit set of Γ equals $\tilde{M}(\infty)$ it follows that Fix $(N) = \tilde{M}$ and $N = \{1\}$. □

Lemma 5.3b Let Γ be an abstract group with subgroups A, Γ^*. Then $[A : \Gamma^* \cap A] \le [\Gamma : \Gamma^*]$.

Proof The map $x(\Gamma^* \cap A) \to x\,\Gamma^*$, $x \in A$, is a well defined injective map of left cosets. □

Lemma 5.3c Let $\Gamma \subseteq I(\tilde{M})$ be an L-subgroup that admits an irreducible decomposition $\Gamma = \Gamma_0 \times \Gamma_1 \times \dots \times \Gamma_r$, and let $\Gamma^* \subseteq \Gamma$ be a subgroup of finite index. Then Γ^* contains a finite index subgroup Γ^{**} with an irreducible decomposition $\Gamma^{**} = \Gamma_0^{**} \times \Gamma_1^{**} \times \dots \times \Gamma_r^{**}$, where Γ_i^{**} has finite index in Γ_i for every i, $0 \le i \le r$.

Proof For $0 \le i \le r$ set $A = \Gamma_i$, $\Gamma_i^{**} = \Gamma^* \cap A$ and apply Lemma 5.3b. □

We are ready to prove 2c) of the theorem. Let Γ^*, Γ^{**} be finite index subgroups of Γ that admit irreducible decompositions $\Gamma^* = \Gamma_0^* \times \Gamma_1^* \times \dots \times \Gamma_r^*$ and $\Gamma^{**} = \Gamma_0^{**} \times \Gamma_1^{**} \times \dots \times \Gamma_s^{**}$. If $G = \Gamma^* \cap \Gamma^{**}$, then G has finite index in both Γ^* and Γ^{**}. By Lemma 5.3c we can find finite index subgroups G^*, G^{**} of G that admit irreducible decompositions $G^* = G_0^* \times G_1^* \times \dots \times G_r^*$ and $G^{**} = G_0^{**} \times G_1^{**} \times \dots \times G_s^{**}$, where G_i^* has finite index in Γ_i^*, $0 \le i \le r$, and G_j^{**} has finite index in Γ_j^{**}, $0 \le j \le s$. Assertion 2c) is now contained in the following.

Lemma 5.3d Let $\Gamma \subseteq I(\tilde{M})$ be an L-subgroup with an irreducible decomposition $\Gamma = \Gamma_0 \times \Gamma_1 \times \dots \times \Gamma_s$. Let Γ^* be a finite index subgroup of Γ with an irreducible decomposition $\Gamma^* = \Gamma_0^* \times \Gamma_1^* \times \dots \times \Gamma_r^*$. Then

a) $\Gamma_0^* \subseteq \Gamma_0$ and $[\Gamma_0 : \Gamma_0^*] \le [\Gamma : \Gamma^*]$.

b) $r = s$

c) After a suitable reordering, $\Gamma_i^* \subseteq \Gamma_i$ for every i, $1 \le i \le r$, and $[\Gamma_i : \Gamma_i^*] \le [\Gamma : \Gamma^*]$.

d) The irreducible decomposition of Γ is unique up to the order of the factors Γ_i, $1 \le i \le r$.

Proof We observe that it suffices to prove the lemma in the case that $s \le r$. Suppose that the lemma has been proved in this case and let $s \ge r$. By Lemma 5.3c there exists a finite index subgroup Γ^{**} of Γ^* of the form $\Gamma^{**} = \Gamma_0^{**} \times \Gamma_1^{**} \times \dots \times \Gamma_s^{**}$, where Γ_i^{**} has finite index in Γ_i for $0 \le i \le s$. Apply the lemma to Γ^{**} and Γ^* to conclude that $s = r$. We now consider the case that $s \le r$.

Note that the factors Γ_i of Γ are all infinite groups by Lemma 5.3a. Note also that d) follows from a), b) and c) by setting $\Gamma^* = \Gamma$. By assertion 2b) of the theorem the groups Γ_0^*, Γ_0 are the Clifford subgroups of Γ^*, Γ and hence $\Gamma_0^* = \Gamma^* \cap \Gamma_0 \subseteq \Gamma_0$ since $\Gamma^* \subseteq \Gamma$. We obtain

i) $[\Gamma_0 : \Gamma_0^*] = [\Gamma_0 : \Gamma^* \cap \Gamma_0] \le [\Gamma : \Gamma^*]$

by Lemma 5.3b. This proves a) of the lemma. If Γ', $\Gamma^{*\prime}$ denote the factor groups Γ/Γ_0, Γ^*/Γ_0^* respectively, then $\Gamma^{*\prime}$ may be regarded as a subgroup of Γ' by means of the well defined injective homomorphism $x\,\Gamma_0^* \to x\,\Gamma_0$ for $x \in \Gamma^*$. It follows immediately that

ii) $[\Gamma' : \Gamma^{*\prime}] \leq [\Gamma : \Gamma^*]$

since $\Gamma^{*\prime}$ is identified with $\{x\,\Gamma_0 : x \in \Gamma^*\}$ and $\Gamma' = \{x\,\Gamma_0 : x \in \Gamma\}$.

We now restrict our attention to the groups Γ', $\Gamma^{*\prime}$ which we identify with the direct products $\Gamma_1 \times \ldots \times \Gamma_s$, $\Gamma_1^* \times \ldots \times \Gamma_r^*$ respectively. We assume that $s \geq 1$ for otherwise $\Gamma = \Gamma_0$ is abelian and the result of the lemma is trivially true. For $1 \leq i \leq s$ let $p_i : \Gamma' \to \Gamma_i$ denote the projection homomorphism. The next step is the assertion

iii) Let $1 \leq j \leq r$ be given. If $p_i(\Gamma_j^*) \neq \{1\}$ for some $1 \leq i \leq s$, then $p_i(\Gamma_\alpha^*) = \{1\}$ for all $\alpha \neq j$, $1 \leq \alpha \leq r$, and $p_i(\Gamma_j^*) = p_i(\Gamma^{*\prime})$. Moreover, $[\Gamma_i : p_i(\Gamma_j^*)] \leq [\Gamma' : \Gamma^{*\prime}] \leq [\Gamma : \Gamma^*]$.

We begin the proof of iii) by noting that the groups $\{p_i(\Gamma_\alpha^*) : 1 \leq \alpha \leq r\}$ commute pairwise and generate $p_i(\Gamma^{*\prime})$. If $\hat{\Gamma}_\alpha^*$ denotes the direct product of the subgroups $\Gamma_\beta^* : \beta \neq \alpha$, $1 \leq \beta \leq r$, then any element in $p_i(\Gamma_\alpha^*) \cap p_i(\hat{\Gamma}_\alpha^*)$ must be central in $p_i(\Gamma_\alpha^*)$ and hence central in $p_i(\Gamma^{*\prime})$, which has finite index in $\Gamma_i = p_i(\Gamma')$. It follows that $p_i(\Gamma_\alpha^*) \cap p_i(\hat{\Gamma}_\alpha^*) = \{1\}$ for all α with $1 \leq \alpha \leq r$ by assertion 2a) of Theorem 5.3, and we conclude that $p_i(\Gamma^{*\prime})$ is the direct product of the subgroups $p_i(\Gamma_\alpha^*)$, $1 \leq \alpha \leq r$. By hypothesis $p_i(\Gamma_j^*) \neq \{1\}$ and hence the irreducibility of Γ_i implies that $p_i(\Gamma_j^*) = p_i(\Gamma^{*\prime})$ and $p_i(\Gamma_\alpha^*) = \{1\}$ if $\alpha \neq j$, $1 \leq \alpha \leq r$. Finally $[\Gamma_i : p_i(\Gamma_j^*)] = [p_i(\Gamma') : p_i(\Gamma^{*\prime})] \leq [\Gamma' : \Gamma^{*\prime}] \leq [\Gamma : \Gamma^*]$ by ii). This proves iii).

We now assert

iv) $r = s$ and for each j with $1 \leq j \leq r$ there exists a unique integer i with $1 \leq i \leq r$ such that $p_i(\Gamma_j^*) \neq \{1\}$.

For each integer j with $1 \leq j \leq r$ we define the set $S_j = \{i : 1 \leq i \leq s$ and $p_i(\Gamma_j^*) \neq \{1\}\}$. Each set S_j is nonempty since $\{1\} \neq \Gamma_j^* \subseteq \Gamma^{*\prime} \subseteq \Gamma'$, and it is clear that $\{1,2,\ldots,s\}$ is the union of the sets S_j, $1 \leq j \leq r$. It follows from iii) that the sets S_j, $1 \leq j \leq r$, are all disjoint. Hence

$$s = \sum_{j=1}^{r} |S_j| \geq r .$$

However at the beginning of the proof of the lemma we reduced to the case $s \leq r$. We conclude that $s = r$ and each set S_j contains exactly one element, which proves iv).

We now conclude the proof of Lemma 5.3d. It follows from iv) that given an integer j with $1 \le j \le r$ there exists a unique integer i with $1 \le i \le r$ such that $\Gamma_j^* \subseteq \Gamma_i$. By iii) $p_i(\Gamma^{*\prime}) = p_i(\Gamma_j^*) = \Gamma_j^*$ and hence $[\Gamma_i : \Gamma_j^*] = [p_i(\Gamma') : p_i(\Gamma^{*\prime})] \le [\Gamma' : \Gamma^{*\prime}] \le [\Gamma : \Gamma^*]$ by ii). After a reordering of the factors Γ_j^* we may assume that $\Gamma_j^* \subseteq \Gamma_j$ for all $1 \le j \le r$. This completes the proof of Lemma 5.3d and also of Theorem 5.3. □

(5.5) Relationship with the de Rham decomposition of $\tilde{M}$

We now relate the irreducible decomposition $\Gamma^* = \Gamma_0^* \times \Gamma_1^* \times \ldots \times \Gamma_s^*$ of a finite index subgroup Γ^* of an L-subgroup $\Gamma \subseteq I(\tilde{M})$ to the de Rham decomposition $\tilde{M} = \tilde{M}_0 \times \tilde{M}_1 \times \ldots \times \tilde{M}_r$ of $\tilde{M}$. By (1.3) and Theorem 5.3 we lose no generality in assuming that Γ leaves invariant the de Rham splitting of $\tilde{M}$. Let $p_i : \Gamma \to I(\tilde{M}_i)$ denote the projection homomorphisms for $0 \le i \le r$. In Theorem 5.3 we have already seen that Γ_0^* is isomorphic to $\mathbb{Z}^k$, where $k = \dim \tilde{M}_0$, and that s is independent of the finite index subgroup Γ^*. We shall decompose the set $A = \{1,2,\ldots,r\}$ uniquely into a disjoint union of "admissible" subsets and shall show that s is the number of these admissible subsets. In particular it will follow that $s \le r$ and equality holds if and only if $p_i(\Gamma)$ is a discrete subgroup of $I(\tilde{M}_i)$ for every integer i with $1 \le i \le r$.

We first consider the case that $\tilde{M} = \tilde{M}_1 \times \ldots \times \tilde{M}_r$ has no Euclidean de Rham factor. In this case the Euclidean factor Γ_0^* is the identity by Theorem 5.3. If $X = \{i_1, i_2, \ldots i_m\}$ is a subset of $A = \{1,2,\ldots,r\}$ with $i_1 < i_2 < \ldots < i_m$, then we define $\tilde{M}_X$ to be the Riemannian product $\tilde{M}_{i_1} \times \ldots \times \tilde{M}_{i_m}$ and we define $p_X = p_{i_1} \times \ldots \times p_{i_m} : \Gamma \to I(\tilde{M}_X)$ to be the corresponding projection homomorphism. Define a subset X of A to be <u>admissible</u> if $p_X(\Gamma)$ is a discrete subgroup of $I(\tilde{M}_X)$ but $p_{X^*}(\Gamma)$ is not a discrete subgroup of $I(\tilde{M}_{X^*})$ for any proper subset X^* of X.

We decompose A into a disjoint union of admissible subsets. Since $\Gamma = p_A(\Gamma)$ is discrete it follows that A contains an admissible subset X_1. Assume that $X_1 \ne A$ for otherwise we are done. If $X \subseteq A$ is any proper subset such that $p_X(\Gamma)$ is a discrete subgroup of $I(\tilde{M}_X)$ and if X' denotes the complement of X in A, then $p_{X'}(\Gamma)$ is a discrete subgroup of $I(\tilde{M}_{X'})$ by Proposition 2.4 and the fact that $\tilde{M}$ has no Euclidean de Rham factor. Setting $X = X_1$ we obtain an admissible subset $X_2 \subseteq X'$. Setting $X = X_1 \cup X_2$ and assuming that $X \ne A$ we obtain an admissible subset $X_3 \subseteq X'$. After a finite number of repetitions of this procedure we obtain a decomposition of A into a disjoint union of admissible subsets $X_1, \ldots, X_s$. We shall see that this decomposition is unique.

Let $A = \bigcup_{i=1}^{s} X_i$ be any decomposition of $A = \{1, 2, \ldots, r\}$ into a disjoint union of admissible subsets $\{X_i\}$. If a set X_i has at least two elements, then the corresponding space $\tilde{M}_i^* = \tilde{M}_{X_i}$ is a symmetric space of noncompact type by the equivalence of 2) and 3) in Proposition 2.7 since $\Gamma_i = p_{X_i}(\Gamma)$ is an L-subgroup of $I(M_i^*)$ by Proposition 2.3 and since $p_{X^*}(\Gamma)$ is a non-discrete subgroup of $I(\tilde{M}_{X^*})$ for each proper subset $X^* \subseteq X_i$ by the admissibility of X_i. This latter fact also implies that Γ_i is an irreducible subgroup of $I(M_i^*)$ by the equivalence of 1) and 2) in Proposition 2.7. If an admissible subset X_i consists of a single element of A, then $\Gamma_i = p_{X_i}(\Gamma)$ is an irreducible subgroup of $I(\tilde{M}_{X_i})$ by Theorem 1.8.1 since $\tilde{M}_{X_i}$ has no Euclidean de Rham factor and is irreducible. Hence Γ_i is irreducible for all i, $1 \leq i \leq s$.

The group $\tilde{\Gamma} = \Gamma_1 \times \ldots \times \Gamma_s$ is a discrete subgroup of $I(\tilde{M})$ since each of the groups Γ_i is discrete, and it follows from Proposition 2.3 that Γ has finite index in $\tilde{\Gamma}$ since $\Gamma \subseteq \tilde{\Gamma}$. By Lemma 5.3c there exists a finite index subgroup Γ^* of Γ such that Γ^* is a direct product $\Gamma_1^* \times \ldots \times \Gamma_s^*$, where Γ_i^* is a finite index subgroup of Γ_i for $1 \leq i \leq s$. The groups $\{\Gamma_i^*\}$ are irreducible since the groups $\{\Gamma_i\}$ are irreducible. Hence $\Gamma^* = \Gamma_1^* \times \ldots \times \Gamma_s^*$ is an irreducible decomposition of Γ^*. The proof shows that $s \leq r$ with equality if and only if each admissible subset contains exactly one element, which happens if and only if $p_i(\Gamma)$ is a discrete subgroup of $I(\tilde{M}_i)$ for $1 \leq i \leq r$.

The discussion above shows that each decomposition of $A = \{1,2,\ldots,r\}$ into a union of disjoint admissible subsets determines an irreducible decomposition of the finite index subgroups of Γ. The uniqueness of irreducible decompositions proved in Theorem 5.3 now shows that the decomposition of A into a union of disjoint admissible subsets is unique.

Finally we consider the case that $\tilde{M}$ has a nontrivial Euclidean factor $\tilde{M}_0$. In this case we assume that Γ is finitely generated, following Theorem 5.3. If $M' = \tilde{M}_1 \times \ldots \times \tilde{M}_r$ denotes the Riemannian product of all non Euclidean de Rham factors of $\tilde{M}$, then the projection of Γ onto $I(M')$ is a discrete subgroup Γ' by Lemma A of [E2]. Hence Γ' is an L-subgroup of $I(M')$ by Proposition 2.3. The proofs of Proposition 3.2 and Theorem 4 above show that Γ admits a finite index subgroup Γ^* that is isomorphic to $\mathbb{Z}^k \times \Gamma^{*\prime}$, where $k = \dim \tilde{M}_0 \geq 1$ and $\Gamma^{*\prime}$ is the projection of Γ^* onto $I(M')$, a finite index subgroup of Γ'. Hence $\Gamma^{*\prime}$ is an L-subgroup of $I(M')$ by Proposition 2.3, and we may now apply the discussion above to $\Gamma^{*\prime} \subseteq I(M')$ since M' has no Euclidean de Rham factor.

Section 6 Algebraic rank of a lattice

In this section we consider lattices $\Gamma \subseteq I(\tilde{M})$ and explain how one may assign to each lattice Γ an integer $k \geq 1$, called the (algebraic) rank of Γ that equals the rank of $\tilde{M}$ as defined in (1.4) when the sectional curvature of $\tilde{M}$ is bounded below. For further details see [BBE] and [BE]. The notion of rank helps to identify the irreducible non Euclidean factors of Γ in the irreducible decomposition of Γ. In fact, one has the following consequence (Theorem C of [BE]) of a striking result of W. Ballmann [B2] and K. Burns and R. Spatzier [BuS 2].

Theorem 6.1 Let M have sectional curvature satisfying $-a^2 \leq K \leq 0$ for some positive constant a, and let $\Gamma \subseteq I(\tilde{M})$ be a lattice that is irreducible, has rank $k \geq 2$ and is either finitely generated or has no nonidentity normal abelian subgroup. Then $\tilde{M}$ is isometric to a symmetric space of noncompact type and rank k.

Before defining the rank of a lattice Γ we shall motivate the definition by considering uniform lattices $\Gamma \subseteq G = I_0(\tilde{M})$, where $\tilde{M}$ is a symmetric space of noncompact type and rank $k \geq 2$. The group G is known to be a connected, semisimple Lie group with trivial center and with no compact factors. Moreover, we shall assume that the uniform lattice Γ has no elements of finite order by passing to an appropriate subgroup of finite index if necessary. If $\phi \in \Gamma$ is an arbitrary element then

$$\tilde{M}_\phi = \{p \in \tilde{M} : d(p,\phi p) \leq d(q,\phi q) \quad \text{for all } q \in \tilde{M}\}$$

is a nonempty, complete, totally geodesic submanifold of $\tilde{M}$. Moreover, ϕ translates the unique geodesic from p to ϕp for every $p \in \tilde{M}$ so that $\tilde{M}_\phi$ is the union of all geodesics in $\tilde{M}$ that are translated by ϕ.

The submanifold $\tilde{M}_\phi$ is itself a symmetric space since any complete, totally geodesic submanifold M* of $\tilde{M}$ is clearly left invariant by all geodesic symmetries at points of M*. Hence we may write

$$\tilde{M}_\phi = \tilde{M}_0(\phi) \times \tilde{M}_1(\phi)$$

where $\tilde{M}_0(\phi)$ is the Euclidean de Rham factor of $\tilde{M}_\phi$ and $\tilde{M}_1(\phi)$, a symmetric space of noncompact type, is the Riemannian product of all non Euclidean de Rham factors of $\tilde{M}_\phi$. If $G_\phi = \{g \in G : g\phi = \phi g\}$, then it is easy to see that G_ϕ leaves $\tilde{M}_\phi$ invariant and acts transitively on $\tilde{M}_\phi$. It is also routine to show that the coset space G_ϕ/Γ_ϕ is compact, where $\Gamma_\phi = \Gamma \cap G_\phi$; see for example Lemma 8.1 of [M]. It follows that the quotient space $\tilde{M}_\phi/\Gamma_\phi$ is compact. By Theorem 4 we conclude that Γ_ϕ admits a finite index subgroup isomorphic to $\mathbb{Z}^r \times \Gamma_1(\phi)$, where $r = \dim \tilde{M}_0(\phi)$ and $\Gamma_1(\phi)$ is a uniform lattice in $\tilde{M}_1(\phi)$. If $\phi \in \Gamma$ translates a geodesic γ such that $\gamma'(0) = v \in R$, the set of regular vectors in $S\tilde{M}$, then one can show that

$\tilde{M}_\phi$ is a k-flat in $\tilde{M}$, where $k = \text{rank } \tilde{M}$. In this case $\tilde{M} = \tilde{M}_0(\phi)$ and $\tilde{M}_1(\phi)$ is absent, using the notation above. It then follows from the discussion above that Γ_ϕ contains a finite index subgroup isomorphic to $\mathbb{Z}^k$.

A general principle asserts that a lattice Γ in a Lie group G reflects the structure of G as much as possible. For example, if $G = I_0(\mathbb{R}^n)$ and $A \approx \mathbb{R}^n$ denotes the normal abelian subgroup of translations in $\mathbb{R}^n$, then G/A is isomorphic to the compact group SO(n). If $\Gamma \subseteq G$ is a lattice, then $\Gamma/\Gamma \cap A$ is finite by one of the Bieberbach theorems [W3, p. 100]. If $\tilde{M}$ is a symmetric space of noncompact type and if $\Gamma \subseteq G = I_0(\tilde{M})$ is a uniform lattice without elements of finite order, then this principle suggests that the geodesics translated by a "generic" element ϕ of Γ should have tangent vectors lying in R since R is a dense open subset of $S\tilde{M}$ (cf. (1.4)). Hence, by the discussion above one would expect that for a generic element ϕ of Γ the centralizer Γ_ϕ contains a finite index subgroup isomorphic to $\mathbb{Z}^k$, where k is the rank of $\tilde{M}$.

The preceding paragraph suggests that one can define the rank of uniform lattice $\Gamma \subseteq I_0(\tilde{M})$, where $\tilde{M}$ is a symmetric space of noncompact type, in terms of the structure of the centralizer Γ_ϕ of a generic element ϕ, provided that one can find an appropriate definition of "generic" element in this context. In Theorem 3.9 of [PR] G. Prasad and M. Raghunathan found a solution to this problem and we rephrase their result as follows:

Theorem 6.2 [PR] Let $\tilde{M}$ be a symmetric space of noncompact type, and let Γ be a uniform lattice in $G = I_0(\tilde{M})$. For each integer $j \geq 1$ let $A_j(\Gamma) = \{\phi \in \Gamma : \Gamma_\phi$ contains a finite index subgroup isomorphic to $\mathbb{Z}^s$ for some integer s with $1 \leq s \leq j$. Then the rank of $\tilde{M}$ equals $r(\Gamma) = \min\{j : \Gamma = \bigcup_{i=1}^{m} \phi_i \cdot A_j(\Gamma)$ for some finite set $\{\phi_1, \ldots, \phi_m\} \subseteq \Gamma\}$.

For an alternate and earlier definition of the rank of Γ see [W1].

Theorem 6.2 can be generalized to arbitrary lattices $\Gamma \subseteq I(\tilde{M})$, where $\tilde{M}$ is any complete, simply connected space whose sectional curvature satisfies $-a^2 \leq K \leq 0$ for some positive constant a. In this setting one first defines in [BE] an integer rank (Γ), the algebraic rank of Γ, by modifying slightly the definition of [PR]. Let $A_j(\Gamma)$ and $r(\Gamma)$ be defined as above. Then define rank $(\Gamma) = \sup\{r(\Gamma^*) : \Gamma^*$ is a finite index subgroup of $\Gamma\}$. We then obtain the following generalization of Theorem 6.2, which is Theorem B of [BE].

Theorem 6.3 Let $\tilde{M}$ be a complete, simply connected manifold with sectional curvature satisfying $-a^2 \leq K \leq 0$ for some positive constant a. Let $\Gamma \subseteq I(\tilde{M})$ be an arbitrary lattice. Then

$$\text{rank } (\Gamma) = r(\tilde{M})$$

where $r(\tilde{M})$ is the rank of $\tilde{M}$ as defined in (1.4).

Remark: This result is false if $r(\Gamma)$ is substituted for rank (Γ). For example, if $\tilde{M} = \mathbb{R}^2$ with its standard flat metric and if $\Gamma \subseteq I(\mathbb{R}^2)$ is a lattice such that $\mathbb{R}^2/\Gamma$ is a Klein bottle, then $r(\tilde{M}) = \text{rank}(\Gamma) = 2$. However, $r(\Gamma) = 1$ since $A_1(\Gamma)$ consists of the orientation reversing isometries of $\mathbb{R}^2$ that lie in Γ and consequently $\Gamma = A_1(\Gamma) \cup \phi \cdot A_1(\Gamma)$ for any element $\phi \neq 1$ in $A_1(\Gamma)$.

Using Theorem 6.3 and the main result of [B2] and [BuS 2] we now obtain Theorem 6.1 stated at the beginning of this section.

Section 7 Centralizer complexes and the Karpelevic boundary of $\tilde{M}$

Let $\tilde{M}$ denote a symmetric space of noncompact type, and let $G = I_0(\tilde{M})$. Let $\Gamma \subseteq G$ be a lattice. To avoid technical complications we shall further assume in the discussion that follows that Γ is a uniform lattice. The result of Prasad and Raghunathan described in the previous section shows that one can determine the rank of $\tilde{M}$, which equals the real rank of G, by a knowledge of the structure of the centralizers Γ_ϕ, $\phi \in \Gamma$. The real rank of G is a crude invariant of G and not surprisingly is determined by rather crude information about the centralizers Γ_ϕ. It is reasonable to ask if more refined information about the structure of G can be determined by a more refined knowledge of the structure of the centralizers Γ_ϕ, $\phi \in \Gamma$. Of course, one must assume that $\tilde{M}$ has rank at least 2 for this program to have any hope of being successful since all centralizers Γ_ϕ are infinite cyclic if $\tilde{M}$ has rank 1. We believe, in fact, that if $\tilde{M}$ has rank at least 2, then G can be identified entirely from the structure of the centralizers Γ_ϕ. Moreover we propose that this identification can be carried out by studying a certain "boundary" $K\Gamma(\infty)$ of the lattice Γ whose elements consist of "centralizer complexes" $\Gamma_\phi(\infty)$, one for each element ϕ in Γ. Each centralizer complex $\Gamma_\phi(\infty)$ is itself the set of all "admissible" finite sequences $(\phi_0, \phi_1, \ldots, \phi_m)$, with $\phi_i \in \Gamma$ for all i, such that $\phi_0 = \phi$. As we shall see the boundary $K\Gamma(\infty)$ of Γ is a discrete analogue of the Karpelevic boundary $K\tilde{M}(\infty)$ of $\tilde{M}$ as defined in [Ka].

We begin with a brief description of the Karpelevic boundary $K\tilde{M}(\infty)$ of a symmetric space $\tilde{M}$ of noncompact type and rank $k \geq 2$. We shall not go into much detail here but instead shall try to convey a flavor of the boundary $K\tilde{M}(\infty)$. For more information see section 13 of [Ka]. The Karpelevic boundary $K\tilde{M}(\infty)$ is a refinement of the standard boundary $\tilde{M}(\infty)$ that leaves regular points of $\tilde{M}(\infty)$ unchanged but replaces each singular point with an entire complex of points. We define a point x in $\tilde{M}(\infty)$ to be regular or singular if the tangent vectors to a geodesic γ that belongs to x are regular (in R) or singular (in $S\tilde{M} - R$). Since R is an open dense subset of $S\tilde{M}$ it follows that the regular points in $\tilde{M}(\infty)$ form an open dense subset of $\tilde{M}(\infty)$. The singular points of $\tilde{M}(\infty)$ are further subdivided into strata corresponding to different degrees of singularity.

We now define $K\tilde{M}(\infty)$. Fix a point $p \in \tilde{M}$. For each point x in $\tilde{M}(\infty)$ we recall that γ_{px} denotes the geodesic belonging to x that begins at p, and we let $F(\gamma_{px})$ denote the union of all geodesics in $\tilde{M}$ that are biasymptotic to γ_{px}. The set $F(\gamma_{px})$ is, in fact, a complete, totally geodesic submanifold of $\tilde{M}$. We write

$$F(\gamma_{px}) = F(\gamma_{px})^e \times F(\gamma_{px})^s$$

where $F(\gamma_{px})^e$ denotes the Euclidean de Rham factor of $F(\gamma_{px})$ and $F(\gamma_{px})^s$ denotes the Riemannian product of all non Euclidean de Rham factors of $F(\gamma_{px})$. The manifold $F(\gamma_{px})^s$ is a symmetric space of noncompact type, and both $F(\gamma_{px})^e$ and $F(\gamma_{px})^s$ are isometric to complete, totally geodesic submanifolds of $\tilde{M}$ that pass through p. The rank and dimension of $F(\gamma_{px})^s$ are strictly smaller than the rank and dimension of $F(\gamma_{px})$ since γ_{px} itself is contained in $F(\gamma_{px})^e$.

For each point x of $\tilde{M}(\infty)$ consider the complex $\tilde{M}_x(\infty)$ of equivalence classes of admissible finite sequences $(x = x_0, x_1, \ldots, x_m)$, with $x_i \in \tilde{M}(\infty)$ for all i, that is defined as follows: a sequence $(x = x_0, x_1, \ldots, x_m)$ is <u>admissible</u> if $x_{i+1} \in F(\gamma_{px_i})^s(\infty) \subseteq M(\infty)$ for every i. We define two admissible sequences $x^* = (x_0, x_1, \ldots, x_N)$ and $y^* = (y_0, y_1, \ldots, y_M)$ to be equivalent if $x_0 = y_0$, $N = M$ and $x_i = \phi(y_i)$ for all $1 \leq i \leq N$ and for some $\phi \in G$ that fixes x_0. The equivalence relation removes the dependence of $x^* = (x_0, x_1, \ldots, x_N)$ upon the base point in $\tilde{M}$. The set of all equivalence classes of admissible sequences x^* is called the <u>Karpelevic boundary</u> of $\tilde{M}$ and is denoted here by $K\tilde{M}(\infty)$. With a suitable topology the space $\tilde{M} \cup K\tilde{M}(\infty)$ becomes a compactification of $\tilde{M}$, and $G = I_0(\tilde{M})$ acts by homeomorphisms on $K\tilde{M}(\infty)$. If x is a regular point in $\tilde{M}(\infty)$, then $F(\gamma_{px})^s$ has dimension zero. Hence the complex $\tilde{M}_x(\infty) \subseteq K\tilde{M}(\infty)$ determined by x is the single point sequence $\{x\}$.

Now let $\Gamma \subseteq I_0(\tilde{M})$ be a uniform lattice without elements of finite order. We construct a boundary $K\Gamma(\infty)$ for Γ that is analogous to the boundary $K\tilde{M}(\infty)$ for $\tilde{M}$. Recall the notation $\Gamma_\phi = \Gamma \cap G_\phi$, where $G_\phi = \{g \in G : g\phi = \phi g\}$, for any element ϕ of Γ. The <u>non Euclidean factor</u> of Γ_ϕ, denoted by Γ_ϕ^s, is defined to be the direct product $\Gamma_1 \times \ldots \times \Gamma_r$ of the irreducible non Euclidean factors of Γ_ϕ^*, where $\Gamma_\phi^* = \Gamma_0 \times \Gamma_1 \times \ldots \times \Gamma_r$ is an irreducible decomposition of some finite index subgroup Γ_ϕ^* of Γ_ϕ. By Theorem 5.3 the group Γ_ϕ^s is well defined up to a subgroup of finite index, and by Theorem 4, Γ_ϕ^s is isomorphic to a uniform lattice in $(\tilde{M}_\phi)^s$, the Riemannian product of all non Euclidean de Rham factors of the symmetric space $\tilde{M}_\phi = \{p \in \tilde{M} : d(p,\phi p) \leq d(q,\phi q) \text{ for all } q \in \tilde{M}\}$.

Given an arbitrary element $\phi = \phi_0$ of $\Gamma = \Gamma_0$ we define Γ_1 to be the non Euclidean factor of $\Gamma_0 \cap G_{\phi_0} = \Gamma_{\phi_0}$, which is a uniform lattice in $\tilde{M}_{\phi_0}$. The group Γ_1 is itself isomorphic to a uniform lattice in $(\tilde{M}_{\phi_0})^s$. Choose an arbitrary element ϕ_1 of Γ_1 and define Γ_2 to be the non Euclidean factor of $\Gamma_1 \cap G_{\phi_1}$, which is isomorphic to a uniform lattice in $(\tilde{M}_{\phi_0})^s \cap \tilde{M}_{\phi_1}$. Hence Γ_2 is isomorphic to a uniform lattice in $M_2{}^* = [(\tilde{M}_{\phi_0})^s \cap \tilde{M}_{\phi_1}]^s$, the product of all non Euclidean de Rham factors of $(\tilde{M}_{\phi_0})^s \cap \tilde{M}_{\phi_1}$. Now choose an arbitrary element ϕ_2 of Γ_2 and continue in this fashion. We obtain a decreasing sequence of groups $\Gamma = \Gamma_0 \supseteq \Gamma_1 \supseteq \Gamma_2 \supseteq \ldots \supseteq \Gamma_i \supseteq \ldots$ and a sequence $(\phi = \phi_0, \phi_1, \ldots, \phi_i, \ldots)$ where for each i we have $\phi_i \in \Gamma_i$, Γ_i is the non Euclidean factor of $\Gamma_{i-1} \cap G_{\phi_{i-1}}$ and Γ_i is isomorphic to a uniform lattice in $M_i{}^* = [M_{i-1}^* \cap \tilde{M}_{\phi_{i-1}}]^s$. Clearly the sequences terminate after a finite number of steps since the rank and dimension of $M_i{}^*$ decrease strictly with i. The fact that the groups $\{\Gamma_i\}$ are uniquely defined only up to a finite index subgroup causes no serious difficulties.

We define $K\Gamma(\infty)$ to be the set of all sequences $(\phi_0, \phi_1, \ldots, \phi_N)$, with $\phi_i \in \Gamma$ for all i, that are constructed in the manner above. Given an element $\phi \in \Gamma$ the centralizer complex $\Gamma_\phi(\infty) \subseteq K\Gamma(\infty)$ that is determined by ϕ is defined to be the set of all such sequences $(\phi_0, \phi_1, \ldots, \phi_N)$ with $\phi = \phi_0$. If ϕ translates a geodesic of $\tilde{M}$ whose tangent vectors are regular, then by the discussion in section 6 the space $\tilde{M}_\phi$ is a k-flat, where k is the rank of $\tilde{M}$, and hence the centralizer complex $\Gamma_\phi(\infty)$ is the single point sequence $\{\phi\}$. In general the size and complexity of the centralizer complex $\Gamma_\phi(\infty)$ determined by an element ϕ of Γ depend directly on the dimension and rank of $(\tilde{M}_\phi)^s$. In our opinion the semisimple group $G = I_0(\tilde{M})$ from which the lattice Γ arises can be identified from the structure of $K\Gamma(\infty)$, but it is unclear at this time how to make this assertion more precise.

Appendix

In this appendix we give the proofs of Propositions 1.5.4, 1.6.4, 2.4 and 2.6, which were omitted in the text.

Proof of Proposition 1.5.4 We shall need the following two lemmas that we state here and prove later after the proof of the Proposition.

Lemma 1.5.4a Let $F_1, \ldots, F_N$ be complete, totally geodesic submanifolds of $\tilde{M}$. Given $p \in \tilde{M}$ and $R > o$ there exists $q \in B_R(p) = \{p^* \in \tilde{M} : d(p,p^*) < R\}$ such that

1) $B_{(R/4^N)}(q) \subseteq B_R(p)$

2) If $q' \in B_{(R/4^N)}(q)$, then $d(q', F_i) \geq R/4^N$ for every $1 \leq i \leq N$.

Lemma 1.5.4b Given $\varepsilon > o$ and a positive integer $N \geq 2$ let $\delta = 2\,\varepsilon \sin(\pi/N)$. If $\phi \in I(\tilde{M})$ has finite order $\leq N$ and if $d(p,\phi p) < \delta$ for some point $p \in \tilde{M}$, then $d(p, \mathrm{Fix}(\phi)) < \varepsilon$, where $\mathrm{Fix}(\phi)$ denotes the set of points in $\tilde{M}$ fixed by ϕ.

Assuming for the moment that these lemmas have been proved, we prove the Proposition. Let $a > o$ and $\varepsilon > o$ be given. It suffices to prove the result in the case that $\varepsilon < a/5$. By Lemma 1.5.4b we may choose $\delta > o$ such that if $\phi \in I(\tilde{M})$ has finite order $\leq n_o$ and if $d(q,\phi q) < \delta$ for some point $q \in \tilde{M}$, then $d(q, \mathrm{Fix}(\phi)) < \varepsilon' = \varepsilon/4^{n_o}$. In addition we require that δ be so small that $2\varepsilon + n_o\,\delta < a$ and $4\varepsilon + 2\delta < a$. We assert that a number δ chosen in this manner satisfies the conditions of the Proposition.

Now let $p \in \tilde{M}$ be a point such that $d_{\Gamma^*}(p) \geq a$. We outline how to locate a point $q \in B_\varepsilon(p)$ such that $\delta_\Gamma(q) \geq \delta$. Let $S = \{\phi \in \Gamma : d_\phi(q) = d(q,\phi q) < \delta$ for some $q \in B_\varepsilon(p)\}$. The set S is clearly finite since Γ is discrete. Let $S_o = \{\phi \in S : \phi$ has finite order $\leq n_o\}$. If S_o is empty then we show that $d_\Gamma(q) \geq \delta$ for all $q \in B_\varepsilon(p)$. If S_o is nonempty and if $\{\phi_1, \ldots, \phi_N\}$ is an enumeration of S_o, then for $1 \leq i \leq N$ we define $F_i = \mathrm{Fix}(\phi_i)$, which is a nonempty, complete, totally geodesic submanifold of $\tilde{M}$. By Lemma 1.5.4a there exists a point $q \in B_\varepsilon(p)$ such that $d(q, F_i) \geq \varepsilon/4^N$ for every i with $1 \leq i \leq N$. We show that $d_\Gamma(q) \geq \delta$, which completes the proof.

Suppose first that S_o is empty and let $q \in B_\varepsilon(p)$ be given arbitrarily. If $d_\Gamma(q) < \delta$ then there exists $\phi \in S$ such that $d_\phi(q) = d_\Gamma(q) < \delta$. Choose an integer k with $1 \leq k \leq n_o$ such that $\phi^k \in \Gamma^*$. Note that $\phi^k \neq 1$ since S_o is empty. Next,

$$
\begin{aligned}
d(p,\phi^k p) &\leq d(p,q) + d(q,\phi^k q) + d(\phi^k q,\phi^k p) \\
&\leq 2\, d(p,q) + k\, d(q,\phi q) \\
&< 2\,\varepsilon + k\delta \leq 2\,\varepsilon + n_o\delta < a
\end{aligned}
$$

by the choice of δ. Hence $d_{\Gamma^*}(p) \leq d_{\phi^k}(p) < a$, contradicting the hypothesis that $d_{\Gamma^*}(p) \geq a$. Therefore $d_\Gamma(q) \geq \delta$ for all $q \in B_\varepsilon(p)$.

Next suppose that $S_o = \{\phi_1, \ldots, \phi_N\}$ is nonempty. We show first that S has at most n_o elements and hence that $N \leq n_o$. If S had more than n_o elements then we could find distinct elements $\phi, \psi \in S$ such that $\phi\Gamma^* = \psi\Gamma^*$ and hence $\psi^{-1}\phi = \gamma^* \neq 1$ in Γ^*. By the triangle inequality and the definition of S it follows that $d(p, \phi p) < 2\varepsilon + \delta$ for any $\phi \in S$, and hence

$$d(p, \gamma^* p) \leq d(p, \psi^{-1}p) + d(\psi^{-1}p, \psi^{-1}\phi p)$$

$$= d(p, \psi p) + d(p, \phi p) < 4\varepsilon + 2\delta < a$$

by the choice of δ. This contradicts the hypothesis that $d_{\Gamma^*}(p) \geq a$. Hence S has at most n_o elements.

Now, by Lemma 1.5.4a we may choose $q \in B_\varepsilon(p)$ so that for $1 \leq i \leq N$ we have $d(q, F_i) \geq \varepsilon/4^N \geq \varepsilon/4^{n_o} = \varepsilon'$, where $F_i = \mathrm{Fix}(\phi_i)$. We show that $d_\Gamma(q) \geq \delta$. Suppose that this is not the case and choose $\phi \in S$ so that $d(q, \phi q) = d_\Gamma(q) < \delta$. Choose an integer k with $1 \leq k \leq n_o$ such that $\phi^k \in \Gamma^*$. If $\phi^k = 1$, then $\phi = \phi_i \in S_o$ for some $1 \leq i \leq N$, and it follows by the definition of δ that $d(q, F_i) < \varepsilon'$, a contradiction to the way in which q was chosen. On the other hand, if $\phi^k \neq 1$ then $d_{\Gamma^*}(p) \leq d_{\phi^k}(p) < a$, exactly as in a previous argument of this proof. This contradiction shows that $d_\Gamma(q) \geq \delta$ and completes the proof of the Proposition. □

We now prove the Lemmas 1.5.4a and 1.5.4b.

<u>Proof of Lemma 1.5.4a</u> We proceed by induction on N, beginning with $N = 1$. If p lies on F_1, then let γ be any unit speed geodesic of $\tilde{M}$ with $\gamma(o) = p$ and $\gamma'(o)$ orthogonal to F_1 at p. The point $q = \gamma(R/2)$ satisfies conditions 1) and 2) of Lemma 1.5.4a. If p does not lie on F_1 let $c = d(p, F_1) > o$ and let $\gamma : [o, c] \to \tilde{M}$ be the unique unit speed geodesic such that $\gamma(o) = p$ and $\gamma(c) = p'$, where p' denotes the unique point on F_1 that is closest to p. The point $q = \gamma(-R/2)$ now satisfies conditions 1) and 2) of the Lemma.

Let $N \geq 2$ be any positive integer and suppose that the Lemma has been proved for $N - 1$. By induction we may choose $q \in B_R(p)$ such that a) $B_{(R/4^{N-1})}(q) \subseteq B_R(p)$ and b) $d(q', F_i) \geq R/4^{N-1}$ for $1 \leq i \leq N-1$ and $q' \in B_{(R/4^{N-1})}(q)$. By the previous paragraph there exists a point $q^* \in B_{R'}(q)$, where $R' = R/4^{N-1}$, such that a') $B_{(R'/4)}(q^*) \subseteq B_{R'}(q)$ and b') If $q' \in B_{(R'/4)}(q^*) = B_{(R/4^N)}(q^*)$, then $d(q', F_N) \geq R'/4 = R/4^N$.

From a') and a) we see that $B_{(R/4^N)}(q^*) = B_{(R'/4)}(q^*) \subseteq B_{R'}(q) \subseteq B_R(p)$.

If $q' \in B_{(R/4^N)}(q^*) \subseteq B_{R'}(q)$ then from b) we obtain $d(q',F_i) \geq R/4^{N-1} > R/4^N$ for $1 \leq i \leq N-1$, and from b') we obtain $d(q', F_N) \geq R/4^N$. Hence the point q^* satisfies conditions 1) and 2) of Lemma 1.5.4a. □

Proof of Lemma 1.5.4b Let $F = \text{Fix}(\phi)$ and suppose that $d(p,\phi p) < \delta$ for some point $p \in \tilde{M}$. If p lies in F then there is nothing to prove so we may assume that $p \in \tilde{M} - F$. The set F is a nonempty, complete, proper, totally geodesic submanifold of $\tilde{M}$ since F contains the entire maximal geodesic joining any two points of F. Let q be the unique point on F that is closest to p. If we identify $T_q\tilde{M}$ with $\mathbb{R}^n$, $n = \dim \tilde{M}$, by means of a fixed orthonormal basis of $T_q\tilde{M}$, then we may identify $A = (d\phi)_q : T_q\tilde{M} \to T_q\tilde{M}$ with an element of the orthogonal group $O(n)$. Let V denote T_qF and $V^\perp$ the orthogonal complement to V in $T_q\tilde{M} \approx \mathbb{R}^n$.

We assert that $\sphericalangle(v,Av) \geq 2\pi/N$ for all $v \in V^\perp$, where $N \geq 2$ is the integer occurring in the statement of the lemma. Decompose $V^\perp$ into an orthogonal direct sum $E_1 \oplus \ldots \oplus E_m$ of irreducible A-invariant subspaces, each of dimension 1 or 2. If $\dim E_i = 2$ then A acts on E_i by rotation through an angle θ_i. If $\dim E_i = 1$, then $A = -I$ on E_i or equivalently A rotates all vectors in E_i by an angle $\theta_i = \pi$. If $\theta = \min\{\theta_1,\ldots,\theta_m\}$ then $\theta \geq 2\pi/N$ since by hypothesis A has order $\leq N$. It follows that $\langle v, Av\rangle \leq \cos\theta \leq \cos(2\pi/N)$ or $\sphericalangle(v,Av) \geq 2\pi/N$ for all unit vectors v in $V^\perp$.

Now let γ be the unique unit speed geodesic with $\gamma(o) = q$ and $\gamma(R) = p$, where $R = d(p,q) = d(p,F)$. If $v = \gamma'(o)$ then $v \in V^\perp$ and from an application of the Rauch comparison theorem to $\tilde{M}$ and $\mathbb{R}^n$ with its flat metric we obtain the inequality

$$d(p,\phi p) \geq 2\,R\,\sin(\tfrac{1}{2}\sphericalangle(v,Av)) \geq 2\,R\,\sin(\pi/N).$$

If $d(p,\phi p) < \delta = 2\,\varepsilon\,\sin(\pi/N)$, then it follows from the inequality above that $d(p,F) = R < \varepsilon$. □

Proof of Proposition 1.6.4 Let $\Gamma \subseteq I(\tilde{M})$ be a group that satisfies the three conditions of the statement. If $\gamma(t)$ is any unit speed geodesic of $\tilde{M} = \tilde{M}_1 \times \tilde{M}_2$, then we may write

(1) $\gamma(t) = (\gamma_1(at), \gamma_2(bt))$

where γ_1,γ_2 are unit speed geodesics of $\tilde{M}_1$, $\tilde{M}_2$ and a,b are numbers $\geq o$ such that $a^2 + b^2 = 1$. We wish to show that the points $\gamma(\infty)$ and $\gamma(-\infty)$ are dual relative to Γ for an arbitrary geodesic γ of $\tilde{M}$. It suffices to consider the case that γ is tangent to neither of the factors $\tilde{M}_1$, $\tilde{M}_2$ (i.e. $a > o$ and $b > o$) since the unit vectors tangent to such geodesics are dense in $S\tilde{M}$; it is easy to verify that if $\{x_n\}$, $\{y_n\}$ are sequences in $\tilde{M}(\infty)$ converging to x,y, then x is Γ-dual to y if x_n is Γ-dual to y_n for every n.

Let a unit speed geodesic $\gamma(t) = (\gamma_1(at), \gamma_2(bt))$ be given in $\tilde{M}$, where $a > 0$, $b > 0$ and γ_1, γ_2 are unit speed geodesics in $\tilde{M}_1$, $\tilde{M}_2$. By the first hypothesis on Γ there exists a sequence $\{\psi_n\} \subseteq \Gamma_2 = p_2(\Gamma)$ such that

(2) $\psi_n(q_2) \to \gamma_2(\infty)$ and $\psi_n^{-1}(q_2) \to \gamma_2(-\infty)$

for any point $q_2 \in \tilde{M}_2$. Let $q_2 = \gamma_2(o) \in \tilde{M}_2$ and define

(3) $t_n = (1/b)\ d(q_2, \psi_n q_2)$

Let β_n, $\beta_n{}^*$ be unit speed geodesics in $\tilde{M}_2$ such that

(4) $\beta_n(o) = q_2$, $\beta_n(bt_n) = \psi_n(q_2)$

$\beta_n{}^*(o) = q_2$, $\beta_n{}^*(bt_n) = \psi_n{}^{-1}(q_2)$

The assertion (2) implies

(5) $\beta_n{}'(o) \to \gamma_2{}'(o)$ and $\beta_n{}^{*\prime}(o) \to -\gamma_2{}'(o)$ as $n \to \infty$

Now let $q_1 = \gamma_1(o) \in \tilde{M}_1$. By the third hypothesis on Γ we can find a sequence $\{\xi_n\} \subseteq N = \text{kernel } (p_2)$ and a positive number R such that

(6) $d(\xi_n q_1, \gamma_1(at_n)) \leq R$ for all n.

If we regard q_1 as the origin of coordinates in the Euclidean space $\tilde{M}_1$, then since the elements of N (and, more generally, the elements of $\Gamma_1 = p_1(\Gamma)$) act as translations on $\tilde{M}_1$ it follows that $\xi_n{}^{-1}(q_1)$ may be identified with $-\xi_n(q_1)$. From (6) and Euclidean geometry we obtain

(7) $d(\xi_n{}^{-1}(q_1), \gamma_1(-at_n)) \leq R$ for all n.

If we set

(8) $a_n = (1/t_n)\ d(q_1, \xi_n q_1)$

then from (6) we obtain

(9) $|a_n - a|\ |t_n| = |a_n t_n - at_n| \leq R$ for every n.

Now let σ_n, $\sigma_n{}^*$ be the unit speed geodesics in $\tilde{M}_1$ such that

(10) $\sigma_n(o) = q_1$, $\sigma_n(a_n t_n) = \xi_n(q_1)$

$\sigma_n{}^*(o) = q_1$, $\sigma_n{}^*(a_n t_n) = \xi_n{}^{-1}(q_1)$

From (6), (7) and the law of cosines we conclude

(11) $\sigma_n{}'(o) \to \gamma_1{}'(o)$ and $\sigma_n{}^{*\prime}(o) \to -\gamma_1{}'(o)$ as $n \to \infty$

Now choose a sequence $\phi_n \subseteq \Gamma$ in the form $\phi_n = T_n \times \psi_n \in I(\tilde{M}_1) \times I(\tilde{M}_2)$ such that

$$(12)\quad \begin{array}{l} T_n \text{ is a translation of } \tilde{M}_1 \text{ with } d(q_1, T_nq_1) \leq R \\ \{\psi_n\} \subseteq \Gamma_2 = p_2(\Gamma) \text{ satisfies (2) above} \end{array}$$

Here R is the positive constant from (6). We can choose $\{\phi_n\}$ in this manner since by hypothesis $\tilde{M}_1$ is compact modulo N, where N = kernel (p_2). Next we define

$$(13)\quad \phi_n^* = \xi_n\, \phi_n = T_n\, \xi_n \times \psi_n$$

where $\{\xi_n\}$ is the sequence of translations in $\tilde{M}_1$ defined by (6). Let γ_n, γ_n^* be the unit speed geodesics defined by

$$(14)\quad \begin{aligned} \gamma_n(t) &= (\sigma_n(at), \beta_n(bt)) \\ \gamma_n^*(t) &= (\sigma_n^*(at, \beta_n^*(bt)) \end{aligned}$$

where β_n, β_n^*, σ_n, σ_n^* are the geodesics defined in (4), (10) respectively. We assert that

$$(15a)\quad \begin{aligned} &d(\phi_n^*q, \gamma_n\,(t_n)) \leq 2R \\ &d((\phi_n^*)^{-1}q, \gamma_n^*(t_n)) \leq 2R \end{aligned}$$

$$(15b)\quad \gamma_n'(o) \to \gamma'(o) \quad \text{and } \gamma_n^{*\prime}(o) \to -\gamma'(o) \quad \text{as } n \to \infty$$

where $q = (q_1, q_2) = \gamma(o)$. Clearly the assertions (15a) and (15b) will imply that $\phi_n^*(q) \to \gamma(\infty)$ and $(\phi_n^*)^{-1}(q) \to \gamma(-\infty)$, completing the proof of the lemma.

We shall prove only the first assertion in (15a) and (15b) since the proof of the second assertion in both parts is entirely similar. By (4), (13) and (14) we have $\phi_n^*(q) = (T_n\xi_n(q_1), \psi_n(q_2))$ and $\gamma_n(t_n) = (\sigma_n(at_n), \beta_n(bt_n)) = (\sigma_n(at_n), \psi_n(q_2))$. Hence from (9), (10) and (12) we obtain

$$\begin{aligned} d(\gamma_n(t_n), \phi_n^*(q)) &= d(\sigma_n(at_n), T_n\xi_n(q_1)) \\ &\leq d(\sigma_n(at_n), \sigma_n(a_nt_n)) + d(\xi_n(q_1), T_n\xi_n(q_1)) \\ &= |a - a_n| \;\; |t_n| + d(q_1, T_nq_1) \leq 2\,R \end{aligned}$$

which is the first assertion of (15a). From (14), (11), (5) and (1) we obtain

$$\gamma_n'(o) = a\; \sigma_n'(o) + b\; \beta_n'(o) \to a\; \gamma_1'(o) + b\; \gamma_2'(o) = \gamma'(o)$$

as $n \to \infty$, which is the first assertion of (15b). $\square$

<u>Proof of Proposition 2.4</u> To prove the second part of assertion 1) we shall use the following result that is somewhat stronger than we

actually need.

Lemma 2.4 Let $\Gamma \subseteq I(\mathbb{R}^n)$ be an arbitrary discrete group of isometries. Then there exists a complete totally geodesic submanifold W of $\mathbb{R}^n$ such that

a) $\Gamma(W) \subseteq W$

b) W is compact modulo Γ (cf. (1.6.3))

Proof Following the proof of Theorem 3.2.8 of [W3, p. 102] we let G_0 denote the component of $G = \Gamma \cdot \mathbb{R}^n \subseteq I(\mathbb{R}^n)$ that contains the identity. Here $\mathbb{R}^n$ denotes also the group of translations of $\mathbb{R}^n$. By Lemma 3.2.7 of [W3, p. 101] the commutator group $[G_0, G_0]$ consists of translations and hence G_0 is solvable. The group $\Gamma_0 = \Gamma \cap G_0$ is also solvable and has finite index in Γ since G_0 has finite index in G. Every element of $I(\mathbb{R}^n)$ is semisimple in the sense of [GW] and it follows from Theorem 1 of [GW] that Γ_0 leaves invariant some complete, totally geodesic submanifold W of $\mathbb{R}^n$ such that W is compact modulo Γ_0. (An elementary proof of the existence of W is also contained in the proof of Theorem 3.2.8 of [W3, p. 102]). Since Γ_0 is a normal subgroup of Γ it follows that for each $\phi \in \Gamma$ the submanifold $\phi(W)$ is invariant under Γ_0 and compact modulo Γ_0. The submanifolds $\{\phi(W) : \phi \in \Gamma\}$ are all parallel to each other in $\mathbb{R}^n$ (that is, they differ by a translation) since all of them are compact modulo Γ_0. Moreover there are only finitely many such submanifolds since Γ_0 has finite index in Γ. We assume that there are at least two such submanifolds for otherwise we are done.

Let S denote the collection of submanifolds of $\mathbb{R}^n$ of the form $W + v$, $v \in \mathbb{R}^n$. Define a metric on S by letting $d(W_1, W_2)$ denote the perpendicular distance between W_1 and W_2. For $W^* \in S$ we define $f(W^*) = \max\{d(W^*, \phi(W)) : \phi \in \Gamma\}$. It is easy to see that there exists $W' \in S$ such that $f(W') \leq f(W^*)$ for all $W^* \in S$. We assert that the minimum point W' is unique. It will then follow that W' is invariant under Γ and compact modulo Γ since $f(\phi(W^*)) = f(W^*)$ for all $\phi \in \Gamma$ and all $W^* \in S$. This will complete the proof.

Suppose that W_1, W_2 are distinct submanifolds in S such that $R = f(W_1) = f(W_2) \leq f(W^*)$ for all W^* in S. Let γ be a line segment that runs from W_1 to W_2 and is perpendicular to both W_1 and W_2. Let W_3 be an element of S that passes through an interior point of γ. We show that $f(W_3) < R$, a contradiction.

Let $\phi \in \Gamma$ and a point p in $\phi(W)$ be given. Let p_1, p_2 be the orthogonal projections of p onto W_1, W_2 respectively, and let σ be the line segment joining p_1 to p_2. If A denotes the closed convex hull in $\mathbb{R}^n$ of $W_1 \cup W_2$, then A contains W_3 and σ. Moreover, W_3 separates W_1 from W_2 in A, and hence σ meets W_3 in an interior point p_3 of σ. It follows that $d(W_3, \phi(W)) \le d(p, p_3) < \max\{d(p,p_1), d(p,p_2)\} = \max\{d(\phi(W), W_1), d(\phi(W), W_2)\} \le R = f(W_1) = f(W_2)$. Hence $d(W_3, \phi(W)) < R$ for each of the finitely many submanifolds $\phi(W)$, $\phi \in \Gamma$, and it follows that $f(W_3) < R$, a contradiction. This completes the proof of the Lemma. □

Assuming that N is an L-subgroup the remaining part of assertion 1) of Proposition 2.4 follows directly from the lemma above and a theorem of Bieberbach [W3, p. 100]. We prove assertion 2) and the fact that N is an L-subgroup simultaneously, but break the proof into 2 cases a) $\Gamma_1 = p_1(\Gamma)$ is discrete and b) $\Gamma_1 = p_1(\Gamma)$ is not discrete.

We consider first the case a) where Γ_1 is discrete. Then Γ is a subgroup of the discrete group $\Gamma_1 \times \Gamma_2 \subseteq I(\tilde{M}_1) \times I(\tilde{M}_2) \subseteq I(\tilde{M})$, and it follows from Proposition 2.3 that Γ has finite index in $\Gamma_1 \times \Gamma_2$. Therefore the groups $\Gamma_1^* = \Gamma_1 \cap \Gamma \subseteq I(\tilde{M}_1) \times \{1\}$ and $\Gamma_2^* = \Gamma_2 \cap \Gamma \subseteq \{1\} \times I(\tilde{M}_2)$ have finite index in $\Gamma_1 = \Gamma_1 \cap (\Gamma_1 \times \Gamma_2)$ and $\Gamma_2 = \Gamma_2 \cap (\Gamma_1 \times \Gamma_2)$ respectively by Lemma 5.3b. If $\Gamma^* = \Gamma_1^* \times \Gamma_2^*$, then $\Gamma^* \subseteq \Gamma \subseteq \Gamma_1 \times \Gamma_2$, and it follows that Γ^* has finite index in Γ. This proves 2a). By assertions 2) and 3) of Proposition 2.3 the group Γ_1^* is an L-subgroup since it has finite index in $\Gamma_1 = p_1(\Gamma)$. Finally $N = \text{kernel } (p_2) = \Gamma_1^*$ is an L-subgroup of $I(\tilde{M}_1)$.

Next, we consider case b) where Γ_1 is not discrete. The proof in this case requires a strengthening of the arguments used in the proof of Theorem 4.1 of [E4]. The proof of that result already shows that N is nonempty and $L(N) = \tilde{M}_1(\infty)$. (One must make minor changes in the proof of Lemma 4.1c in arguments involving the displacement function d_N. See a) following (1.5.3)). Clearly N is discrete since Γ is discrete. The definition of L-subgroup requires that we show

$$\tilde{M}_1^{\,a} = \{p \in \tilde{M}_1 : d_N(p) \ge a\} \text{ is compact modulo N for every } a > 0$$

and

N satisfies the duality condition in $\tilde{M}_1$.

The remaining assertions in 2b) will be established in the course of the proof.

We prove that $\tilde{M}_1^{\,a}$ is compact modulo N. Let $a > 0$ be given and let $\{p_n\} \subseteq \tilde{M}_1$ be a sequence such that $d_N(p_n) \ge a$ for every n. It suffices to construct a sequence $\{\rho_n\} \subseteq N$ such that $\{\rho_n(p_n)\}$ is bounded. Fix $q_2 \in \tilde{M}_2$ and let $q_n = (p_n, q_2) \in \tilde{M}_1 \times \tilde{M}_2 = \tilde{M}$. Observe that $d_\Gamma(q_n) \ge c > 0$ for all sufficiently large n and some positive constant c. If this were not the case, then we would have $d_\Gamma(q_n) \to 0$, passing to a subsequence if necessary. It would then follow as in the proof of Lemma 4.1b of [E4] that $d_N(q_n) = d_N(p_n) \to 0$ as $n \to +\infty$, contradicting our assumption.

Choose $c > 0$ so that $d_\Gamma(q_n) \ge c$ for all sufficiently large n. Since Γ is an L-subgroup we may choose $\{\xi_n\} \subseteq \Gamma$ so that $\xi_n(q_n)$ is a bounded sequence in $\tilde{M}$. It follows as in the proof of Lemma 4.1a of [E4] that by passing to a suitable subsequence $p_2(\xi_n)$ is constant in $I(\tilde{M}_2)$ and hence $\rho_n = \xi_1^{-1}\xi_n$ lies in N for every n. Therefore $p_n^* = \rho_n(p_n)$ is a bounded sequence in $\tilde{M}_1$, and we conclude that $\tilde{M}_1^{\,a}$ is compact modulo N.

We prove that N satisfies the duality condition in $\tilde{M}_1$. Since $\Gamma_1 = p_1(\Gamma)$ is assumed to be not discrete the arguments of [E4, pp. 464-466] apply and show that there exists a Γ-invariant splitting

$$(*) \quad \tilde{M} = \tilde{M}_\alpha \times \tilde{M}_\beta \times \tilde{M}_2$$

with projection homomorphisms $p_\alpha : \Gamma \to I(\tilde{M}_\alpha)$, $p_\beta : \Gamma \to I(\tilde{M}_\beta)$ and $p_2 : \Gamma \to I(\tilde{M}_2)$ such that the following conditions are satisfied:

i) $\tilde{M}_\alpha$ is a Euclidean space of positive dimension.

ii) $\tilde{M}_1 = \tilde{M}_\alpha \times \tilde{M}_\beta$, Riemannian product

iii) $\Gamma_\beta = p_\beta(\Gamma)$ is a discrete subgroup of $I(\tilde{M}_\beta)$

iv) If we write $\tilde{M} = M_1^* \times M_2^*$, where $M_1^* = \tilde{M}_\alpha$ and $M_2^* = \tilde{M}_\beta \times \tilde{M}_2$, and if we let $p_1^* : \Gamma \to I(M_1^*)$ and $p_2^* : \Gamma \to I(M_2^*)$ denote the corresponding projection homomorphisms, then

a) $\Gamma_2^* = p_2^*(\Gamma)$ is a discrete subgroup of $I(M_2^*)$

b) $N^* = \text{kernel } (p_2^*)$ is a uniform lattice of translations in M_1^*. In particular N^* is a subgroup of $N = \text{kernel } (p_2)$ that consists of Clifford translations.

Note that assertion 2b) of the Proposition follows from assertions i) and ivb) above.

We outline the proof that $N = \text{kernel } (p_2)$ satisfies the duality condition. We consider the group $N_\alpha = p_\alpha(N) \subseteq I(\tilde{M}_\alpha)$ and we let N_α^* denote $C(N_\alpha)$, the subgroup of (Clifford) translations in N_α. We shall show that N_α^* has finite index in N_α. It will then follow that $N^{**} = N \cap p_\alpha^{-1}(N_\alpha^*)$ has finite index in N and $N_\alpha^* = p_\alpha(N^{**})$ consists of translations of $\tilde{M}_\alpha$. Applying Proposition 1.6.4 to the splitting $\tilde{M}_1 = \tilde{M}_\alpha \times \tilde{M}_\beta$ in ii) above we shall conclude that N^{**} satisfies the duality condition and consequently N does also.

We show first that N_α^* has finite index in N_α. Suppose first that N_α is a discrete subgroup of $I(\tilde{M}_\alpha)$. Since $N^* \subseteq N_\alpha$ is a uniform lattice of translations in $\tilde{M}_\alpha$ by ivb) above it follows that the quotient space M_α/N_α is compact and hence N^* has finite index in N_α. It follows that $N_\alpha^* = C(N_\alpha)$ has finite index in N_α since $N^* \subseteq N_\alpha^* \subseteq N_\alpha$. Next, suppose that N_α is not discrete and let $G = \overline{N_\alpha}$, the closure of N_α in $I(\tilde{M}_\alpha)$. G is a Lie group of positive dimension. Let G_0 denote the connected component of G that contains the identity. The fact that $N^* = \text{kernel } (p_2^*)$ is discrete and is normalized by N_α implies that N^* is normalized by G and hence centralized by G_o. It follows either directly or from Proposition 2.3 of [CE] that G_o consists of translations in $\tilde{M}_\alpha$ since N* is a uniform lattice of translations in $\tilde{M}_\alpha$.

Let $S \subseteq N_\alpha$ be any set such that the right cosets $\{N_\alpha{*}\phi : \phi \in S\}$ are all distinct in N_α. We must prove that S is finite. Fix a point $p_\alpha \in N_\alpha$ and choose $R > o$ so that the diameter of $\tilde{M}_\alpha/N^*$ is at most R. Since $N^* \subseteq N_\alpha^*$ we may assume that the elements of S have been chosen so that $d(p_\alpha, \phi(p_\alpha)) \leq R$ for all $\phi \in S$. It follows from Proposition 1.5.1 that S lies in some compact subset C of G. Let $U \subseteq G$ be a neighborhood of the identity such that $U \cdot U \subseteq G_o$ and $U = U^{-1}$. The compact set C is covered by finitely many translates $U \cdot \phi_i$, $1 \leq i \leq N$, with $\phi_i \in G$. If S were an infinite set then infinitely many elements of S would lie in one of the sets $U \cdot \phi_i$, and hence we could find distinct elements ϕ,ψ in S such that $\phi \cdot \psi^{-1} \in N_\alpha \cap (U \cdot U) \subseteq N_\alpha \cap G_o$. Since G_o consists of translations of $\tilde{M}_\alpha$ it would then follow that $N_\alpha{*}\phi = N_\alpha{*}\psi$, a contradiction to the definition of S. Hence S is finite and N_α^* has finite index in N_α.

If we define $N^{**} = N \cap p_\alpha^{-1}(N_\alpha^*)$, then $p_\alpha(N^{**}) = N_\alpha^*$ consists of translations in $\tilde{M}_\alpha$ and N** has finite index in N since N_α^* has finite index in $N_\alpha = p_\alpha(N)$. We wish to apply Proposition 1.6.4 to the Riemannian splitting $\tilde{M}_1 = \tilde{M}_\alpha \times \tilde{M}_\beta$ from ii) above and the group $N^{**} \subseteq I(\tilde{M}_1)$. Since $N^* \subseteq$ kernel $(p_\beta | N^{**})$ it follows from iv b) that $\tilde{M}_\alpha$ is compact modulo kernel $(p_\beta | N^{**})$. To prove that N** and hence also N satisfy the duality condition it suffices by Proposition 1.6.4 to show that $p_\beta(N^{**})$ satisfies the duality condition in $\tilde{M}_\beta$.

Consider the group $\Gamma_2^* = p_2^*(\Gamma) = \{\{1\} \times p_\beta(\phi) \times p_2(\phi) : \phi \in \Gamma\}$, using the notation of (*) above. By a slight abuse of notation we let p_β, p_2 also denote the projections $\Gamma_2^* \to I(\tilde{M}_\beta)$ and $\Gamma_2^* \to I(\tilde{M}_2)$ respectively. By assertion iv a) of (*) above we know that Γ_2^* is discrete and hence by Proposition 2.3 Γ_2^* is an L-subgroup of $I(M_2^*)$, where $M_2^* = \tilde{M}_\beta \times \tilde{M}_2$.

Now $\Gamma_2 = p_2(\Gamma_2^*) = p_2(\Gamma)$ is a discrete subgroup of $I(\tilde{M}_2)$ by one of the hypotheses of Proposition 2.4, and $\Gamma_\beta = p_\beta(\Gamma) = p_\beta(\Gamma_2^*)$ is discrete by assertion iii) of (*). Applying assertion 2a) of Proposition 2.4 to the L-subgroup $\Gamma_2^* \subseteq I(M_2^*)$ we conclude that N_β = kernel $(p_2 \mid \Gamma_2^*) = \Gamma_2^* \cap \Gamma_\beta$ has finite index in Γ_β. Now by definition $N_\beta = \{p_\beta(\phi) : \phi \in \Gamma$ and $p_2(\phi) = 1\} = p_\beta(N)$ since N = kernel (p_2). It follows that $p_\beta(N^{**})$ has finite index in N_β and hence in Γ_β since N** has finite index in N. Since $\Gamma_\beta = p_\beta(\Gamma)$ satisfies the duality condition in $\tilde{M}_\beta$ by assertion 3) of Proposition 1.6.2 it follows from assertion 2) of Proposition 1.6.2 that $p_\beta(N^{**})$ satisfies the duality condition in $\tilde{M}_\beta$. Finally N** satisfies the duality condition in $\tilde{M}_1$ by Proposition 1.6.4, which proves that N satisfies the duality condition and concludes the proof that N is an L-subgroup of $I(\tilde{M}_1)$. □

Proof of Proposition 2.6 We break the proof into a series of lemmas. The result follows from Lemmas 2.6b, 2.6c and 2.6d, which show (in the notation of Lemmas 2.6a and 2.6b) that $\pi_A(A)$ is an L-subgroup of $I(\tilde{M}_A)$ that is isomorphic to A, where $\tilde{M}_A = V_A \times E_A$. We shall assume that E_A has positive dimension for otherwise $\tilde{M}_A = V_A$ and $\pi_A(A) = p_A(\Gamma)$ is an L-subgroup of $I(\tilde{M}_A)$ by assertions 2) and 3) of Lemma 2.6a.

Lemma 2.6a Let $\Gamma \subseteq I(\tilde{M})$ be an L-subgroup whose center Z is nontrivial. Then there exists a Γ-invariant splitting

$$\tilde{M} = V_A \times E_A \times V_B \times E_B$$

such that

1) The spaces E_A, E_B are Euclidean spaces.

2) $\phi = \phi^A \times T^A \times \{1\} \times T^B$ if $\phi \in A$

$\phi = \{1\} \times \tilde{T}^A \times \phi^B \times \tilde{T}^B$ if $\phi \in B$

where T^A, $\tilde{T}^A$ and T^B, $\tilde{T}^B$ are translations in E_A, E_B respectively and ϕ^A, ϕ^B are arbitrary elements of $I(\tilde{M}_A)$, $I(\tilde{M}_B)$.

3) The groups $p_A(\Gamma)$ and $p_B(\Gamma)$ are discrete L-subgroups with trivial center of $I(V_A)$ and $I(V_B)$, where $p_A : \Gamma \to I(V_A)$ and $p_B : \Gamma \to I(V_B)$ are the projection homomorphisms.

4) If $\phi \in Z_B$, the center of B, then

$$\phi = \{1\} \times \{1\} \times \{1\} \times \tilde{T}^B$$

where $\tilde{T}^B$ is a translation of E_B, and moreover $Z_B \subseteq I(E_B)$ is a uniform lattice in E_B.

Lemma 2.6b In terms of the Γ-invariant splitting from Lemma 2.6a let $\tilde{M}_A = V_A \times E_A$ and let $\pi_A : \Gamma \to I(\tilde{M}_A)$ be the corresponding projection homomorphism. Then $\Gamma^* = \pi_A(A)$ is a discrete subgroup of $I(\tilde{M}_A)$ and $\pi_A : A \to \Gamma^*$ is an isomorphism.

Lemma 2.6c For each positive number a the set $\tilde{M}_A^a = \{p \in \tilde{M}_A : d_{\Gamma^*}(p) \geq a\}$ is compact modulo $\Gamma^* = \pi_A(A)$.

Lemma 2.6d $\Gamma^* = \pi_A(A)$ satisfies the duality condition in $\tilde{M}_A$.

We now prove the lemmas 2.6a through 2.6d.

Proof of Lemma 2.6a Let $Z(\Gamma) = \{\phi \in I(\tilde{M}) : \phi\psi = \psi\phi$ for all $\psi \in \Gamma\}$ denote the centralizer of Γ in $I(\tilde{M})$, and let $Z = \Gamma \cap Z(\Gamma)$ denote the nontrivial center of Γ. By Proposition 3.1 there exists a Γ-invariant splitting $\tilde{M} = \tilde{M}_1 \times \tilde{M}_2$ and projection homomorphisms $p_i : \Gamma \to I(\tilde{M}_i)$, i=1,2, that satisfy the following properties:

1) $\tilde{M}_1$ is a Euclidean space of positive dimension.

2) $p_1(\Gamma)$ consists of translations in $\tilde{M}_1$ and $Z(\Gamma) \subseteq I(\tilde{M}_1) \times \{1\}$ is the subgroup consisting of all translations in $\tilde{M}_1$.

3) Z = kernel (p_2) and Z is a uniform lattice in $\tilde{M}_1$.

4) $\Gamma_2 = p_2(\Gamma)$ is an L-subgroup of $I(\tilde{M}_2)$ whose center is trivial.

We decompose $\tilde{M}_2$. Since $p_2(\Gamma)$ has trivial center and the groups $p_2(A)$, $p_2(B)$ commute it follows that $p_2(A) \cap p_2(B) = \{1\}$. Hence $p_2(\Gamma)$ is a direct product $p_2(A) \times p_2(B)$ and since $p_2(\Gamma)$ is an L-subgroup of $I(\tilde{M}_2)$ it follows from Proposition 2.5 that we may write $\tilde{M}_2$ as a Riemannian product $V_A \times V_B$, where $p_2(A) \subseteq I(V_A) \times \{1\}$ and $p_2(B) \subseteq \{1\} \times I(V_B)$ are L-subgroups of $I(V_A)$ and $I(V_B)$ respectively. It may be the case that either $p_2(A) = \{1\}$ or $p_2(B) = \{1\}$, and under these circumstances the decomposition of $\tilde{M}_2$ is trivial and the statement of Lemma 2.6a becomes correspondingly simpler.

Next we decompose $\tilde{M}_1$ into a $p_1(\Gamma)$-invariant splitting $E_A \times E_B$. For each point $p \in \tilde{M}_1$ let $E_B(p)$ denote the closed convex hull in $\tilde{M}_1$ of $Z_B{}^*(p) = \{p_1(\phi)(p) : \phi \in Z_B = Z \cap B\}$. Since $\tilde{M}_1$ is a Euclidean space and since the elements of $p_1(\Gamma)$ are translations on $\tilde{M}_1$ the set $E_B(p)$ may also be described as the linear span in $\tilde{M}_1$ of the orbit $Z_B{}^*(p)$ if one identifies p with the origin in $\tilde{M}_1$. Hence if ε_B is the distribution in $\tilde{M}_1$ whose value at p is the tangent space at p of $E_B(p)$, then ε_B is an integrable parallel foliation in $\tilde{M}_1$. If ε_A denotes the orthogonal complementary foliation to ε_B in $\tilde{M}_1$, then since $p_1(\Gamma)$ is abelian we obtain a $p_1(\Gamma)$-invariant Riemannian splitting

$$\tilde{M}_1 = E_A \times E_B$$

where the factors E_A, E_B induce the foliations ε_A, ε_B. Again, if $Z \cap B = \{1\}$ then the decomposition becomes $\tilde{M}_1 = E_A$.

Finally by combining the decompositions $\tilde{M}_2 = V_A \times V_B$ and $\tilde{M}_1 = E_A \times E_B$ we obtain the decomposition of $\tilde{M}$ that is stated in Lemma 2.6a. The first two assertions of Lemma 2.6a have already been verified in the course of the discussion above. To prove 3) of Lemma 2.6a we observe that $p_A(\Gamma) = p_2(A)$ and $p_B(\Gamma) = p_2(B)$ by construction. Hence $p_A(\Gamma)$ and $p_B(\Gamma)$ are discrete centerless groups since $p_2(\Gamma) = p_2(A) \times p_2(B)$ is a discrete centerless group by 4) above. To prove 4) of Lemma 2.6a we observe that if $\phi \in Z_B = Z \cap B$, then by an earlier observation ϕ lies in kernel (p_2) and hence $Z_B = p_1(Z_B) \subseteq p_1(\Gamma)$. It follows that Z_B regarded as a subgroup of $p_1(\Gamma)$ leaves invariant each leaf $E_B(p)$ of the foliation ε_B and acts as a uniform lattice of translations on $E_B(p)$. This completes the proof of 4).

<u>Proof of Lemma 2.6 b</u> We show first that $\pi_A(A)$ is isomorphic to A by showing that $A \cap$ kernel $(\pi_A) = \{1\}$. Let $a^* \in A \cap$ kernel (π_A) be given. By assertion 2) of Lemma 2.6 a the element a^* has the form

$$a^* = \{1\} \times \{1\} \times \{1\} \times T^B$$

$\in \{1\} \times I(E_B)$ for some translation T^B of E_B. Let Γ_B denote the discrete subgroup of Γ generated by a* and $Z_B = Z \cap B$, where Z denotes the center of Γ. By assertion 4) of Lemma 2.6 a the group Γ_B leaves invariant each leaf $E_B(p)$ of the foliation ε_B of $\tilde{M}$ induced by E_B and acts by translations on $E_B(p)$. If $a^* \neq \{1\}$ then since $A \cap B = \{1\}$ it follows that rank $(\Gamma_B) = 1 + \text{rank}\ (Z_B) = 1 + \dim E_B(p)$ since Z_B is a uniform lattice on each submanifold $E_B(p)$ by 4) of Lemma 2.6 a. Hence rank $(\Gamma_B) > \dim E_B(p)$, but this contradicts the fact that Γ_B acts as a discrete group of translations on the Euclidean space $E_B(p)$. Therefore $a^* = \{1\}$ and $\pi_A : A \to \pi_A(A)$ is an isomorphism.

Next we show that $\pi_A(A)$ is a discrete subgroup of $I(\tilde{M}_A)$. Choose a sequence $\{a_n\} \subseteq A$ such that $\pi_A(a_n) \to 1$ in $I(\tilde{M}_A)$ as $n \to +\infty$. It suffices to show that $\pi_A(a_n) = 1$ for large n. By Lemma 2.6 a we can write

(1) $a_n = \phi_n^A \times T_n^A \times \{1\} \times T_n^B$

where T_n^A, T_n^B are translations of E_A, E_B and ϕ_n^A is an arbitrary element of $I(V_A)$. Hence $a_n^* = \pi_A(a_n) = \phi_n^A \times T_n^A \to 1$ as $n \to \infty$. By 3) of Lemma 2.6 a the group $p_A(\Gamma)$ is discrete and hence

(2) $\phi_n^A = 1$ for sufficiently large n

since $\phi_n^A = p_A(a_n^*) \to 1$ as $n \to \infty$. Here $p_A : \Gamma \to I(V_A)$ denotes the projection homomorphism. It now follows from (1) and (2) above and assertion 2) of Lemma 2.6 a that a_n lies in Z, the center of Γ, for sufficiently large n.

Now fix a point $p = (p^A, q^A, p^B, q^B) \in \tilde{M} = V_A \times E_A \times V_B \times E_B$. Since $Z_B = Z \cap B$ is a uniform lattice in E_B by 4) of Lemma 2.6 a, we can find a number $R^* > o$ and elements $b_n \in Z_B$ of the form

$$b_n = \{1\} \times \{1\} \times \{1\} \times \tilde{T}_n^B$$

where $\tilde{T}_n^B$ is a translation of E_B, such that $d(T_n^B(q^B), \tilde{T}_n^B(q^B)) \leq R^*$ for every n, where T_n^B is the translation of E_B defined in (1). If $\xi_n = a_n b_n^{-1} = \{1\} \times T_n^A \times \{1\} \times T_n^B(\tilde{T}_n^B)^{-1}$, then for sufficiently large n the elements ξ_n lie in the center Z of Γ and

$$d^2(p, \xi_n p) = d^2(q^B, T_n^B(\tilde{T}_n^B)^{-1}(q^B)) + d^2(q^A, T_n^A(q^A))$$

$$\leq R^* + 1 \quad \text{for large n}$$

since $T_n^A \to 1$ as $n \to \infty$ by hypothesis. By the discreteness of Γ there are only finitely many distinct elements in the sequence $\{\xi_n\}$ and hence

(3) $T_n^A = 1$ for sufficiently large n

It follows from (1), (2) and (3) that $\pi_A(a_n) = \phi_n^A \times T_n^A = 1$ for sufficiently large n, and this proves that $\pi_A(A)$ is a discrete subgroup of $I(\tilde{M}_A)$. □

Proof of Lemma 2.6 c Recall that $p_B(\Gamma) \subseteq I(V_B)$ is discrete by assertion 3) of Lemma 2.6 a. We shall need the following result, stated in the notation of Lemma 2.6 a.

Sublemma Let $p = (p^A, q^A, p^B, q^B) \in V_A \times E_A \times V_B \times E_B = \tilde{M}$ be a point such that $p^B \in V_B$ is fixed by no element of $p_B(\Gamma)$ except the identity. Let F_p denote the leaf through p of the foliation of $\tilde{M}$ that corresponds to $\tilde{M}_A$. Let Γ^* denote $\pi_A(A) \subseteq I(\tilde{M}_A)$. Then for every $a > o$ there exists $c > o$ such that if $d_{\Gamma^*}(q^*) \geq a$ for some point $q^* = ((p^A)^*, (q^A)^*) \in V_A \times E_A = \tilde{M}_A$, then $d_\Gamma(q) \geq c$ where $q = ((p^A)^*, (q^A)^*, p^B, q^B) \in F_p$.

Proof We prove the sublemma by contradiction. Suppose that the sublemma is false and choose $p \in \tilde{M}$ as above, $a > o$ and a sequence $q_n^* = (p_n^A, q_n^A) \subseteq \tilde{M}_A$ such that $d_{\Gamma^*}(q_n^*) \geq a$ for all n but $d_\Gamma(q_n) \to o$ as $n \to \infty$, where $q_n = (p_n^A, q_n^A, p^B, q^B)$. Let $\{\phi_n\} \subseteq \Gamma$ be a sequence such that $\phi_n \neq 1$ and $d_\Gamma(q_n) = d(q_n, \phi_n q_n)$ for every n. From Lemma 2.6a and the fact that $\Gamma = A \times B$ we may write

$$(1) \quad \phi_n = a_n b_n \qquad \text{with } a_n \in A,\ b_n \in B$$

$$a_n = \phi_n^A \times T_n^A \times \{1\} \times T_n^B$$

$$b_n = \{1\} \times \tilde{T}_n^A \times \phi_n^B \times \tilde{T}_n^B$$

where ϕ_n^A, ϕ_n^B are elements of $I(V_A)$, $I(V_B)$ and where T_n^A, $\tilde{T}_n^A$ and T_n^B, $\tilde{T}_n^B$ are translations in E_A and E_B. It follows that $d(p^B, \phi_n^B(p^B)) \leq d(q_n, \phi_n q_n) \to o$ as $n \to +\infty$. Since $p_B(\Gamma)$ is discrete and p^B is not fixed by any non-identity element of $p_B(\Gamma)$ it follows that $\phi_n^B = \{1\}$ for all sufficiently large n. Hence $b_n \in Z \cap B = Z_B$ for large n by (1) above and assertion 2) of Lemma 2.6a. From assertion 4) of Lemma 2.6a it follows that $\tilde{T}_n^A = \{1\}$ for all sufficiently large n and hence we obtain

$$(2) \quad \phi_n = \phi_n^A \times T_n^A \times \{1\} \times T_n \qquad \text{for large n}$$

where $T_n = T_n^B \tilde{T}_n^B$, a translation of E_B. If $\phi_{n_k}^A \times T_{n_k}^A = \pi_A(\phi_{n_k}) \neq \{1\}$ for some subsequence $\{\phi_{n_k}\}$, then $d(q_{n_k}, \phi_{n_k}(q_{n_k})) \geq d(q_{n_k}^*, \pi_A(\phi_{n_k})(q_{n_k}^*) \geq d_{\Gamma^*}(q_{n_k}^*) \geq a$ for every k, which contradicts the assumption that $d(q_n, \phi_n q_n) \to o$ as $n \to \infty$. Hence we conclude that $\phi_n^A \times T_n^A = \{1\}$ in $\Gamma^* = \pi_A(A)$ for all sufficiently large n, and from (2) we obtain

$$(3) \quad \phi_n = \{1\} \times \{1\} \times \{1\} \times T_n \qquad \text{for large n}$$

where T_n is a translation of E_B. The fact that $d(q_n, \phi_n q_n) \to o$ implies that $T_n \to \{1\}$ in $I(E_B)$ as $n \to \infty$ and hence $\phi_n \to \{1\}$ in $I(\tilde{M})$. The discreteness of Γ implies that $\phi_n = 1$ for all sufficiently large n, which contradicts the definition of $\{\phi_n\}$ and completes the proof of the sublemma. □

We now prove Lemma 2.6c. Let $a > 0$ be given. To prove that $\tilde{M}^a_A = \{p^* \in \tilde{M}_A : d_{\Gamma^*}(p^*) \geq a\}$ is compact modulo $\Gamma^* = \pi_A(A)$ it suffices to show that for any sequence $\{p_n^*\} \subseteq \tilde{M}^a_A$ and any point $p^* \in \tilde{M}_A$ there exists a subsequence $\{p_{n_k}^*\}$, a sequence $\{a_k\} \subseteq A$ and a number $R' > 0$ such that $d(p^*_{n_k}, \pi_A(a_k)(p^*)) \leq R'$ for every k. Choose a point $p = (p^A, q^A, p^B, q^B)$ as in the sublemma above. Let $F_p \subseteq \tilde{M}$ correspond to $\tilde{M}_A$ and choose $c > 0$ to correspond to $a > 0$ as in the sublemma. Let $p^* = (p^A, q^A) \in \tilde{M}_A$.

Let $p_n^* = (p_n^A, q_n^A) \in \tilde{M}^a_A$ be any sequence, and let $p_n = (p_n^A, q_n^A, p^B, q^B) \in F_p$. By the sublemma $d_\Gamma(p_n) \geq c > 0$ for every n and since Γ is an L-subgroup of $I(\tilde{M})$ there exists a sequence $\{\phi_n\} \subseteq \Gamma$ and a positive number R such that $d(p_n, \phi_n p) \leq R$ for every n. From the fact that $\Gamma = A \times B$ we may choose sequences $\{a_n\} \subseteq A$ and $\{b_n\} \subseteq B$ as in (1) of the sublemma such that

(1) $\phi_n = a_n b_n$

$a_n = \phi_n^A \times T_n^A \times \{1\} \times T_n^B$

$b_n = \{1\} \times \tilde{T}_n^A \times \phi_n^B \times \tilde{T}_n^B$

It follows that

(2) $\phi_n(p) = (\phi_n^A(p^A),\ T_n^A \tilde{T}_n^A(q^A),\ \phi_n^B(p^B),\ T_n^B \tilde{T}_n^B(q^B))$

for every n. The condition $d(p_n, \phi_n p) \leq R$ for all n implies that for all n we have

(3a) $d(p_n^A, \phi_n^A(p^A)) \leq R$

(3b) $d(q_n^A, T_n^A \tilde{T}_n^A(q^A)) \leq R$

(3c) $d(p^B, \phi_n^B(p^B)) \leq R$

(3d) $d(q^B, T_n^B \tilde{T}_n^B(q^B)) \leq R$

By assertion 3) of Lemma 2.6a the group $p_B(\Gamma) \subseteq I(V_B)$ is discrete and hence there are only finitely many distinct elements in the sequence $\{\phi_n^B\}$. By passing to a subsequence we may assume that $\phi_n^B = \phi_1^B \in I(V_B)$ for every n. We may therefore write

(4) $b_n = \xi_n b_1$

where $\xi_n = b_n b_1^{-1} \in$ kernel (p_B) for every n and $p_B : \Gamma \to I(V_B)$ is the projection homomorphism. From assertion 2) of Lemma 2.6a it is clear that $B \cap \text{kernel } (p_B) \subseteq Z_B = B \cap Z$, and hence $\xi_n \in Z_B$ for every n. It now follows from (1) and (4) above and from assertion 4) of Lemma 2.6a that

(5) $\tilde{T}_n^A = \tilde{T}_1^A$ for every n.

Hence we obtain from (3b) above the inequality

(3b)' $d(q_n^A, T_n^A(q^A)) \leq R^*$ for every n

where $R^* = R + d(q^A, \tilde{T}_1^A(q^A))$. Finally if $a_n^* = \pi_A(a_n) = \phi_n^A \times T_n^A \times \{1\} \times \{1\}$, then the inequalities (3a) and (3b)' together imply

$d(p_n^*, a_n^*(p^*)) \leq R'$ for every n

where $p^* = (p^A, q^A)$ and $R' = [R^2 + (R^*)^2]^{\frac{1}{2}}$. This proves that $\tilde{M}_A^a$ is compact modulo $\Gamma^* = \pi_A(A)$ as explained above. □

Proof of Lemma 2.6 d We shall apply Proposition 1.6.4 to the space $\tilde{M}^* = \tilde{M}_1^* \times \tilde{M}_2^*$, where $\tilde{M}^* = \tilde{M}_A$, $\tilde{M}_1^* = V_A$ and $\tilde{M}_2^* = E_A$ in the notation of Lemma 2.6a. In this case the group $\Gamma^* \subseteq I(\tilde{M}^*)$ is the group $\Gamma^* = \pi_A(A)$. The splitting of $\tilde{M}^*$ is Γ^*-invariant since the splitting of $\tilde{M}$ in Lemma 2.6a is Γ-invariant. Let $p_i^* : \Gamma^* \to I(\tilde{M}_i^*)$ denote the projection homomorphisms for i=1,2, and let $p_A : \Gamma \to I(V_A)$ and $q_A : \Gamma \to I(E_A)$ denote the projection homomorphisms arising from the splitting of $\tilde{M}$ in Lemma 2.6a.

By assertion 2) of Lemma 2.6a we observe that $p_1^*(\Gamma^*) = p_A(\Gamma)$ and $p_2^*(\Gamma^*) = q_A(A)$ consists of translations in $\tilde{M}_2^* = E_A$. Hence by assertion 3) of Proposition 2.3 it follows that $p_1^*(\Gamma^*)$ satisfies the duality condition in $\tilde{M}_1^* = V_A$ since Γ satisfies the duality condition in $\tilde{M}$. Now consider $N = \text{kernel } (p_1^*) \subseteq p_2^*(\Gamma^*)$. It will follow from Proposition 1.6.4 that $\Gamma^* = \pi_A(A)$ satisfies the duality condition in $\tilde{M}^* = \tilde{M}_A$ once we have shown that $\tilde{M}_2^*$ is compact modulo N.

From assertions 2) and 3) of Lemma 2.6a we see that Z, the center of Γ, consists of elements of the form

$\{1\} \times T^A \times \{1\} \times T^B$

in $I(V_A) \times I(E_A) \times I(V_B) \times I(E_B)$, where T^A, T^B are translations in the Euclidean spaces E_A, E_B. It follows from assertion 2) of Lemma 2.6a that if $\phi \in A$, then $\pi_A(\phi) \in N = \text{kernel } (p_1^*)$ if and only if $\phi \in Z \cap A$. Hence $N = q_A(Z \cap A)$. Assertion 3) in the proof of Lemma 2.6a shows that Z is a uniform lattice in $\tilde{M}_1 = E_A \times E_B$, and hence Z satisfies the duality condition in $E_A \times E_B$. It follows from Proposition 1.6.2 that $q_A(Z)$ satisfies the duality condition in E_A. However, Z is the direct product of $Z \cap A$ and $Z \cap B$ since $\Gamma = A \times B$, and hence $q_A(Z) = q_A(Z \cap A) \cdot q_A(Z \cap B) = q_A(Z \cap A) \cdot \{1\} =$ N by assertion 4) of Lemma 2.6a. Therefore N satisfies the duality condition in E_A and since N consists of translations in $E_A = \tilde{M}_2^*$ it must contain a uniform lattice N^* in $\tilde{M}_2{}^*$ generated by k linearly independent translations, where $k = \dim \tilde{M}_2{}^*$. Therefore $\tilde{M}_2{}^*$ is compact modulo N and it follows from Proposition 1.6.4 that $\Gamma^* = \pi_A(A)$ satisfies the duality condition in $\tilde{M}_A$. This completes the proof of Lemma 2.6d and hence of Proposition 2.6. □

REFERENCES

[A] M. Anderson, "On the fundamental group of nonpositively curved manifolds", preprint.

[AS] M. Anderson and V. Schroeder, "Existence of flats in manifolds of nonpositive curvature", to appear in Invent. Math.

[B1] W. Ballmann, "Axial isometries of manifolds of nonpositive curvature", Math. Ann. 259 (1982), 131-144.

[B2] W. Ballmann, "Nonpositively curved manifolds of higher rank", to appear in Annals of Math.

[BBE] W. Ballmann, M. Brin and P. Eberlein, "Structure of manifolds of nonpositive curvature, I", Annals of Math. 122 (1985), 171-203.

[BBS] W. Ballmann, M. Brin and R. Spatzier, "Structure of manifolds of nonpositive curvature, II", Annals of Math. 122 (1985), 205-235.

[BE] W. Ballmann and P. Eberlein, "Fundamental group of manifolds of nonpositive curvature", to appear in J. Diff. Geom.

[BGS] W. Ballmann, M. Gromov and V. Schroeder, Manifolds of Nonpositive Curvature, Birkhäuser, 1985.

[BMS] H. Bass, J. Milnor and J.P. Serre, "Solution of the congruence subgroup problem for $SL_n (n \geq 3)$ and $Sp_{2n} (n \geq 2)$, Publ. IHES, Number 33 (1967), 59-137.

[Bo] A. Borel, "Compact Clifford-Klein forms of symmetric spaces", Topology 2 (1963), 111-122.

[BuS 1] K. Burns and R. Spatzier, "Classification of a class of topological Tits buildings", preprint.

[BuS 2] K. Burns and R. Spatzier, "Manifolds of nonpositive curvature and their buildings", preprint.

[CE] S. Chen and P. Eberlein, "Isometry groups of simply connected manifolds of nonpositive curvature", Ill. J. Math. 24 (1980), 73-103.

[E 1] P. Eberlein, "A canonical form for compact nonpositively curved manifolds whose fundamental groups have nontrivial center", Math. Ann. 260 (1982), 23-29.

[E 2] P. Eberlein, "Euclidean de Rham factor of a lattice of nonpositive curvature", J. Diff. Geom. 18 (1983), 209-220.

[E 3] P. Eberlein, "Isometry groups of simply connected manifolds of nonpositive curvature, II", Acta Math. 149 (1982), 41-69.

[E 4] P. Eberlein, "Lattices in spaces of nonpositive curvature", Annals of Math. 111 (1980), 435-476.

[E 5] P. Eberlein, "Structure of manifolds of nonpositive curvature" in Global Differential Geometry and Global Analysis 1984, Lecture Notes in Mathematics, vol. 1156, Springer-Verlag, Berlin, pp. 86-153.

[E 6] P. Eberlein, "Survey of manifolds of nonpositive curvature", Surveys in Geometry, Tokyo, August 1985, preprint.

[EO] P. Eberlein and B. O'Neill, "Visibility manifolds", Pac. J. Math. 46 (1973), 45-109.

[GW] D. Gromoll and J. Wolf, "Some relations between the metric structure and the algebraic structure of the fundamental group in manifolds of nonpositive curvature", Bull. Amer. Math. Soc. 77 (1971), 545-552.

[G] M. Gromov, "Manifolds of negative curvature", J. Diff. Geom. 13 (1978), 223-230.

[H] S. Helgason, Differential Geometry and Symmetric Spaces, Academic Press, New York, 1962.

[Ka] F. Karpelevic, "The geometry of geodesics and the eigenfunctions of the Laplace-Beltrami operator on symmetric spaces", Trans. Moscow Math. Soc. (AMS Translation) Tom 14 (1965), 51-199.

[KN] S. Kobayashi and K. Nomizu, Foundations of Differential Geometry, vol. 1, J. Wiley and Sons, New York, 1963, pp. 179-193.

[Ku] A. Kurosh, Theory of Groups, vol. 2, second English edition, Chelsea, New York, 1960.

[LY] H. B. Lawson and S. -T. Yau, "Compact manifolds of nonpositive curvature", J. Diff. Geom. 7 (1972), 211-228.

[M] G.D. Mostow, Strong Rigidity of Locally Symmetric Spaces, Annals of Math Studies Number 78, Princeton University Press, Princeton, New Jersey 1973.

[PR] G. Prasad and M. Raghunathan, "Cartan subgroups and lattices in semisimple Lie groups", Annals of Math. 96 (1972), 296-317.

[R] M. Raghunathan, Discrete Subgroups of Lie Groups, Springer, New York, 1972.

[Sch 1] V. Schroeder, "Finite volume and fundamental group on manifolds of negative curvature", J. Diff. Geom. 20 (1984), 175-183.

[Sch 2] V. Schroeder, "On the fundamental group of a Visibility manifold", to appear in Math. Zeit.

[Sch 3] V. Schroeder, "A splitting theorem for spaces of nonpositive curvature", Invent. Math 79 (1985), 323-327.

[Sh] H. Shimizu, "On discontinuous groups operating on the product of upper half planes", Annals of Math. 77 (1963), 33-71.

[W1] J. Wolf, "Discrete groups, symmetric spaces and global holonomy", Amer. J. Math. 84 (1962), 527-542.

[W2] J. Wolf, "Homogeneity and bounded isometries in manifolds of negative curvature", Ill. J. Math. 8 (1964), 14-18.

[W3] J. Wolf, Spaces of Constant Curvature, 2d edition, published by the author, Berkeley, California, 1972.

[Y] S. -T. Yau, "On the fundamental group of compact manifolds of nonpositive curvature", Annals of Math 93 (1971), 579-585.

ON A COMPACTIFICATION OF THE SET OF RIEMANNIAN MANIFOLDS WITH BOUNDED CURVATURES AND DIAMETERS

Kenji Fukaya

Departement of Mathematics
Faculty of Science
University of Tokyo
Hongo, Tokyo 113, Japan

Introduction

In this paper we shall study the following class $\mathfrak{M}(n,D)$ of Riemannian manifolds : we say that an n-dimensional Riemannian manifold M is contained in $\mathfrak{M}(n,D)$ if

(0-1-1) the sectional curvature of M is smaller than 1 and greater than -1,

(0-1-2) the diameter of M is smaller than D.

In [12] Gromov introduced the following notion, the Hausdorff distance, between two metric spaces. The purpose of this paper is to study the closure of $\mathfrak{M}(n,D)$ with respect to the Hausdorff distance.

<u>Definition 0-2</u> Let X and Y be metric spaces and $f : X \to Y$ be a map which is not necessarily continuous. We say that f is an ε-<u>Hausdorff approximation</u> if

(0-3-1) for each p,q X, we have

$$| d(f(p),f(q)) - d(p,q) | < \varepsilon ,$$

(0-3-2) the ε-neighborhood of $f(X)$ contains Y.

The Hausdorff distance $d_H(X,Y)$ between X and Y is the infinimum of all numbers ε such that there exist ε-Hausdorff approximations from X to Y and from Y to X.

The following result of Gromov is a starting point of our study.

Theorem 0-4 ([12] 5.3) The closure of $\mathfrak{M}(n,D)$ with respect to the Hausodrff distance is compact.

Notation 0-5 $C\mathfrak{M}(n,D)$ denotes the closure of $\mathfrak{M}(n,D)$ with respect to the Hausdorff distance.

We shall study the following problem in this paper.

Problem 0-6

(1) Determine the set $C\mathfrak{M}(n,D)$. Namely, what kind of a metric space can be a limit of some sequence of elements of $\mathfrak{M}(n,D)$.

(2) Let $X_i, X \in C\mathfrak{M}(n,D)$. Suppose $\lim_{i\to\infty} d_H(X_i,X) = 0$. Then, study the relation between the topological structures of X and X_i.

We shall decompose $C\mathfrak{M}(n,D)$ into its "interior" $\mathrm{Int}(\mathfrak{M}(n,D))$ and its "boundary" $\partial\mathfrak{M}(n,D)$. This decomposition is based on Theorem 0-9 below. To state it, we need a notion.

Definition 0-7 Let X and Y be metric spaces. The Lipschitz distance, $d_L(X,Y)$, between X and Y is the infinimum of all positive numbers ε such that there exists a homeomorphism $f : X \to Y$ satisfying

$$e^{-\varepsilon} \leqq d(f(p),f(q))/d(p,q) \leqq e^{\varepsilon}$$

for each points p and q of X.

Notation 0-8 An element M of $\mathcal{M}(n,D)$ is contained in $\mathcal{M}(n,D,\mu)$ if the injectivity radius of M is greater than μ.

Theorem 0-9 ([12] 8.25) *On $\mathcal{M}(n,D,\mu)$, the Hausdorff distance and the Lipschitz distance are equivalent.*

Remark that if $d_L(X,Y) < \infty$ then X and Y are homeomorphic. Therefore, Theorem 0-9 implies that all elements of the closure of (n,D) are topological manifolds. This fact justifies the following :

Notation 0-10 The interior of $C\mathcal{M}(n,D)$, $\mathrm{Int}(\mathcal{M}(n,D))$, denotes the union of the closures of $\mathcal{M}(n,D,\mu)$ ($\mu > 0$). And the boundary $\partial\mathcal{M}(n,D)$ denotes the set $C\mathcal{M}(n,D) - \mathrm{Int}(\mathcal{M}(n,D))$.

In Section 1, we shall define a stratification on $C\mathcal{M}(n,D)$ and gives a rigidity theorem on boundary (Theorem 1-1), which is a generalization of Theorem 0-9. Also we shall study the local structure of elements of $\partial\mathcal{M}(n,D)$.

In Section 2, we shall study Problem 0-5 (2). Our result is not yet complete. In Section 2, we shall also study the limit of nonpositively curved manifolds and aspherical manifolds, and give compactness theorems (Theorems 2-12 and 2-20).

In this paper, the author does not try to give proofs of our results. Theorem 2-1 is proved in [7] . Proofs of other results will appear in [8] and [9] .

Section 1 The boundary $\partial\mathcal{M}(n,D)$

§ 1-a We shall study Problem 0-6 (1) in this section. First we define a stratification $\{\mathcal{X}_i\}_{i=0,1,2,\cdots}$ of $C\mathcal{M}(n,D)$.

Definition 1-1 We put

$$\mathcal{X}_k = \left\{ X \in C\mathcal{M}(n,D) \;\middle|\; \begin{array}{l}\text{The Hausdorff dimension of } X \text{ is equal} \\ \text{to or smaller than } n-k.\end{array} \right\}$$

Theorem 0-9 and the following result implies $\mathcal{X}_1 = \partial\mathcal{M}(n,D)$.

Theorem 1-2 (Gromov [12] 8.10) Suppose that $M_i \in \mathcal{M}(n,D)$ and that the injectivity radius of M_i is smaller than $1/i$. Let X be a metric space satisfying $\lim_{i\to\infty} d_H(X,M_i) = 0$. Then the Hausdorff dimension of X is equal to or smaller than $n-1$.

When M_i and X are as in Theorem 1-2, we say that M_i collapse to X.

Theorem 1-2, combined with Theorem 0-9, implies also that the Hausdorff distance and the Lipschitz distance define the same topology on $\mathrm{Int}(\mathcal{M}(n,D))$ $(= \mathcal{X}_0 - \mathcal{X}_1)$.

Theorem 1-3 For each k, the set $\mathcal{X}_k - \mathcal{X}_{k+1}$ is complete with respect to the Lipschitz distance. The Hausdorff distance and the Lipschitz distance define the same topology on $\mathcal{X}_k - \mathcal{X}_{k+1}$. In other words, if $X_i, X \in \mathcal{X}_k - \mathcal{X}_{k+1}$ and if $\lim_{i\to\infty} d_H(X_i,X) = 0$, then $\lim_{i\to\infty} d_L(X_i,X) = 0$. In particular X_i and X are homeomorphic for sufficiently large i.

To prove Theorem 1-3, we use Theorems 1-8 and 1-14 described later

§ 1-b Next, we shall study the topology of elements of $\partial\mathcal{M}(n,D)$. We recall an example of elements of $\partial\mathcal{M}(n,D)$ given in [15].

Example 1-4. Suppose that an m-dimensional torus T^m acts on an n-dimensional manifold M. Assume that, for each point p of M, the isotropy group ($= \{ g \in T^m \mid g(p) = p \}$) does not coincide with T^m. Then there exist a sequence of Riemannian metrics g_i on M and a metric d on M/T^m such that $(M,g_i) \in \mathcal{M}(n,D)$ and that

$$\lim_{i\to\infty} d_H((M,g_i),(M/T^m,d)) = 0.$$

In fact, let g_M be a T^m-invariant Riemannian metric on M, and G a subgroup of T^m which is isomorphic to $\mathbb{R}$ and which is dense in T^m. Put

$$g_i(V,V) = \begin{cases} 1/i\cdot g_M(V,V) & \text{if } V \text{ is tangent to an orbit of } G. \\ g_M(V,V) & \text{if } V \text{ is perpendicular to an orbit of } G. \end{cases}$$

Then, some constant multiple of g_i has required property.

§ 1-c We call an element of $\mathcal{CM}(n,D)$ to be smoooth if it is locally diffeomorphic to M/T^m in Example 1-4. Precisely :

Definition 1-5 For a metric space X, we consider the following condition.

(1-6) For each point p of X, there exist a neighborhood U of p in X, a compact Lie group G_p, and a faithful representation of G_p into $SO(m,\mathbb{R})$, such that the identity component of G_p is isomorphic to a torus and that U is homeomorphic to V/G_p for some neighborhood V of 0 in $\mathbb{R}^m$. Furthermore there exists a G_p invariant smooth Riemannian metric g on V such that U is isometric to $(V/G_p,\bar{g})$, where $\bar{g}$ denotes the quotient metric.

We call an element of $\mathcal{CM}(n,D)$ to be <u>smooth</u> if it satisfies Condition (1-6).

Definition 1-7 Let X satisfy Condition (1-6). We put

$$S(X) = \{ p \in X \mid \text{The group } G_p \text{ in (1-6) is nontrivial} \}.$$

Clearly $X - S(X)$ is a smooth Riemannian manifold. Moreover $S(X)$ has a stratification $S(X) = S_1(X) \supseteq S_2(X) \supseteq \cdots$ in an obvious way such that $S_k(X) - S_{k+1}(X)$ is a $(\dim(X)-k)$-dimensional smooth Riemannian manifold.

The manifold M/T^m in Example 1-4 is a smooth element of $\mathcal{CM}(n,D)$.

§ 1-d Now our result on singularities of elements of $\partial\mathcal{M}(n,D)$ is as follows.

Theorem 1-8 For each k, the set of smooth elements are dense in $\mathcal{X}_k - \mathcal{X}_{k+1}$ with respect to the Lipschitz distance. In particular, every element of $\mathcal{CM}(n,D)$ is homeomorphic to a smooth one.

Sketch of the proof The proof is divided into three steps.

Step 1. First we need the following result.

Theorem 1-9 (Bemelmans, Min-Oo, Ruh [1]) For each Positive number ε and Riemannian manifold $M \in \mathcal{M}(n,\infty)$, there exists a Riemannian manifold $M' \in \mathcal{M}(n,\infty)$ such that

(1-10-1) $$d_L(M,M') < \varepsilon$$

(1-10-2) $$\| \nabla^k R(M') \| < C(n,k,\varepsilon).$$

Here the symbol $R(M')$ denotes the curvature tensor, $\| \ \|$ the C^0-

norm, and $C(n,k,\varepsilon)$ the positive number depending only on n,k and ε.

Step 2

Lemma 1-11 Let X_i, Y_i, X, Y be metric spaces such that all bounded subsets are relatively compact. Suppose that $\lim_{i\to\infty} d_H(X_i, X) = 0$, $\lim_{i\to\infty} d_H(Y_i, Y) = 0$, and that $d_L(X_i, Y_i) \leqq \varepsilon$ for each i. Then we have $d_L(X,Y) \leqq \varepsilon$.

The proof of this lemma is an easy exercise of general topology.

Step 3 By Steps 1 and 2, it suffices to show the following : if a sequence M_i of elements of $\mathfrak{M}(n,D)$ and a metric space X satisfies $\lim_{i\to\infty} d_H(M_i, X) = 0$ and $\| \nabla^k R(M_i) \| \leqq C_k$ for each k, then X is a smooth element of $C\mathfrak{M}(n,D)$. To prove this fact, we recall the argument presented in [12] 8.33 $\sim$ 8.38. There it is proved that, for each point p of X, there exists a metric d on B, the n-dimensional ball, and a pseudogroup of isometries G of (B,d) such that B/G is isometric to a neighborhood of p in X. In our case, because of the inequality $\| \nabla^k R(M_i) \| < C_k$, the metric d is a smooth Riemannian metric. It follows that the identity component G_0 of G is a Lie group germ. Put

$$H = \{ g \in G_0 \mid g \text{ has a fixed point on } B \}.$$

It suffices to show that H is a torus and is contained in the center of G_0. Let $\mathfrak{g}$ be the Lie algebra of G_0. Then $\mathfrak{g}$ is a Lie subalgebra of $\Gamma T(B)$, the Lie algebra of all vector fields on B. Put

$$\mathfrak{h} = \{ X \in \mathfrak{g} \mid X(p) = 0 \text{ for some point } p \text{ of } B \}.$$

<u>Lemma 1-12</u> <u>$\mathfrak{h}$ is a Lie subalgebra and is contained in the center of $\mathfrak{g}$.</u>

Lemma 1-12 follows essentially from the following two facts.

(1-13-1) $\mathfrak{g}$ is a nilpotent Lie algebra. (This is a consequence of Margulis' lemma. See [12] 8.51.)

(1-13-2) Every compact subgroup of nilpotent Lie group is contained in its center.

Now Lemma 1-12 implies that H has the required property. Theorem 1-8 follows.

<u>§ 1-e</u> As was remarked in § 1-c, the subspaces $S_k(X) - S_{k+1}(X)$ (especially $X - S(X)$) are Riemannian manifolds if X is a smooth element of $\mathcal{CM}(n,D)$. It is natural to ask whether the sectional curvatures of $S_k(X) - S_{k+1}(X)$ have an upperbound, while X moves on the set of smooth elements contained in $\mathcal{CM}(n,D)$. But the answer is negative.

<u>Example 1-13</u> Let S^2 be the 2-dimensional sphere with the standard metric. SO(3) acts on S^2 as isometries. Put

$$\gamma_t = \begin{pmatrix} \cos t & -\sin t & 0 \\ \sin t & \cos t & 0 \\ 0 & 0 & 1 \end{pmatrix} \in SO(3).$$

Let $\varphi_{n,C}$ be the selfisometry of $S^2 \times \mathbb{R}$ defined by $\varphi_{n,C}(x,r) = (\gamma_{C/n}(x), r+1/n)$. Let $(S^2 \times S^1, g_{n,C})$ be the quotient space of $S^2 \times \mathbb{R}$ by the group generated by $\varphi_{n,C}$ and Y_C be the metric space such that

$$\lim_{n\to\infty} d_H((S^2 \times S^1, g_{n,C}), Y_C) = 0$$

Then, $(S^2 \times S^1, g_{n,C}) \in \mathcal{M}(3,5)$. On the other hand Y_C is isometric to the following rugby ball. Therefore the sectional curvature of Y_C

at the north pole goes to infinity when C goes to 0.

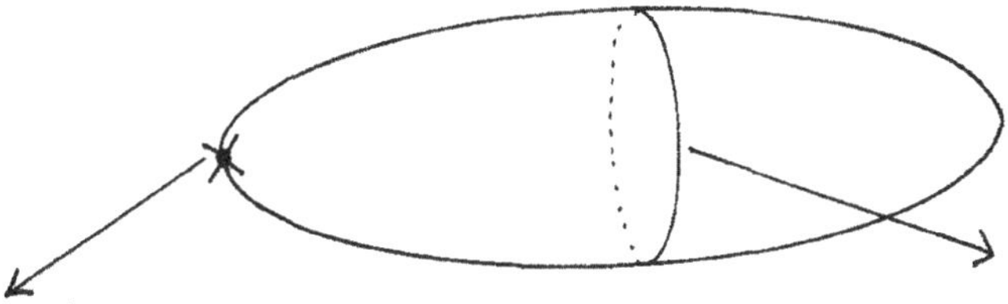

In this example, $\lim_{C\to 0} Y_C = [0,\pi] \in \mathcal{X}_2$. And $Y_C \in \mathcal{X}_1$.

§ 1-f

Theorem 1-14 Let X_i be a sequence of smooth elements of $\mathcal{X}_k - \mathcal{X}_{k+1}$ and X be a metric space. Suppose $\lim_{i\to\infty} d_H(X_i,X) = 0$. Then $X \in \mathcal{X}_{k+1}$, if one of the following three conditions is satisfied.

(1-15-1) There exist a positive number C and a sequence of elements p_i of $S_j(X_i) - S_{j+1}(X_i)$ such that

(1-15-1-a) $$d(p_i,S_{j+1}(X_i)) \geqq C,$$

(1-15-1-b) the sectional curvatures of $S_j(X_i) - S_{j+1}(X_i)$ at p_i are unbounded.

(1-15-2) There exists p_i which satisfies (1-15-1-a) and

(1-15-2-b) the injectivity radius of $S_j(X)$ at p_i converges to 0 when i goes to infinity.

(1-15-3) There exists a sequence of pairs of connected components A_i and B_i of $S_{j_1}(X_i) - S_{j_1+1}(X_i)$ and $S_{j_2}(X_i) - S_{j_2+1}(X_i)$ such that

$$\lim_{i\to\infty} d(A_i,\ B_i - B_C(S_{j+1}(X_i))) = 0$$

for some positive number C. Here $B_C(S_{j+1}(X_j))$ denotes the set

$$\{\ p \in X \mid d(p,S_{j+1}(X_j)) \leqq C\ \}.$$

In this paper, we do not give a proof of Theorem 1-14 and restrict ourselves to state a lemma which plays a key role in the proof. We need a notation. Let M be an element of $\mathcal{M}(n,\infty)$, ψ a selfisometry of M, p an element of M, ℓ a minimal geodesic joining p with $\psi(p)$. Let A denote the composition of $(d\psi)_p$ and the parallel transformation along ℓ. Put $r_p(\psi) = \sup \{ \| A(x) - X \| \mid \|X\| = 1, X \in T_p(M) \}$ Let G be a group of isometries acting freely on M. Set

$$r/t(G) = \sup \{ r_p(g)/d(p,g(p)) \mid p \in M, \ g \in G - \{1\} \}.$$

Lemma 1-16 There exists a constant C depending only on $r/t(G)$ such that the sectional curvature of M/G is greater than -1 and smaller than C.

Remark 1-17 Theorems 1-8 and 1-14, combined with [10] or [12] 8.28, imply that, for each (not necessarily smooth) element X of $C\mathcal{M}(n,D)$, the subspaces $S_k(X) - S_{k+1}(X)$ are Riemannian manifolds with $C^{1,\alpha}$ distance. Here α is an arbitrary number contained in $[0,1)$.

§ 1-g Theorems 1-8 and 1-14 describe how a neighborhood of each point of each element of $C\mathcal{M}(n,D)$ looks like. The following problem on global condition is open.

Problem 1-18 Let X be a space satisfying Condition (1-6). Is X homeomorphic to an element of $C\mathcal{M}(n,D)$ for some n and D ?

In particular, are there any elements of $C\mathcal{M}(n,D)$ which are not homeomorphic to a qutient space M/G of a manifold M by a compact Lie group G whose identity component is isomorphic to a torus ?

Section 2 Fibre bundle theorem and its application

§ 2-a We shall study problem 0-6 (2), in this section.

Theorem 2-1 ([7]) There exists a positive number $\varepsilon(n,\mu)$ depending only on n and μ such that the following holds.

If $M \in \mathcal{M}(n_1,\infty)$, $N \in \mathcal{M}(n_2,\infty,\mu)$, $n_1,n_2 \leq n$, and if the Hausdorff distance between M and N is smaller than $\varepsilon(n,\mu)$, then there exists a map $f : M \to N$ satisfying the conditions below.

(2-2-1) (M,N,f) is a fibre bundle.

(2-2-2) The fibre of f is diffeomorphic to an infranilmanifold. Here a manifold is said to be an infranilmanifold, if its finite covering is diffeomorphic to the quotient of a nilpotent Lie group by its lattice.

(2-2-3) If $V \in T(M)$ is perpendicular to a fibre of f, then we have

$$e^{-o(\varepsilon)} < \|df(V)\| / \|V\| < e^{o(\varepsilon)}.$$

Here $o(\varepsilon)$ denotes a positive number depending only on ε, n, μ and satisfying $\lim_{\varepsilon\to 0} o(\varepsilon) = 0$.

The proof of this theorem is presented in [7]. Here we mention only the construction of the map f. For simplicity we assume M is compact. Let ψ be an ε-Hausdorff approximation Z_N be a finite subset of N such that, for each distinct elements z and z' of Z_N, we have $d(z,z') > 3\varepsilon$, and that the 6ε-neighborhood of Z_N contains N. Let K be the order of N. Now, following [12] or [14], we shall construct an embedding $f_N : N \to \mathbb{R}^K$. Take a C^∞-function $h : \mathbb{R} \to [0,1]$ such that $h(t) = 1$ for $t < 0$ and that $f(t) = 0$ for $t > 1$. Define $f_N : N \to \mathbb{R}^K$ by $f_N(p) = (hd(p,z))_{z \in Z_N}$. The following is proved in [14].

Lemma 2-3

(1) f_N is an embedding.

(2) Put

$B_C(Nf_N(N)) = \{ (p,u) \in$ the normal bundle of $f_N(N) \mid \|u\| \leqq C \}$.

Then, the restriction of the exponential map to $B_{CK^{1/2}}(Nf_N(N))$ is a diffeomorphism to its image. Here C denotes a constant independent of N, μ, K.

Next we construct a map $f_M : M \to \mathbb{R}^K$. For $z \in Z_N$ and $p \in M$, put

$$d_z(p) = \int_{y \in B_\varepsilon(\psi(z))} d(y,p)\, dy \Big/ \mathrm{Vol}(B_\varepsilon(\psi(z))),$$

$$f_M(p) = (hd_z(p))_{z \in Z_N}.$$

Here $B_\varepsilon(\psi(z))$ denotes $\{ y \in M \mid d(\psi(p),y) < \varepsilon \}$. It is easy to see that f_M is of C^1-class. It is also easy to see that $f_M(M)$ is contained in the 10ε-neighborhood of $f_N(N)$. Then we put

$$f = f_N^{-1} \circ \pi \circ \mathrm{Exp}^{-1} \circ f_M.$$

Here π denotes the projection : $Nf_N(N) \to f_N(N)$, Exp the exponential map. The property (2-2-1) follows from the following lemma.

Lemma 2-4 The submanifold $\mathrm{Ext}^{-1}(f_M(M))$ is transversal to each fibre of the normal bundle $Nf_N(N) \to f_N(N)$.

To prove this lemma, we use a Toponogov type comparison theorem on triangles of M and N.

The method used to prove (2-2-2) is a modification of the argument appeared in the proof of the theorem of almost flat manifolds in [2] or [11].

§ 2-b In § 1-a we gave an example of collapsing with help of a torus action. In § 2-b and § 2-c, we shall give examples of collapsing concerning nilmanifolds.

Example 2-5 Let G be a solvable Lie group and Γ its lattice. Put $G_0 = G$, $G_1 = [G,G]$, $G_2 = [G_1,G_1]$, $\cdots$, $G_{i+1} = [G_i,G_1]$. Take a right invariant Riemannian metric g on G and define a right invariant metric g_ε on G by $g_\varepsilon(V,V) = \varepsilon^{i\cdot 2^i}\cdot g(V,V)$ when $V \in T_1(G)$ is tangent to G_i and perpendicular to G_{i+1}. Then $\lim_{\varepsilon\to 0}(G/\Gamma, g_\varepsilon)$ is equal to the flat torus $G_1\backslash G/\Gamma$, and the sectional curvatures of g_ε are uniformly bounded. In this example, the fibre bundle in Theorem 2-1 is $G_1/(G_1 \cap \Gamma) \to G/\Gamma \to G_1\backslash G/\Gamma$.

This example proves the implication (2-7-1) $\Rightarrow$ (2-7-2) in the following conjecture.

Conjecture 2-6 For an n-dimensional compact manifold M, the following two conditions are equivalent.

(2-7-1) There exists a finite covering M' of M which is diffeomorphic to the quotient G/Γ of a solvable Lie group G by its lattice Γ.

(2-7-2) There exists a sequence of Riemannian metrics g_i on M, and a discrete subgroup Γ of the group of all isometries of $\mathbb{R}^n$ such that $(M,g_i) \in \mathfrak{M}(n,D)$ and that

$$\lim_{i\to\infty} d_H((M,g_i),(\mathbb{R}^n/\Gamma,\text{quotient metric})) = 0.$$

§ 2-c Next we shall give an example of collapsing to a metric space which is not a manifold.

Example 2-8 Let (G_i, Γ_i) be a sequence of pairs consisting of nilpotent Lie groups G_i and their lattices Γ_i. Let (M,g) be a compact Riemannian manifold, and ψ_i be a homomorphism from Γ_i to the group of isometries of (M,g). Put $T = \bigcap_i (\overline{\bigcup_{j>i} \psi_j(\Gamma_j)})$. Here the closure $(\overline{\bigcup_{j>i} \psi_j(\Gamma_j)})$ is taken in the sense of compact open topology. It is proved in 4 7.7.2 that there exists a sequence of right invariant Riemannian metrics g_i on G_i such that the sectional curvatures of g_i are uniformly bounded and that $\lim_{i\to\infty} (G_i/\Gamma_i, \bar{g}_i) = \text{point}$. We define an action of Γ_i on $G_i \times M$ by $\gamma(g,x) = (\gamma g, \psi_i(\gamma)(x))$. Let $(G_i \times_{\Gamma_i} M, g_i \times g)$ denote the quotient space with the quotient Riemannian metric. Then it is easy to see that $(G_i \times_{\Gamma_i} M, g_i \times g) \in \mathfrak{M}(n,D)$ and that $\lim_{i\to\infty} (G_i \times_{\Gamma_i} M, g_i \times g) = (M/T, \bar{g})$. Here $\bar{g}$ denotes the quotient metric. In this example, there exists also a map from $G_i \times_{\Gamma_i} M$ to M/T.

The author has a feeling that the above example gives all possible phenomena which can occur at a neighborhood of each point of the limit. Namely :

Conjecture 2-9 *Let M_i be a sequence of elements of $\mathfrak{M}(n,D)$ and X be a smooth element of $\mathcal{C}\mathfrak{M}(n,D)$. Suppose $\lim_{i\to\infty} d_H(M_i, X) = 0$. Then, for sufficiently large i, there exists a map $f : M_i \to X$ such that the following holds.*

(2-10-1) $(f^{-1}(S_k(X)-S_{k+1}(X)), S_k(X)-S_{k+1}(X), f)$ *is a fibre bundle.*

(2-10-2) *Let $p_0 \in X - S(X)$, $p \in S_k(X) - S_{k+1}(X)$. Put $F = f^{-1}(p_0)$. Then the group G_p in Condition (1-6) acts freely on F and $f^{-1}(p)$ is diffeomorphic to F/G_p.*

An affirmative answer to the above conjecture will enable us to know completely what happens at a neighborhood of each point when manifolds collapse keeping their curvatures and diameters bounded. The global problem on collaping is open even in the case of fibre bundle.

Problem 2-11 Let F be an infranilmanifold and (M,N,f) be a fibre bundle with fibre F. Give a necessary and sufficient condition for the existence of a sequence of metrics g_i on M such that (M,g_i) is contained in $\mathcal{M}(n,D)$ and that $\lim_{i\to\infty} (M,g_i)$ is diffeomorphic to N.

§ 2-d In this section we shall give an application of our fibre bundle theorem and the results of Section 1.

Theorem 2-12 Let $\mathcal{NP}(n,D)$ denote the set of all metric spaces X/Γ satisfying the following five conditions.

(2-13-1) X is a simply connected complete Riemannian manifold.

(2-13-2) The sectional curvature of X is equal to or smaller than 0 and greater than -1.

(2-13-3) Γ is a properly discontinuous group of isometries of X.

(2-13-4) The diameter of X/Γ is smaller than D.

(2-13-5) The dimension of X is equal to n.

Then, the closure of $\bigcup_{k=1}^{n} \mathcal{NP}(k,D)$ in the sense of the Lipschitz distance is compact with respect to the Hausdorff distance. In particular, if a sequence of elements of $\mathcal{NP}(n,D)$ converges to a metric space Y with respect to the Hausdorff distance then there exists a metric d on Y such that $(Y,d) \in \mathcal{NP}(k,D)$ for some $k \leqq n$.

Remark 2-14 Theorem 2-12 was suggested by V. Shroeder to the author on May 1985 at Berkeley. I am not certain whether he said that he gave a proof of this fact or not.

Remark 2-15 In order to obtain a compactness theorem like Theorem 2-12, we have to deal with the set of orbifolds and cannot restrict ourselves to manifolds. In other words, there exists a sequence M_i of nonpositively curved Riemannian manifolds with bounded curvatures and diameters, such that the limit of M_i with respect to the Hausdorff distance is not a manifold. In fact, let Γ be a discrete group of isometries of $\mathbb{R}^n$, and Γ_0 be its torsionfree normal subgroup such that Γ has a torsion element and that Γ/Γ_0 is a cyclic group of oder p ($< \infty$). Let Γ act on S^1 so that Γ_0 acts trivially and that Γ/Γ_0 acts freely. Take the skew product $\mathbb{R}^n \times_\Gamma S^1 = M$. Then M is a flat Riemannian manifold. If we let the second factor S^1 shrink to a point, then M collapses to $\mathbb{R}^n/\Gamma$, which is not a manifold.

Proof of Theorem 2-12 Suppose $M_i \in \mathcal{NP}(n,D)$, $Y \in \mathcal{CM}(n,D)$, $\lim_{i\to\infty} d_H(M_i,Y) = 0$. Take the universal covering space X_i of M_i. Let Γ_i denote the fundamental group of M_i and the symbol X the limit of X_i with respect to the pointed Hausdorff distance, (which is defined in [12] 3.14). By taking a subsequence if necessary we may assume that the pair (X_i,Γ_i) converges to a pair (X,G) with respect to the equivariant Hausdorff distance. (See [6] section 1 for the definition and [6] section 3 for the proof of this fact.) Put

$$H = \{ g \in G \mid g \text{ has a fixed point on } X \}.$$

Then, a method similar to the argument in step 3 (§ 1-d) implies that H is a compact group contained in the center of G.

Lemma 2-16 H is a trivial group.

Proof We prove by contradiction. Put

$$X_0 = \{ p \in X \mid g(p) = p \text{ for each } g \in H.\}$$

Since X is a limit of nonpositively curved manifolds, X_0 is a convex subset of X. On the other hand, since H is contained in the center of G, it follows that X_0 is G-invariant and that X_0 is nonempty Therefore, for each poit p of X, we have

$$d(p,X_0) \leqq \operatorname{diam}(X/G) \leqq D.$$

But this is impossible because X_0 is convex.

Lemma 2-16 implies that G acts freely on X. Using this fact, it is easy to show that $Y = X/G$ is a quotient of a contractible manifold by a properly discontinuous action of a group. For simplicity, we assume that Y is an (aspherical) manifold. Then, Theorem 2-1 implies that there exists a fibre bundle $F \to M_i \to Y$ with an infranilmanifold fibre F. Hence, becase of the homotopy exact sequence $\pi_2(Y) \to \pi_1(F) \to \pi_1(M_i)$, the virtually nilpotent group $\pi_1(F)$ is a normal subgroup of $\pi_1(M_i)$. Therefore 5 chapter 9 implies that X_i is a Riemannian product $\mathbb{R}^k \times Y_i'$ such that F is a quotient of $\mathbb{R}^k$ and that Y_i' converges to a space Y' such that Y is a quitient of Y' by a discrete group. Therefore, Y is diffeomorphic to an element of (n,D). This completes the proof of Theorem 2-12.

Theorem 2-12 gives an alternative proof of the following result due to Buyalo.

<u>Theorem 2-17</u> (Buyalo [3]) <u>For each n and D, the number of diffeomorphism classes of elements of $\mathcal{NP}(n,D)$ is finite.</u>

<u>§ 2-e</u> In the proof of Theorem 2-12, we used convexity of the set X_0. We can avoid this and can prove the following result, by making use of the notion $\operatorname{Diam}_{n-1}$ defined in [13].

Definition 2-18 Let $\mathcal{AS}(n,D,C)$ be the set of metric spaces X/Γ satisfying (2-13-1),(2-13-3),(2-13-4),(2-13-5) and

(2-19-1) the sectional curvature of X is smaller than C and greater than $-C$,

(2-19-2) X is contractible.

Theorem 2-20 For each positive number C_n, there exist positive numbers $C_1, C_2, \cdots$ such that, for every $D > 0$, the closure of $\bigcup_{k=1}^{n} \mathcal{AS}(k,D,C_k)$ in the sense of the Lipschitz distance is compact with respect to the Hausdorff distance.

Using Theorem 2-20 and an orbifold version of Theorem 2-1, we can prove the following :

Theorem 2-21 For each n and D, there exists a finite number of elements $N_1, N_2, \cdots, N_k$ of $\bigcup_{k=1}^{n} \mathcal{AS}(k,D,\infty)$ such that the following holds.

For each $M \in \mathcal{AS}(n,D,1)$ there exist a sequence of elements $M_0 = M$, $M_1, \cdots, M_j$ of $\bigcup_{k=1}^{n} \mathcal{AS}(k,D,\infty)$ and fibre bundles $f_i : M_i \to M_{i+1}$ such that

(2-22-1) M_j is diffeomorphic to N_i for some i.

(2-22-2) The fibre of f_i is diffeomorphic to an infranilmanifold.

Added on September 26 ; The author verified that Conjecture 2-9 is varied. The proof will appear in [8].

References

[1] Bemelmans,J. Min-Oo, and Ruh,E.A., Smoothin Riemannian metrics Math Z., 188 (1984),69-74.

[2] Buser,P. and Karcher,H., Gromov's almost flat manifolds, Astérisque 81 (1981).

[3] Buyalo,S.,V., Volume and the fundamental group of a manifold of nonpositive curvature, Math. U.S.S.R. Sbornik 50 (1985),137-150.

[4] Cheeger,J. and Ebin,D.G., Comparison theorems in Riemannian geometries, North-Holland, 1975.

[5] Cheeger,J. and Gromov,M., Collapsing Riemannian manifolds while keeping their curvatures bounded, to appear.

[6] Fukaya,K., Theory of convergence for Riemannian orbifolds, Preprint.

[7] Fukaya,K., Collapsing Riemannian manifolds to a lower dimensional one, Preprint.

[8] Fukaya,K., A boundary of the set of Riemannian manifolds with bounded curvatures and diameters, in preparation.

[9] Fukaya,K., A compactness theorem of a set of aspherical Riemannian orbifolds, in preparation.

[10] Green,R.E. and Wu,H., Lipschitz convergence of Riemannian manifolds, Preprint.

[11] Gromov,M., Almost flat manifolds, J. of Differential Geometry, 13 (1978) 231-241.

[12] Gromov,M., Lafontaine,J., and Pansu,P., Structure métrique pour les variétés riemanniennes, Cedic/Fernand Nathan, 1981.

[13] Gromov,M., Large Riemannian manifolds, in this preceeding.

[14] Katsuda,A., Gromov's convergence theorem and its application, to appear in Nagoya J. Math.

[15] Pansu,P., Effondrement des variétés riemannienne d'après J. Cheeger and M.Gromov, Seminaire Bourbaki 36 année 1983/84 n°618.

Large Riemannian Manifolds

M. Gromov

Institut des Hautes Études Scientifiques
35, Route de Chartres, 91440 Bures-Sur-Yvette
France

We want to discuss here several unsolved problems concerning metric invariants of a Riemannian manifold $V = (V, g)$ which mediate between the curvature and topology of V.

1. VOLUME OF BALLS $B_v(\rho)$ IN LARGE MANIFOLDS V.

Assume V is complete and define for all $\rho \geq 0$.

$$\sup \mathrm{Vol}(V; \rho) = \sup_{v \in V} \mathrm{Vol}\, B_v(\rho)$$

for the balls $B_v(\rho) \subset V$. If V has bounded geometry (e.g. compact), then the behavior of $\sup \mathrm{Vol}(V; \rho)$ for $\rho \longrightarrow 0$ is controlled by the lower bound of the scalar curvature of V, called

$$\inf S(V) = \inf_{v \in V} S(V, v).$$

On the other hand, the asymptotic behaviour of $\sup \mathrm{Vol}$ for $\rho \longrightarrow \infty$ has a topological meaning if, for example, V is metrically covers some compact manifold.

1.A. Vague Conjecture.

If V is large compared to $\mathbb{R}^n$ for $n = \dim V$, then

$$\sup \mathrm{Vol}(V; \rho) \geq \sup \mathrm{Vol}(\mathbb{R}^n; \rho) = A_n \rho^n,$$

where A_n is the volume of the unit ball in $\mathbb{R}^n$. Furthermore, $\inf S(V) = 0$ for large manifolds V (compare [GL] and [S]).

To make sense of 1.A, we give several precise notions of largeness.

$\mathcal{L}_1$. Contractible almost homogeneous manifolds (CAH).

This means that V is contractible and that the action of the isometry group $Is(V)$ is cocompact on V. For example the universal coverings of compact aspherical manifolds are CAH.

$\mathcal{L}_2$. Geometrically contractible manifolds (GC).

Define $GC_k(V, \rho)$ for all $\rho \geq 0$ to be the lower bound of the numbers $r \geq \rho \geq 0$, such that the inclusion of the concentric balls in V

$$B_v(\rho) \hookrightarrow B_v(r)$$

is a k-contractible map for all $v \in V$.

Recall, that a continuous map $f: X \longrightarrow Y$ is called k-degenerate, if there exist a k-dimensional polyhedron P and continuous maps $f_1: X \longrightarrow P$ and $f_2: P \longrightarrow Y$, such that $f = f_2 \circ f_1$. Then, f is called k-contractible if it is homotopic to a k-degenerate map.

A manifold V is called GC if $GC_0(V, \rho) < \infty$ for all $\rho \geq 0$.

Obviously, CAH $\Longrightarrow$ GC. (Compare [G]$_2$ P.43.)

$\mathcal{L}_3$. Manifolds with $\mathrm{Diam}_{n-1} = \infty$.

Define $\mathrm{Diam}_k V$ to be the lower bound of those $\delta > 0$ for which there exists a continuous map of V into some k-dimensional polyhedron, say $f: V \longrightarrow P$, such that

$$\mathrm{Diam}\, f^{-1}(p) \leq \delta,$$

for all $p \in P$ (compare [G]$_2$ P.127).

It is not hard to prove the following relation between Diam_k and GC for $k + \ell = n - 1 = \dim V - 1$ (compare [G]$_2$ P.143).

There exists a function $\rho_n(\delta)$ for $\delta \geq 0$, such that

$$\mathrm{Diam}_k V \leq \delta \Longrightarrow GC_\ell(V, \rho) = \infty \quad \text{for} \quad \rho \geq \rho_n(\delta).$$

In particular, GC $\Longrightarrow \mathrm{Diam}_{n-1} = \infty$.

$\mathcal{L}_4$. Manifolds with $\mathrm{Cont}_{n-1}\mathrm{Rad} = \infty$.

Imbed V into the space of functions $L_\infty(V)$ by $v \longmapsto \mathrm{dist}(v, *)$. If V is compact, define $\mathrm{Cont}_k\mathrm{Rad}\ V$ to be the lower bound of the numbers $\varepsilon \geq 0$, such that the inclusion map of V into the ε-neighborhood $U_\varepsilon(V) \subset L_\infty(V)$ is k-contractible, where the function space $L_\infty(V)$ is equipped with the L_∞-norm:

$$\|f(v)\| = \sup_{v \in V} |f(v)|$$

(compare $[G]_2$ P.P.41, 138).

If V is noncompact, one modifies this definition by restricting to proper k-contracting homotopies which keep pull-backs of bounded subsets in $U_\varepsilon(V)$ bounded in V.

It is easy to see that

$$\mathrm{Cont}_k\mathrm{Rad}\ V \leq \tfrac{1}{2}\ \mathrm{Diam}_k V$$

and that

$$\mathrm{GC} \Longrightarrow \mathrm{Cont}_{n-1}\mathrm{Rad} = \infty .$$

Furthermore, (see $[G]_2$ P.138).

$$\mathrm{Cont}_{n-1}\mathrm{Rad}\ V \leq C_n(\mathrm{Vol}\ V)^{1/n}$$

for some universal constant $C_n > 0$. In particular

$$\mathrm{CG} \Longrightarrow \mathrm{Vol}\ V = \infty.$$

$\mathcal{L}_5$. Manifolds with Fill Rad $= \infty$.

Define Fill Rad V to be the minimal ε for which V is $\mathbb{Z}_2$-homologous to zero in the ε-neighborhood $U_\varepsilon(V) \subset L_\infty(V)$ (compare $[G]_2$ P.41). Clearly, Filling Rad $\leq \mathrm{Cont}_{n-1}\mathrm{Rad}$. Yet Fill Rad > 0 for all manifolds V. (See $[G]_2$ for applications of Fill Rad and [K] for a computation of Fill Rad of some symmetric spaces). It is also clear that

$$GC \Longrightarrow \text{Fill Rad} = \infty.$$

Also notice that Fill Rad decreases under proper distance decreasing maps $V_1 \longrightarrow V_2$ of degree one (mod 2) (see $[G]_2$ P.8).

$\mathcal{L}_6$. Hyperspherical manifolds.

Assume V is oriented and define $\text{HS Rad}_k V$ to be the upper bound of those numbers $R \geq 0$ for which there exists a proper Λ^k-contracting map of V onto the sphere $S^n(R) \subset \mathbb{R}^{n+1}$ of radius R, say

$$f: V \longrightarrow S^n(R),$$

such that $\deg f \neq 0$. Here "proper" means that the complement of some compact subset in V goes to a single point in S^n and "Λ^k-contracting" signifies that f decreases the k-dimensional volumes of all k-dimensional submanifolds in V (compare [GL]). One says that V is HS if $\text{HS Rad}_1 V = \infty$.

Remark. One can modify the definition of HS Rad by restricting to maps f with $\deg f \equiv 1 \pmod 2$. Then modified HS clearly implies Fill Rad $= \infty$.

Stable classes $\mathcal{L}_i^+$ and $\mathcal{L}_i^-$.

Given a class $\mathcal{L}$ of n-dimensional manifolds. One defines $V \in \mathcal{L}^+$ iff V admits a proper distance decreasing map of degree one onto some manifold $V' \in \mathcal{L}$. One also defines $V \in \mathcal{L}^-$ iff the existence of a proper distance decreasing map $V' \longrightarrow V$ of degree one implies $V' \in \mathcal{L}$. The stabilization $\mathcal{L}^+$ looks interesting for the classes $\mathcal{L}_2$, $\mathcal{L}_3$ and $\mathcal{L}_4$. Furthermore, it is logical to allow an arbitrary pseudo-manifold V' in the definition of $\mathcal{L}_3^-$ and $\mathcal{L}_4^-$ and to stabilize (in an obvious way) the invariants Diam_k and Cont_kRad in order to match the classes $\mathcal{L}_3^-$ and $\mathcal{L}_4^-$. Following this line of reasoning, one can define $\text{Diam}_k^- h$ and $\text{Cont}_k^- h$ for an arbitrary homology class h in V by representing h by distance decreasing maps $V' \longrightarrow V$ for $\dim V' = \dim h$.

1.B. On the Vague Conjecture.

There is no solid evidence for 1.A for manifolds in the classes $\mathcal{L}_i$ and $\mathcal{L}_i^{\pm}$. One even does not know if

$$\sup \operatorname{Vol}(V;\rho) \geq \pi\rho^2$$

for CAH surfaces. However it is easy to see that

$$\sup \operatorname{Vol}(V;\rho) \geq 3\rho^2$$

for GC surfaces (compare $[G]_2$ P.40). This suggests relaxing 1.A to the inequality

$$\sup \operatorname{Vol}(V;\rho) \geq C_n\rho^n \tag{1}$$

for some universal constant in the interval $0 < C_n < A_n$. In fact, (a quantitative version of) (1) is proven on P.130 in $[G]_2$ for manifolds V with $\operatorname{Diam}_{n-1} V = \infty$, provided $\operatorname{Ricci} V \geq -1$.

Finally, a non-sharp version of 1.A is known to be true asymptoticly for $\rho \longrightarrow \infty$ for CAH manifolds. Namely, the polynomial growth theorem for abstract groups reduces the problem to the universal covering $V \longrightarrow T^n$ of the homotopy n-torus T^n and the argument on P.100 in $[G]_2$ yields the bound

$$\liminf_{\rho\to\infty} \rho^{-n} \operatorname{Vol} B(\rho) \geq C_n > 0$$

for the concentric balls $B(\rho)$ in the universal covering of T^n.

Now, we turn to the inequality $\inf S(V) \leq 0$ for large manifolds V. One is able to prove (see [GL] and $[G]_2$ P.129) that

$$\inf S(V) \leq (\pi 6\sqrt{2})/\operatorname{Diam}_1 V)^2 \tag{2}$$

for complete simply connected 3-manifolds. In particular $\operatorname{Diam}_1 V = \infty$ implies $\inf S(V) \leq 0$ for these V. Next one believes that

$$\inf S(V) \leq C_n(\text{HS Rad}_2 V)^{-2}. \tag{3}$$

This is proven for spin manifold V in [GL] and a similar inequality is anounced in [S] for the general case. Yet, one does not know the best constant C_n in (3). For example, let a metric g on S^n satisfy $g \geq g_0$ for the standard metric g_0 on S^n. One does not know if $\inf S(g) \leq S(g_0)$.

Many CAH manifolds V are shown to be HS (see [GL] and references therein) and no counterexample to CAH $\Longrightarrow$ HS is known. More generally, let V' be a closed manifold whose classifying map to the Eilenberg Maclain space $K(\pi, 1)$ for $\pi = \pi_1(V')$ sends the fundamental class $[V]$ (here, V is assumed oriented) to a non-zero class in $H_n(K(\pi, 1); \mathbb{Q})$. Then, one asks if the universal covering V of V' is HS. (The HS property of V does not depend on the metric in V'). If so, the manifold V' admits no metric with $S(V) > 0$ as it follows from (3).

If $V \in \mathcal{L}_i$, $i = 1,\dots,6$, then, clearly, $V \times \mathbb{R}^N \in \mathcal{L}_i$ for all N. In particular, if V is HS then $V \times R^N$ also is HS. The converse is unlikely to be true but no counter example is known. On the other hand, the largeness of $V \times \mathbb{R}^N$ has roughly the same effect on $S(V)$ as that of V itself. Namely,

$$\inf S(V) \leq C'_{n+N}(\text{HS Rad}_2 V \times \mathbb{R}^N)^{-2}, \qquad (3')$$

provided V is spin (compare [S] for non-spin manifolds).

2. MANIFOLDS WITH $K \geq 0$.

Let V be a complete connected manifold with non-negative sectional curvature. Then one can show that the largeness conditions $\mathcal{L}_i$ are equivalent for $i = 3,4,5,6$, and V is $\mathcal{L}_i$-large for $i = 3,\dots,6$ if and only if

$$\sup \text{Vol}(V; \rho) = \sup \text{Vol}(\mathbb{R}^n; \rho) = A_n \rho^n \qquad (4)$$

for all $\rho \geq 0$. Furthermore, if

$$\sup \text{Vol}(V; 1) \leq A' < A_n,$$

then

$$\sup \mathrm{Vol}(V;\rho) \le C\rho^{n-1} \tag{5}$$

for all $\rho \ge 1$ and for some universal constant $C = C(n, A')$.

If in addition to $K(V) \ge 0$ one assumes $S(V) \ge \sigma^2 > 0$, then one can strengthen (5) by

$$\sup \mathrm{Vol}(V;\rho) \le C'_{n,\sigma}\rho^{n-2} \tag{5'}$$

and show that

$$\mathrm{Diam}_{n-2} V \le C''_{n,\sigma}/\sigma. \tag{6}$$

2.A. Open Questions.

(a) It seems likely, that complete hyperspherical manifolds with $K(V) > 0$ are geometrically contractible.

(b) The relating (4), (5) and (5') may generalize to the case Ricci $V \ge 0$. This seems quite realistic if $|K(V)| \le 1$ and Inj Rad $V \ge 1$.

(c) It is unknown if (6) holds true for all complete manifolds with $S(V) \ge \sigma^2$.

2.B. Idea of the Proof of (4) - (6).

For certain sequences of points $v_i \in V$ the sequences of the pointed metric spaces (V, v_i) converge in the Hausdorff topology to isometric products $\mathbb{R}^d \times V'$ for (possibly singular) spaces V' with $K \ge 0$. If d is the largest possible, then V' with is compact and $\mathrm{Diam}_d V \le \text{const} \sup_{V'} \mathrm{diam}\, V'$. In particular, if V is large, then (the maximal) $d = n$ and $\lim_{i\to\infty} \mathrm{Vol}\, B_{V_i}(g) = A_n\rho^n$. This proves (4); the inequalities (5), (5') and (6) follow by a similar argument.

2.C.

To grasp the geometric meaning of the invariants $\mathrm{diam}_k V$, consider the Euclidean solid

$$V' = \{(x_0, \ldots, x_{n-1}) \mid |x_k| \le \mathrm{Diam}_k V,\ k = 0,\ldots,n-1\} \subset \mathbf{R}^n.$$

One believes that every compact manifold V with (possibly empty) convex boundary and with $K(V) \ge 0$ roughly looks like V'. For example, the volume of V' seems a good approximation to Vol V and the spectrum of the Laplace operator on V' might approximate that on V. Namely, the corresponding numbers of eigenvalues $\le \lambda$ are conjectured to satisfy,

$$N'(C_n\lambda) \ge N(\lambda) \ge N'(C_n^{-1}\lambda).$$

A similar rough approximation is expected for small balls in manifolds with $K(V) \le 1$. Here the case $|K(V)| \le 1$ looks easy.

2.D. <u>Manifolds with</u> $S_k(V) \ge \alpha$ <u>and</u> $R_k(V) \ge \alpha$.

Write $S_k(V) \ge \alpha$ if the average of the sectional curvatures over the 2-planes in every tangent k-dimensional surface in $T(V)$ is $\ge \alpha$. Write $R_k(V) \ge \alpha$ if the sum of the first k eigenvalues of Ricci on $T_v(V)$ is $\ge \alpha$ for all $v \in V$. One does not know the geometric significanse of the inequalites $S_k > 0$ for $3 \le k \le n - 1$ and $R_k > 0$ for $2 \le k \le n - 1$, unless some additional conditions are imposed on V. What one wishes is an upper bound like $\mathrm{Diam}_\ell \le C/\sigma$ for $S_{\ell+2} \ge \sigma^2$. Here is a simple fact supporting this conjecture.

<u>Let</u> V <u>be a complete manifold with</u> Ricci ≥ 0 <u>and</u> $R_k \ge \sigma^2$ <u>for some fixed</u> $k \le n$. <u>Then</u> $\sup \mathrm{Vol}(V; \rho) \le C\rho^{k-1}/\sigma$ <u>provided</u> $|K(V)| \le$ const $< \infty$ <u>and</u> Inj Rad $V \ge \varepsilon > 0$.

This is shown by a limit argument as in 2.B.

Observe, that the inequality $R_k \ge \alpha$ defines a <u>convex</u> subset in the space of the curvature tensors on every space $T_v(V)$. This insures the stability of this inequality under certain (weak) limits of metrics.

3. VERY LARGE MANIFOLDS.

Define $\mathrm{Vol}_k(V)$ as the lower bound of those $s \geq 0$ for which there exists a simplicial map $f: V \longrightarrow P$ for some smooth triangulation of V and some $(n-k)$-dimensional polyhedron P, such that the k-dimensional volume of the pull-back $f^{-1}(p) \subset V$ is $\leq s$ for all $p \in P$. It is known that

$$(\mathrm{Vol}_k V)^{1/k} \geq C_n \text{ Fill Rad } V,$$

for all complete manifolds V (see $[G]_2$ P.134), but a similar inequality with C_k instead of C_n (here $n = \dim V$) is unknown.

Next, let

$$h_k(V; \rho) = \inf_{v \in V} \log \mathrm{Vol}_k B_v(\rho)$$

for the ball $B_v(\rho) \subset V$ and define the entropy $h_k(V)$ by

$$h_k(V) = \liminf_{\rho \to \infty} \rho^{-1} h_k(V, \rho).$$

The most interesting is the entropy of the universal coverings $\tilde{V}$ of compact manifolds V. Here one expect the ratios such as $h_k(\tilde{V})/(\mathrm{Vol}\ V)^{1/n}$ or as $h_k(V)/\mathrm{Diam}_\ell V$ to bound some topological invariants of V. It is known, for instance, that

$$(h_n(\tilde{V}))^n/\mathrm{Vol}\ V \geq C_n \|V\| \tag{7}$$

where $\|V\|$ denotes the <u>simplicial volume</u> of V, that is, roughly speaking, the minimal number of simplices needed to trianglate the fundamental classes of V (see $[G]_1$ P.245).

If $\tilde{V}$ is contractible, then one expects a similar bound for Pontryagin numbers and for the L_2-Betti numbers of V (see $[G]_1$ P.293 for related results).

A complementary problem is to bound h_k by some curvature condition on V. For example, does the inequality $S(V) \geq -\sigma^2$ implies $h_2(V) \leq C\sigma$? Here is a closely related.

3.A. Conjecture.

Every closed manifold V with $S(V) \geq -\sigma^2$ satisfies

$$\|V\| \leq C_n \sigma^n \operatorname{Vol} V. \tag{8}$$

Remarks.

(A) The inequality (8) for $\operatorname{Ricci} V \geq -\sigma^2$ follows from (7), but the best constant C_n is unknown for $n \geq 3$.

(B) One can imagine a stronger version of (8), namely

$$\|V\| \leq C_n \int_V |S_v^-(v)|^{n/2} dv \tag{8'}$$

where $S_v^- = \min(0, S_v)$. But this is unknown even with $K(V)$ in place of $S(V)$. In fact, the only known lower bound for the total curvature $\int_V |K|_v^{n/2} dv$ comes from characteristic numbers of V. One does not know, for example, if every hyperbolic 3-manifold admits a sequence of metrics such that $\int_V |S_v|^{3/2} dv \longrightarrow 0$, even if one insists on $K < 0$ for these metrics.

3.B. Specific Entropy $sh_k V$.

Let $sh_k(V; \rho)$ be the upper bound of the numbers $\ell \geq 0$ with the following property. There exists a C^1-map $f: V \longrightarrow V$, such that $\operatorname{dist}(f, \operatorname{Id}) \leq \rho$ and every k-dimensional submanifold V' in V satisfies

$$\log \operatorname{Vol}_k V' - \log \operatorname{Vol}_k f(V') \geq \ell .$$

Then set

$$sh_k V = \liminf_{\rho \to \infty} \rho^{-1} sh_k(V; \rho).$$

Observe, for the universal covering $\tilde{V}$ of a compact manifold V, that $sh_k \tilde{V} = 0$ iff the fundamental group $\pi_1(V)$ is amenable and that $sh_2 \tilde{V} > 0$ iff $\pi_1(V)$ is hyperbolic (e.g. V admits a metric with $K < 0$). Furthermore, every symmetric space with $K \leq 0$ and rank $= 2$ has $h_k > 0$ and $sh_k > 0$ if and only if $k > 2$.

<u>Conjecture</u>. <u>Let</u> V <u>be a complete geometrically contractible manifold with</u> $S(V) \geq -\sigma^2$. <u>Then</u>

$$sh_2 V \leq C_n |\sigma| .$$

A related question is as follows. Let V be a compact manifold with $S(V) \geq \sigma^2$. Does there exist a (possibly singular) 2-dimensional surface (or a varifold) $V' \subset V$, such that Area $V' \leq C_n \sigma^{-2}$? In fact, one expects that

$$\mathrm{Vol}_2 V \leq C_n \sigma^{-2} .$$

4. NORMS ON THE COHOMOLOGY AND ON THE K-FUNCTOR.

The L_∞-<u>norm</u> on $H^*(V; \mathbb{R})$ is obtained by minimizing the L_∞-norm $= \sup_{v \in V} \|\omega\|_v$ of closed forms ω representing classes in H^* (see §7.4 in $[G]_2$ for details and references). Next, for an isomorphism class α of an orthogonal or unitary vector bundle $X \longrightarrow V$ we define $\|\alpha\|$ by minimizing the L_∞-norm of the curvature forms of (orthogonal or unitary) connections on X. An alternative "norm", called $\|\alpha\|^+$, is obtained by minimizing the Lipschitz constant of classifying maps of V into the pertinent Grassmann manifold G. Clearly

$$\|\alpha\| \leq C \|\alpha\|^+$$

for $C = C(n, \dim \alpha)$. Furthermore, if α is the class of a complex line bundle, then $\|\alpha\| = \|c_1(\alpha)\|$ for the first Chern class $c_1(\alpha)$. In fact, every closed 2-form ω on V in an <u>integral</u> cohomology class is the curvature form of some line bundle with curvature $= \omega$.

4.A. <u>Theorem</u> (see $[G_1 L]$, $[G]_1$ P.294 and references therein).

<u>Denote by</u> $s = s(V)$ <u>the minimal norm</u> $\|\gamma\|$ <u>for all orthogonal bundles with</u> $w_2(\gamma) = w_2(V)$ <u>for the second Stiefel Whitney class</u> w_2. <u>Then every unitary</u> β <u>satisfies</u>

$$|\{ch\, \beta \cdot \hat{A}(V)\}[V]| \leq C_n N(C'_n(s + \|\beta\|) - C'_n \sigma) \qquad (9)$$

where $\sigma = \inf S(V)$, where V is assumed compact and oriented, and where C_n, C'_n and C''_n are some universal positive constants. (Recall that $N(\lambda)$ denotes the number of eigenvalues $\leq \lambda$ of the Laplace operators on functions on V).

Corollaries.

(a) No metric g on V with $S(V, S) \geq \sigma > 0$ can be too large.

Proof. Take some β for which the left hand side of (9) does not vanish and observe that $s \longrightarrow 0$ and $\|\beta\| \longrightarrow 0$ as g is getting large. If n is odd, apply the above to $V \times S^1$ for a long circle S^1.

(b) Let (V, g) be a closed oriented manifold, such that, for a fixed metric g_0 on V, one has $g \wedge g \geq g_0 \wedge g_0$, that is the identity map $(V, g) \longrightarrow (V, g_0)$ decrease areas of the surfaces in V. Then the Laplace operator on $(V, \dot{g})$ satisfies for all $\lambda \geq 0$

$$N^{2/n}(\lambda) \geq C_n\lambda + C'_n\sigma - C'', \tag{9'}$$

where $\sigma = \dot{i}nf\ S(V, g)$ and where the constant C'' depends on (V, g_0). Furthermore, if V is spin, then

$$N^{2/n}(\lambda) \geq C_n\lambda + C'_n\sigma - C'''_n\rho^{-2},$$

where $\rho = \text{HS Rad}_2(V, g)$.

Proof. Apply (9) with appropriate β and γ.

Remarks.

(1) The inequalities (9') and (9") can be applied to the universal covering of V where the dimension $N(\lambda)$ is understood in the sense of Von Neumann algebras.

(2) The best constants C'' in (9') seems an interesting invariant of (V, g_0).

The norm of an appropriate β (as well as of $s(V)$) can be often made arbitrary small by passing to the universal covering $\tilde{V}$ of V where some version of (9) still holds true (see [GL]). This is so, for instance, if $\tilde{V}$ is a hyperspherical manifold with $w_2(\tilde{V}) = 0$. In this case (9) implies $\inf S(V) \leq 0$ for every metric on V. Furthermore, the norm $\|\beta\|^+$ also becomes arbitrary small in the hyperspherical case. Thus, by combining [GL]-twisting with [VW]-untwisting (see [VW]), one gets the following result.

4.B.

Let the universal covering $\tilde{V}$ *of a compact manifold* V *be spin and hyperspherical. Then the spectrum of the Dirac operator on* $\tilde{V}$ *contains zero.*

Remark. A similar argument applies to the Laplace operator on *forms* on $\tilde{V}$. However, the Laplace on functions on $\tilde{V}$ contains zero in the spectrum iff $sh_n \tilde{V} = 0$.

Question. Let V be a "large" manifold, e.g. V is contractible and covers a compact manifold V'. Does the spectra of Dirac and Laplace (on forms!) contain zero ? This is likely if $\pi_1(V')$ satisfies the strong Novikov conjecture.

4.C. *Symplectic Forms.*

Let ω be a symplectic (i.e. closed and nonsingular) 2-form on a closed manifold V. Write $g \geq \omega$ if the L_∞-norm of ω with respect to (the metric) g is ≤ 1 and set

$$\|\omega\|_S = \sup_{g \geq \omega} \sigma_g$$

for $\sigma_g = \inf S(V, g)$. If V is spin and if some real multiple of ω represents an *integral* class in $H^*(V; \mathbf{R})$ then (9) implies $\|\omega\|_S < \infty$. Furthermore all metrics $g \geq \omega$ on V satisfy

$$N^{2/n}(\lambda) \geq C_n \lambda + C_n' \sigma_y - C'' \tag{10}$$

for some (interesting ?) constant $C'' = C''(V, \omega)$ (compare (9')).

Question. Are the spin and the integrality conditions essential ?

How can one evaluate $\|\omega\|_S$ for known examples of symplectic manifolds ?

Observe the following useful property of the L_∞-norm on the image $I^* = f^*(H^*(K; \mathbf{R})) \subset H^*(V; \mathbf{R})$ for an arbitrary continuous map $f: V \longrightarrow K$ where $K = K(\Gamma/1)$ for a *residually finite* group Γ.

4.C'.

For every $\alpha \in I^*$ and every $\varepsilon > 0$, there exists a finite covering $\tilde{V} \longrightarrow V$ and some integral classes $\tilde{\alpha}_1, \ldots, \tilde{\alpha}_p$ in $H^*(\tilde{V}; \mathbb{Z}) \subset H^*(V; \mathbb{R})$ such that $\|\alpha_i\| \leq \varepsilon$ for $i = 1, \ldots, p$ and the pull-back $\tilde{\alpha} \subset H^*(\tilde{V}; \mathbb{R})$ of α is representible by some real combination of $\tilde{\alpha}_i$.

4.C". Corollary.

If a closed even dimensional spin manifold V possesses a 2-dimensional class $\alpha \in I^*$, such that $\alpha^{n/2} \neq 0$ (for $n = \dim V$), then V admits no metric with $S > 0$, provided the implied group Γ is residually finite.

Proof. Apply (9) to some line bundles $\tilde{\beta}_i$ on $\tilde{V}$ with $c_1(\tilde{\beta}_i) = \tilde{\alpha}_i$.

Probably, one can drop the residual finiteness condition by elaborating on non-compact thechniques in [GL]. It also would be interesting to eliminate spin by Schoen-Yau minimal manifolds techniques (see [S] and references therein).

References

$[G]_1$ M. Gromov, Volume and bounded cohomology, Publ. Math. IHES, #56, P.P.213-307 (1983).

$[G]_2$ M. Gromov, Filling Riemannian manifolds, J. of Differential Geometry, #18, P.P.1-147 (1983).

[GL] M. Gromov and B. Lawson, Positive scalar curvature and the Dirac operator on complete Riemannian manifolds, Publ. Math. IHES, #58, P.P.295-408 (1983).

[K] M. Katz, The filling radius of two point homogeneous spaces, J. of Differential Geometry, #18, P.P.505-511 (1983).

[S] R. Schoen, Minimal manifolds and positive scalar curvature, Proc. ICM 1982, Warsaw, P.P.575-579, North Holland 1984.

[VW] C. Vafa and E. Witten, Eigenvalues inequalities for fermions in Gauge theories, Comm. Math. Physics 95:3 P.P.257-277 (1984).

ANALYTIC INEQUALITIES, AND ROUGH ISOMETRIES BETWEEN NON-COMPACT RIEMANNIAN MANIFOLDS

Masahiko KANAI

Department of Mathematics
Faculty of Science and Technology
Keio University
Yokohama 223, Japan

1. Introduction

For a non-compact riemannian manifold, how it spreads at infinity is one of the most interesting problems we have to study, and, in this pont of view, its local geometry and topology are of no matter to us. The notion of rough isometry was introduced in [**K1**] in this spirit:

Definition. A map $\varphi : X \to Y$, *not necessarily continuous*, between metric spaces X and Y, is called a *rough isometry*, if the following two conditions are satisfied:

(i) for a sufficiently large $\epsilon > 0$, the ϵ-neighborhood of the image of φ in Y coincides with Y itself;

(ii) there are constants $a \geq 1$ and $b \geq 0$ such that

$$a^{-1}d(x_1, x_2) \; - \; b \; \leq d(\varphi(x_1), \varphi(x_2)) \; \leq a\, d(x_1, x_2) \; + \; b$$

for all $x_1, x_2 \in X$.

We say that X is *roughly isometric* to Y if there is a rough isometry of X into Y.

It is quite easy to see that being roughly isometric is an equivalence relation. In fact, (1) the composition $\psi \circ \varphi : X \to Z$ of two rough isometries $\varphi : X \to Y$ and $\psi : Y \to Z$ is again a rough isometry: (2) For a rough isometry $\varphi : X \to Y$, an "inverse" rough isometry $\varphi^- : Y \to X$ is constructed as follows; for $y \in Y$ take $x \in X$ so that $d(\varphi(x), y) < \epsilon$, where ϵ is the constant in the definition above, and set $\varphi^-(y) = x$. Here, we should note that the above construction of the inverse rough isometry φ^- is possible because we do not assume that a rough isometry is to be continuous: In general, φ^- is not continuous even if φ is continuous. This is a remarkable feature of rough isometries, and by virtue of it, we can identify some spaces of different topological types by rough isometries. For example, the inclusion map of the complete "periodic" surface in Fig.1 into the euclidean 3-space $\mathbf{R}^3$ is a rough isometry, and therefore the surface is roughly isometric to $\mathbf{R}^3$.

As we have just seen, a rough isometry does not, in general, preserve the topological structures of spaces, but in the preceding papers [**K1**] and [**K2**] we have exhibited that some geometric invariants and properties of non-compact riemannian manifolds

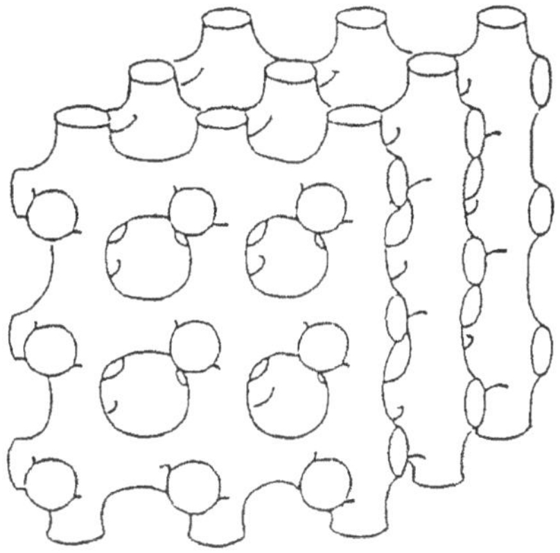

Fig.1

are inherited through rough isometries. One of them is the validity of isoperimetric inequalities. First of all, recall the classical isoperimetric inequality: It suggests that, for a bounded domain Ω in the euclidean space $\mathbf{R}^n$ with smooth boundary, the inequality

$$(\mathrm{vol}\,\Omega)^{1/n} \leq c_n \cdot (\mathrm{area}\,\partial\Omega)^{1/(n-1)}$$

alway holds with a constant c_n depending only on the dimension n. This leads us to the following definition of the isoperimetric constant $I_m(X)$ for a general complete riemannian manifold X with $\dim X \leq m \leq \infty$:

$$(1.1) \qquad I_m(X) = \inf_{\Omega} \frac{\mathrm{area}\,\partial\Omega}{(\mathrm{vol}\,\Omega)^{(m-1)/m}},$$

where Ω ranges through all the non-empty bounded domains in X with smooth boundaries, and, in the definition, we adopt the natural convention that $(m-1)/m = 1$ for $m = \infty$; in other words, $I_\infty(X)$ is, so-called, Cheeger's isoperimetric constant (for a non-compact riemannian manifold). Now the classical isoperimetric inequality is nothing but $I_n(\mathbf{R}^n) > 0$. And a theorem in the previous paper [**K1**] says that the validity of the isoperimetric inequality $I_m(X) > 0$ is preserved by rough isometries, under the additional condition that

(*) the Ricci curvature is bounded below, and the injectivity radius is positive,

which ensures the uniformness of local geometry: More precisely we proved

Theorem 1.1. *Let X and Y be complete riemannian manifolds satisfying the condition (*) and roughly isometric to each other. Then, for $\max\{\dim X, \dim Y\} \leq m \leq \infty$, the inequality $I_m(X) > 0$ is equivalent to the inequality $I_m(Y) > 0$.*

A reason why isoperimetric inequalities have a lot of applications is that they are closely related to analytic inequalities. An application of this kind was, in fact, done in [**K1**], where we proved the Liouville theorem generalized in terms of rough isometries:

Theorem 1.2. *Let X be a complete riemannian manifold satisfying the condition $(*)$ and roughly isometric to the euclidean m-space with $m \geq \dim X$. Then any positive harmonic function on X is constant.*

One of the crucial steps of the proof of the above theorem is to translate Theorem 1.1 into an assertion concerned with a Sobolev inequality, and, to state it in a more concrete form, we should introduce the analytic constants $S_{l,m}(X)$ for a complete riemannian manifold X:

$$(1.2) \quad S_{l,m}(X) = \inf_{u \in C_0^\infty(X)} \frac{\left\{\int_X |\nabla u|^l \, dx\right\}^{1/l}}{\left\{\int_X |u|^{m/(m-1)} dx\right\}^{(m-1)/m}}, \qquad l \geq 1,\ 1 < m \leq \infty,$$

where $C_0^\infty(X)$ denotes the space of C^∞ functions on X with compact supports, and we again assume that $(m-1)/m = m/(m-1) = 1$ provided $m = \infty$. A fundamental fact relating isoperimetric inequalities to Sobolev inequalities is the identity

$$(1.3) \qquad I_m(X) = S_{1,m}(X)$$

due to Federer-Fleming [**FF**] and Maz'ya [**M**] (see also Osserman [**O**]), and, by virtue of it, Theorem 1.1 have

Corollary 1.3. *Let X and Y be as in Theorem 1.1. Then, for* $\max\{\dim X, \dim Y\} \leq m \leq \infty$, $S_{1,m}(X) > 0$ *if and only if* $S_{1,m}(Y) > 0$.

For the euclidean space $\mathbf{R}^m$, we have the Sobolev inequality $S_{1,m}(\mathbf{R}^m) > 0$, and, by the corollary above, we obtain $S_{1,m}(X) > 0$ for X as in Theorem 1.2, and this is one of the inequalities we need to prove Theorem 1.2.

Also, there are relations between an isoperimetric constant and an analytic constant other than the Federer-Fleming-Maz'ya identity (1.3). In fact, Cheeger [**Ch**] established the inequality

$$(1.4) \qquad \frac{1}{2} I_\infty(X) \leq S_{2,2}(X)$$

for an arbitrary complete riemannian manifold X (cf. Yau [**Y**]), while Buser proved in [**B**] that if X is a complete riemannian manifold with Ricci curvature bounded below then

$$(1.5) \qquad S_{2,2}(X)^2 \leq \text{const} \cdot I_\infty(X),$$

where the constant in the inequality depends only on the dimension of X and on the infimum of the Ricci curvature. Note that $S_{2,2}(X)^2$ appears as the best constant in the Poincaré inequality $\int_X |\nabla u|^2 dx \geq \text{const} \cdot \int_X u^2 dx$ for $u \in C_0^\infty(X)$, and therefore

$S_{2,2}(X)^2$ is equal to the infimum of the spectrum of $-\Delta$, the Laplace operator multiplied by -1, acting on L^2-functions on X. In particular, the inequalities of Cheeger (1.4) and of Buser (1.5) imply that

$$I_\infty(X) > 0 \quad \text{if and only if} \quad S_{2,2}(X) > 0 \tag{1.6}$$

for a complete riemannian manifold X with Ricci curvature bounded below, and consequently, together with Theorem 1.1, we have

Corollary 1.4. *Let X and Y be as in Theorem 1.1. Then $S_{2,2}(X) > 0$ is equivalent to $S_{2,2}(Y) > 0$.*

Now these two corollaries of Theorem 1.1 lead us to the natural question: To what extent are the analytic constants $S_{l,m}(X)$ preserved by rough isometries? In the present article, we will prove the following generalization of Corollary 1.4 without use of isoperimetric inequalities or Theorem 1.1.

Theorem 1.5. *Suppose that X and Y are complete riemannian manifolds satisfying the condition $(*)$ and roughly isometric to each other. Then, for $1 < m \leq \infty$ and $l \geq m/(m-1)$, $S_{l,m}(X) > 0$ if and only if $S_{l,m}(Y) > 0$.*

This theorem is also motivated by author's previous work [**K2**], in which he showed that the parabolicity is preserved by rough isometries: By definition, a riemannian manifold X is said to be parabolic if there is no positive superharmonic function on X other than constants, and the main result obtained in [**K2**] is

Theorem 1.6. *Let X and Y be complete riemannian manifolds which satisfy the condition $(*)$ and are roughly isometric to each other. Then X is parabolic if and only if so is Y.*

To prove the theorem, we first showed that a complete riemannian manifold X is non-parabolic if and only if $\operatorname{cap}\Omega > 0$ for a non-empty bounded domain Ω in X with smooth boundary, where the capacity $\operatorname{cap}\Omega$ of Ω is defined by

$$\operatorname{cap}\Omega = \inf\left\{\int_X |\nabla u|^2 dx \; : \; u \in C_0^\infty(X), \; u_{|\Omega} = 1\right\}.$$

Then Theorem 1.6 is reduced to the problem of showing that the non-vanishing of the capacity is preserved by rough isometries, and the proof of this fact is almost the same with that of Theorem 1.5, because the behavior of the capacity under rough isometries is quite similar to that of the analytic constant $S_{2,2}(X)$, as is expected from their similarity in the definitions.

The construction of this article is as follows. §§2 and 3 are devoted to the proof of Theorem 1.5, which will be done, as in the preceding works [**K1**] and [**K2**], by approximating "continuous" geometry of a riemannian manifold, say X, by "combinatorial" geometry of a certain discrete subset P of X endowed with a suitable combinatorial structure. We will call P a net in X, and its "intrinsic" aspects are considered in §2. In the next section, we will show that P actually approximates X, and will complete the proof of Theorem 1.5. Finally, in §4, we will discuss another application of the discrete approximation method. In particular we will reveal relationship between the work of Kesten [**Ks2**] on the random walks and the Cheeger-Buser inequalities (1.4) and (1.5): With the aid of our discrete approximation theorems, the latter (in a weaker form) will be followed from the former.

2. Intrinsic Studies of Nets

We begin this section with the precise "intrinsic" definition of nets. A *net* is a countable set P equipped with a family $\{N_p\}_{p\in P}$ indexed by the elements of P itself such that

(i) each N_p is a finite subset of P, and that
(ii) for $p, q \in P$, $p \in N_q$ if and only if $q \in N_p$.

A net is nothing but a kind of 1-dimensional graphs: In fact, each element of P can be considered as a vertex of a graph, and two vertices p and q are considered to be combined by an edge if $p \in N_q$. Now let P be a net. P is said to be *uniform* if $\sup_{p\in P} \#N_p < \infty$, where, for a set S, $\#S$ denotes its cardinality. A sequence $\mathbf{p} = (p_0, \ldots, p_L)$ of elements of P is called a *path from* p_0 *to* p_L *of length* L if $p_k \in N_{p_{k-1}}$ for $k = 1, \ldots, L$, and the net P is said to be *connected* if any two points of P are combined by a path. In the case when P is connected, the *combinatorial metric* δ of P is defined by

$$\delta(p, q) = \min\{\text{the lengths of paths from } p \text{ to } q\}$$

for $p, q \in P$. We always consider a connected net as a metric space with the combinatorial metric δ.

Next we introduce the analytic constants for the nets. Again let P be a net. For real-valued functions u and v on P, put

$$(2.1)\qquad \begin{aligned} \langle Du, Dv\rangle(p) &= \sum_{q\in N_p} \{u(q) - u(p)\}\{v(q) - v(p)\}, \\ |Du|(p) &= \sqrt{\langle Du, Du\rangle(p)}, \qquad p \in P: \end{aligned}$$

The former is a combinatorial analogue of the inner product of the gradients of the functions u and v, and the latter the norm of the gradient of u. Now, for each $l \geq 1$ and $1 < m \leq \infty$, the analytic constant of the net P is defined by

$$(2.2)\qquad S_{l,m}(P) = \inf_u \frac{\left\{\sum_{p\in P} |Du|^l(p)\right\}^{1/l}}{\left\{\sum_{p\in P} |u|^{m/(m-1)}(p)\right\}^{(m-1)/m}},$$

where u ranges over all finitely supported functions on P. Now we have a combinatorial version of Theorem 1.5:

Proposition 2.1. *Let P and Q be uniform connected nets. If P is roughly isometric to Q, then $S_{l,m}(P) > 0$ is equivalent to $S_{l,m}(Q) > 0$ for any $l \geq 1$ and $1 < m \leq \infty$.*

Proof. In the proof, denote by $B_\rho(q)$ the "closed" ρ-ball in Q around $q \in Q$; i.e., $B_\rho(q) = \{r \in Q : \delta(r,q) \leq \rho\}$. Also, let $\varphi : P \to Q$ be a rough isometry such that $Q = \bigcup_{p\in P} B_\tau(\varphi(p))$ for some constant $\tau > 0$.

To begin with, suppose that v is an arbitrary non-negative function on Q with finite support, and define another finitely supported function $\overline{v}$ on Q by

$$\overline{v}(q) = \frac{1}{\#B_\tau(q)} \sum_{r\in B_\tau(q)} v(r).$$

Our first purpose is to derive the inequality (2.3) below. Let q and q' be points of Q with $\delta(q,q') = 1$. Then we have

$$\begin{aligned}
|\overline{v}(q') - \overline{v}(q)| &= \left| \frac{1}{\#B_\tau(q')} \sum_{r'\in B_\tau(q')} v(r') - \frac{1}{\#B_\tau(q)} \sum_{r\in B_\tau(q)} v(r) \right| \\
&= \left| \frac{1}{\#B_\tau(q)\cdot \#B_\tau(q')} \sum_{r\in B_\tau(q), r'\in B_\tau(q')} \{v(r') - v(r)\} \right| \\
&\leq \frac{1}{\#B_\tau(q)\cdot \#B_\tau(q')} \sum_{r\in B_\tau(q), r'\in B_\tau(q')} |v(r') - v(r)|.
\end{aligned}$$

Moreover, for $r \in B_\tau(q)$ and $r' \in B_\tau(q')$, combining them by a length-minimizing path $\mathbf{q} = (q_0, \ldots, q_L)$ with $q_0 = r$, $q_L = r'$, and of length $L \leq 2\tau + 1$, we obtain

$$\begin{aligned}
|v(r') - v(r)| &\leq |v(q_0) - v(q_1)| + \cdots + |v(q_{L-1}) - v(q_L)| \\
&\leq |Dv|(q_0) + \cdots + |Dv|(q_{L-1}) \\
&\leq \sum_{q''\in B_\tau(q)} |Dv|(q'')
\end{aligned}$$

since $\delta(q_i, q) \leq \tau$ for $i = 0, \ldots, L-1$, and therefore we get

$$|\overline{v}(q') - \overline{v}(q)| \leq \sum_{q''\in B_\tau(q)} |Dv|(q'').$$

Thus, for any $q \in Q$, we have

$$|D\overline{v}|(q) = \left\{ \sum_{q'\in N_q} \{\overline{v}(q') - \overline{v}(q)\}^2 \right\}^{1/2} \leq \nu_{Q,1}^{1/2} \sum_{r\in B_\tau(q)} |Du|(r),$$

and

$$|D\overline{v}|^l(q) \leq \nu_{Q,1}^{l/2}\left\{\sum_{r\in B_\tau(q)}|Dv|(r)\right\}^l \leq \nu_{Q,1}^{l/2}\nu_{Q,\tau}^{l-1}\sum_{r\in B_\tau(q)}|Dv|^l(r)$$

by the Hölder inequality, where for any $\rho > 0$, we put $\nu_{Q,\rho} = \sup_{q\in Q}\#B_\rho(q)$ which has a finite value by the assumption of uniformness of Q. This yields

$$\sum_{q\in Q}|D\overline{v}|^l(q) \leq \nu_{Q,1}^{l/2}\nu_{Q,\tau}^{l-1}\sum_{q\in Q}\sum_{r\in B_\tau(q)}|Dv|^l(r) \leq \nu_{Q,1}^{l/2}\nu_{Q,\tau}^{l}\sum_{q\in Q}|Dv|^l(q),$$

i.e.,

$$\left\{\sum_{q\in Q}|D\overline{v}|^l(q)\right\}^{1/l} \leq c_1\cdot\left\{\sum_{q\in Q}|Dv|^l(q)\right\}^{1/l} \tag{2.3}$$

with a suitable constant c_1.

Now define a finitely supported non-negative function u on P by $u = \overline{v}\circ\varphi$. The next purpose of ours is to obtain the estimates (2.4) and (2.5) below. First note that, for $p, p' \in P$ with $\delta(p,p') = 1$, there is a constant L_0 such that $\delta(\varphi(p),\varphi(p')) \leq L_0$ since φ is a rough isometry. Therefore, combining $\varphi(p)$ and $\varphi(p')$ by a path $\mathrm{q} = (q_0,\ldots,q_L)$ in Q with $q_0 = \varphi(p)$, $q_L = \varphi(p')$, and of length $L \leq L_0$, we have

$$\begin{aligned}
|u(p) - u(p')| &= |\overline{v}(q_0) - \overline{v}(q_L)| \\
&\leq |\overline{v}(q_0) - \overline{v}(q_1)| + \cdots + |\overline{v}(q_{L-1}) - \overline{v}(q_L)| \\
&\leq |D\overline{v}|(q_0) + \cdots + |D\overline{v}|(q_{L-1}) \\
&\leq \sum_{q\in B_{L_0-1}(\varphi(p))}|D\overline{v}|(q),
\end{aligned}$$

and this implies, as above,

$$\begin{aligned}
|Du|^l(p) &\leq \nu_{P,1}^{l/2}\left\{\sum_{q\in B_{L_0-1}(\varphi(p))}|D\overline{v}|(q)\right\}^l \\
&\leq \nu_{P,1}^{l/2}\nu_{Q,L_0-1}^{l-1}\sum_{q\in B_{L_0-1}(\varphi(p))}|D\overline{v}|^l(q)
\end{aligned}$$

with $\nu_{P,\rho} = \sup_{p\in P}\#\{p' \in P : \delta(p',p) \leq \rho\} < \infty$ for $\rho > 0$. Hence we obtain

$$\left\{\sum_{p\in P}|Du|^l(p)\right\}^{1/l} \leq c_2\cdot\left\{\sum_{q\in Q}|D\overline{v}|^l(q)\right\}^{1/l} \tag{2.4}$$

with a certain constant c_2. Finally, for each $p \in P$, we have

$$\begin{aligned} u^{m/(m-1)}(p) &= \left\{ \frac{1}{\#B_r(\varphi(p))} \sum_{q \in B_r(\varphi(p))} v(q) \right\}^{m/(m-1)} \\ &\geq \nu_{Q,r}^{-m/(m-1)} \sum_{q \in B_r(\varphi(p))} v^{m/(m-1)}(q), \end{aligned}$$

and consequently we get

$$\sum_{p \in P} u^{m/(m-1)}(p) \geq \nu_{Q,r}^{-m/(m-1)} \sum_{q \in Q} v^{m/(m-1)}(q)$$

since $Q = \bigcup_{p \in P} B_r(\varphi(p))$. This shows

$$\left\{ \sum_{p \in P} u^{m/(m-1)}(p) \right\}^{(m-1)/m} \geq c_3 \cdot \left\{ \sum_{q \in Q} v^{m/(m-1)}(q) \right\}^{(m-1)/m} \tag{2.5}$$

with a constant $c_3 > 0$.

By (2.3), (2.4) and (2.5) we conclude

$$\begin{aligned} &\frac{c_1 c_2}{c_3} \cdot \frac{\left\{ \sum_{q \in Q} |Dv|^l(q) \right\}^{1/l}}{\left\{ \sum_{q \in Q} v^{m/(m-1)}(q) \right\}^{(m-1)/m}} \\ &\qquad \geq \frac{\left\{ \sum_{p \in P} |Du|^l(p) \right\}^{1/l}}{\left\{ \sum_{p \in P} u^{m/(m-1)}(p) \right\}^{(m-1)/m}} \geq S_{l,m}(P) \end{aligned}$$

for an arbitrary non-negative function v on Q with finite support. Moreover because $|Dv| \geq |D|v||$ for any function v on Q, we obtain $(c_1 c_2 / c_3) \cdot S_{l,m}(Q) \geq S_{l,m}(P)$. This completes the proof of the proposition. ▯

3. Discrete Approximation Theorem

In this section, we construct a net P in a complete riemannian manifold X, and show that P indeed approximates X combinatorially. Then Theorem 1.5 will follow immediately.

Now let X be a complete riemannian manifold. A subset P of X is said to be *ϵ-separated* if $d(p,q) \geq \epsilon$ whenever p and q are distinct points of P, and for a maximal ϵ-separated subset of X, a structure of net on it is canonically defined by

$$N_p = \{q \in P : 0 < d(p,q) \leq 3\epsilon\}, \qquad p \in P.$$

We call a maximal ϵ-separated subset P of X with this structure of net, an *ϵ-net in X*. Suppose that P is an ϵ-net in X. Then we have the following facts since P is maximally ϵ-separated in X:

(3.1) The open geodesic balls $B_{\epsilon/2}(p)$ in X of radius $\epsilon/2$ and with centers at $p \in P$ are disjoint;

(3.2) The geodesic balls $B_\epsilon(p)$, $p \in P$, cover X.

Moreover we can easily show that

(3.3) P is connected if X is connected.

So, in the rest of this paper, we assume that all manifolds and nets are connected.

To relate geometry of P to that of X, the following lemma proved in [**K1**] is fundamental.

Lemma 3.1. *Suppose that the Ricci curvature of X is bounded below. Then*

(1) *for $r > 0$ and $x \in X$, we have*

$$\#(P \cap B_r(x)) \leq \nu(r), \tag{3.4}$$

where $\nu(r)$ is a constant independent of x. In particular, P is uniform.

(2) *P is roughly isometric to X: In fact, there are constants $a > 1$ and $b > 0$ such that*

$$a^{-1}d(p,q) \leq \delta(p,q) \leq a\,d(p,q) + b \qquad \text{for } p,q \in P. \tag{3.5}$$

Consequently any two nets in X are roughly isometric to each other.

The most crucial part of the proof of Theorem 1.5 is the following discrete approximation theorem.

Theorem 3.2. *Let P be an ϵ-net in a complete riemannian manifold X satisfying the condition $(*)$. Then, for any $1 < m \leq \infty$ and $l \geq m/(m-1)$, $S_{l,m}(X) > 0$ if and only if $S_{l,m}(P) > 0$.*

We begin the proof of this theorem with referring to volume estimates of geodesic balls. For a complete riemannian manifold X with Ricci curvature bounded below, a standard comparison theorem gives

$$\operatorname{vol} B_r(x) \leq V_+(r) \qquad \text{for } x \in X \text{ and } r > 0. \tag{3.6}$$

On the other hand, for a complete riemannian manifold X with injectivity radius $\operatorname{inj} X > 0$, Croke [Cr] showed the inequality

$$\operatorname{vol} B_r(x) \geq V_-(r) \qquad \text{for } x \in X \text{ and } 0 < r \leq \frac{1}{2}\operatorname{inj} X. \tag{3.7}$$

Note that both $V_+(r)$ and $V_-(r)$ above are constants independent of $x \in X$.

Another necessity for the proof of Theorem 3.2 is

Lemma 3.3 (see, for the proof, [**K2**]). *Let X be a complete riemannian manifold with Ricci curvature bounded below. Then for the geodesic ball $B = B_r(p)$ in X of radius r and with the center p, there is a constant $\beta = \beta(r) > 0$ independent of p for which*

$$\int_B |\nabla u|\, dx \ \geq \ \beta \cdot \int_B |u - u^*|\, dx$$

holds for all $u \in C^\infty(\overline{B})$, where u^* denotes the integral mean of u over B; $u^* = (\operatorname{vol} B)^{-1} \int_B u\, dx$.

And this lemma yields

Corollary 3.4. *Let X and B be as in the lemma. Then for each $1 < m \leq \infty$ and $l \geq m/(m-1)$, there is a constant $\gamma = \gamma(r) > 0$ independent of p such that*

$$u^{*1/(m-1)} \left\{ \int_B |\nabla u|^l dx \right\}^{1/l} \geq \gamma \cdot \int_B \left| |u|^{m/(m-1)} - u^{*m/(m-1)} \right| dx \tag{3.8}$$

for all $u \in C^\infty(\bar{B})$ with

$$u^* = \left\{ \frac{1}{\operatorname{vol} B} \int_B |u|^{m/(m-1)} dx \right\}^{(m-1)/m}.$$

Proof. Apply Lemma 3.3 to $|u|^{m/(m-1)}$: Then with the Hölder inequality and (3.6), we have

$$\begin{aligned}
&\beta \cdot \int_B \left| |u|^{m/(m-1)} - u^{*m/(m-1)} \right| dx \\
&\leq \int_B \left| \nabla |u|^{m/(m-1)} \right| dx \\
&\leq \frac{m}{m-1} \int_B |u|^{1/(m-1)} |\nabla u| dx \\
&\leq \frac{m}{m-1} \left\{ \int_B |u|^{m/(m-1)} dx \right\}^{1/m} \left\{ \int_B |\nabla u|^{m/(m-1)} dx \right\}^{(m-1)/m} \\
&= \frac{m}{m-1} (\operatorname{vol} B)^{1/m} u^{*1/(m-1)} \left\{ \int_B |\nabla u|^{m/(m-1)} dx \right\}^{(m-1)/m} \\
&\leq \frac{m}{m-1} (\operatorname{vol} B)^{(l-1)/l} u^{*1/(m-1)} \left\{ \int_B |\nabla u|^l dx \right\}^{1/l} \\
&\leq \frac{m}{m-1} V_+(r)^{(l-1)/l} u^{*1/(m-1)} \left\{ \int_B |\nabla u|^l dx \right\}^{1/l}.
\end{aligned}$$

∎

Proof of Theorem 3.2. We may prove the theorem only in the case when $\epsilon \leq \operatorname{inj} X/2$, because any two nets in X are uniform and roughly isometric to each other by Lemma 3.1, and consequently, by Proposition 2.1, whether $S_{l,m}(P) > 0$ or not is independent of the choice of the net P in X.

First we prove the "if" part of the theorem, and to do this it is sufficient to show that

$$\frac{\left\{\int_X |\nabla u|^l dx\right\}^{1/l}}{\left\{\int_X |u|^{m/(m-1)} dx\right\}^{(m-1)/m}} \geq c_1 \cdot S_{l,m}(P) \tag{3.9}$$

for all $u \in C_0^\infty(X)$ with a suitable constant $c_1 > 0$. Now take $u \in C_0^\infty(X)$ arbitrarily, and define a finitely supported non-negative function u^* on P by

$$u^*(p) = \left\{\frac{1}{\operatorname{vol} B_{4\epsilon}(p)} \int_{B_{4\epsilon}(p)} |u|^{m/(m-1)} dx\right\}^{(m-1)/m}.$$

Then we immediately have

$$\int_X |u|^{m/(m-1)} dx \leq \sum_{p\in P} \int_{B_{4\epsilon}(p)} |u|^{m/(m-1)} dx \leq V_+(4\epsilon) \sum_{p\in P} u^{*m/(m-1)}(p), \tag{3.10}$$

because $\{B_{4\epsilon}(p) : p \in P\}$ covers X and because of (3.6). On the other hand, for $p, q \in P$ with $\delta(p,q) = 1$, we have

$$\begin{aligned}
&\{u^{*1/(m-1)}(p) + u^{*1/(m-1)}(q)\} \left\{\int_{B_{7\epsilon}(p)} |\nabla u|^l dx\right\}^{1/l} \\
&\geq u^{*1/(m-1)}(p) \left\{\int_{B_{4\epsilon}(p)} |\nabla u|^l dx\right\}^{1/l} \\
&\qquad + u^{*1/(m-1)}(q) \left\{\int_{B_{4\epsilon}(q)} |\nabla u|^l dx\right\}^{1/l} \\
&\geq \gamma(4\epsilon) \cdot \int_{B_{4\epsilon}(p)} \left| |u|^{m/(m-1)}(x) - u^{*m/(m-1)}(p)\right| dx \\
&\qquad + \gamma(4\epsilon) \cdot \int_{B_{4\epsilon}(q)} \left| |u|^{m/(m-1)}(x) - u^{*m/(m-1)}(q)\right| dx \\
&\geq \gamma(4\epsilon) \cdot \int_{B_{4\epsilon}(p)\cap B_{4\epsilon}(q)} \left\{\left| |u|^{m/(m-1)}(x) - u^{*m/(m-1)}(p)\right| \right. \\
&\qquad\qquad \left. + \left| |u|^{m/(m-1)}(x) - u^{*m/(m-1)}(q)\right| \right\} dx \\
&\geq \gamma(4\epsilon) \cdot \int_{B_{4\epsilon}(p)\cap B_{4\epsilon}(q)} \left| u^{*m/(m-1)}(p) - u^{*m/(m-1)}(q)\right| dx \\
&= \gamma(4\epsilon) \operatorname{vol}(B_{4\epsilon}(p) \cap B_{4\epsilon}(q)) \cdot \left| u^{*m/(m-1)}(p) - u^{*m/(m-1)}(q)\right| \\
&\geq \gamma(4\epsilon) \operatorname{vol} B_\epsilon(p) \cdot \left| u^{*m/(m-1)}(p) - u^{*m/(m-1)}(q)\right| \\
&\geq \gamma(4\epsilon) V_-(\epsilon) \cdot \left| u^{*m/(m-1)}(p) - u^{*m/(m-1)}(q)\right| :
\end{aligned} \tag{3.11}$$

The first inequality follows from the fact that $B_{4\epsilon}(p)$, $B_{4\epsilon}(q) \subset B_{7\epsilon}(p)$ because $d(p,q) \leq 3\epsilon$; the second inequality just follows from (3.8); the inequality before last is a consequence of the fact that $B_\epsilon(p) \subset B_{4\epsilon}(p) \cap B_{4\epsilon}(q)$, and the last is by Croke's inequality (3.7). Here note that for any real numbers $\xi, \eta \geq 0$ and $\alpha > 0$ we always have

$$2\,|\xi^{1+\alpha} - \eta^{1+\alpha}| \geq |\xi - \eta|(\xi^\alpha + \eta^\alpha).$$

Applying this to $\xi = u^*(p)$, $\eta = u^*(q)$, $\alpha = 1/(m-1)$ in (3.11), we obtain

$$\left\{\int_{B_{7\epsilon}(p)} |\nabla u|^l dx\right\}^{1/l} \geq \frac{1}{2}\gamma(4\epsilon)V_-(\epsilon)\cdot|u^*(p) - u^*(q)|$$

for $p, q \in P$ with $\delta(p,q) = 1$, and therefore, by (3.4), we get

$$\left\{\int_{B_{7\epsilon}(p)} |\nabla u|^l dx\right\}^{1/l} \geq \frac{1}{2}\gamma(4\epsilon)V_-(\epsilon)\nu(3\epsilon)^{-1/2}\cdot|Du^*|(p),$$

with $\nu(3\epsilon) \geq \sup_{p\in P} \# N_p$. This yields, again by (3.4),

$$\int_X |\nabla u|^l dx \geq \nu(7\epsilon)^{-1}\sum_{p\in P}\int_{B_{7\epsilon}(p)} |\nabla u|^l dx \geq c_2 \sum_{p\in P} |Du^*|^l(p) \tag{3.12}$$

with a suitable constant $c_2 > 0$. Now the inequality (3.9) immediately follows from (3.10) and (3.12).

Next we give the "only if" part of the theorem; i.e., we will show that for any function u^* on P with finite support, the inequality

$$\frac{\left\{\sum_{p\in P} |Du^*|^l(p)\right\}^{1/l}}{\left\{\sum_{p\in P} |u^*|^{m/(m-1)}(p)\right\}^{(m-1)/m}} \geq c_3 \cdot S_{l,m}(X) \tag{3.13}$$

always holds with a certain constant $c_3 > 0$. The proof of this inequality is rather easier than that of (3.9), and is done by "smoothing" u^* by use of a partition of unity of X. So, first of all, we construct a partition of unity associated to a covering of X by geodesic balls around $p \in P$. For each $p \in P$, define a function $\widehat{\eta}_p$ on X with finite support by

$$\widehat{\eta}_p(x) = \begin{cases} 1 - \frac{1}{2\epsilon}d(x,p) & \text{if } x \in B_{2\epsilon}(p) \\ 0 & \text{otherwise,} \end{cases}$$

and then define a partition of unity, $\{\eta_p : p \in P\}$, by

$$\eta_p(x) = \frac{1}{\sum_{q\in P}\widehat{\eta}_q(x)}\,\widehat{\eta}_p(x).$$

It is easy to see that there are constants $c_4 > 0$ and c_5 independent of $p \in P$ such that $\eta_p \geq c_4$ on $B_{\epsilon/2}(p)$ and $|\nabla \eta_p| \leq c_5$. Now let u^* be an arbitrary finitely supported function on P. We may consider only non-negative u^*, because for general u^* we have $|Du^*| \geq |D|u^*||$. Define a non-negative function u on X by

$$u(x) = \sum_{p \in P} \eta_p(x) u^*(p).$$

This function u on X is not smooth but is Lipschitz continuous, and therefore differentiable almost everywhere. So we can treat this function u as a smooth function. Then, with Croke's inequality (3.7), we immediately have

$$\begin{aligned}\int_{B_{\epsilon/2}(p)} u^{m/(m-1)} dx &\geq \int_{B_{\epsilon/2}(p)} \eta_p^{m/(m-1)}(x)\, u^{*m/(m-1)}(p)\, dx \\ &\geq c_4^{m/(m-1)} \operatorname{vol} B_{\epsilon/2}(p) \cdot u^{*m/(m-1)}(p) \\ &\geq c_4^{m/(m-1)} V_-(\epsilon/2) \cdot u^{*m/(m-1)}(p)\end{aligned}$$

and this implies from (3.1) that

$$\text{(3.14)} \qquad \int_X u^{m/(m-1)} dx \geq \sum_{p \in P} \int_{B_{\epsilon/2}(p)} u^{m/(m-1)} dx \geq c_6 \sum_{p \in P} u^{*m/(m-1)}(p)$$

with some constant $c_6 > 0$. On the other hand, at a point $x \in B_\epsilon(p)$,

$$u(x) = \sum_{q \in N_p \cup \{p\}} u^*(q) \eta_q(x)$$

since each η_q is supported on $\overline{B_{2\epsilon}(q)}$, and

$$\nabla u(x) = \sum_{q \in N_p \cup \{p\}} u^*(q) \nabla \eta_q(x) = \sum_{q \in N_p} \{u^*(q) - u^*(p)\} \nabla \eta_q(x)$$

because $\sum_{q \in N_p \cup \{p\}} \nabla \eta_q(x) = 0$ (recall that $\sum_{q \in P} \eta_q = 1$). Thus by the Schwarz inequality, (3.4), and the fact that $|\nabla \eta_q| \leq c_5$, we obtain

$$|\nabla u|(x) \leq c_5 \nu(3\epsilon)^{1/2} \cdot |Du^*|(p)$$

for $x \in B_\epsilon(p)$. This implies with (3.6) that

$$\int_{B_\epsilon(p)} |\nabla u|^l dx \leq c_5^l \nu(3\epsilon)^{l/2} V_+(\epsilon) \cdot |Du^*|^l(p),$$

and therefore we have

$$\text{(3.15)} \qquad \int_X |\nabla u|^l dx \leq \sum_{p \in P} \int_{B_\epsilon(p)} |\nabla u|^l dx \leq c_7 \cdot \sum_{p \in P} |Du^*|^l(p).$$

Finally (3.13) follows from (3.14) and (3.15). ▯

Now we can complete

Proof of Theorem 1.5. Let X and Y be complete riemannian manifolds satisfying the condition $(*)$ which are roughly isometric to each other, and take nets P and Q in X and Y, respectively. Then, by Lemma 3.1 (2) and the assumption that X is roughly isometric to Y, P and Q, both of which are also uniform by Lemma 3.1 (1), are roughly isometric to each other, and consequently $S_{l,m}(P) > 0$ if and only if $S_{l,m}(Q) > 0$ by Proposition 2.1. On the other hand, Theorem 3.2 says that the inequalities $S_{l,m}(P) > 0$ and $S_{l,m}(Q) > 0$ are, respectively, equivalent to $S_{l,m}(X) > 0$ and $S_{l,m}(Y) > 0$. Thus we conclude that $S_{l,m}(X) > 0$ if and only if $S_{l,m}(Y) > 0$. This completes the proof of the theorem. ▯

4. Inequalities of Kesten and of Cheeger-Buser

In the preceding sections as well as our earlier works [**K1**] and [**K2**], we have seen that nets are enriched with combinatorial geometry: They have a lot of geometric notions corresponding to those for riemannian manifolds, such as volume growth rate, isoperimetric and analytic inequalities, and potential-theoretic and probability-theoretic notions. Also nets relate to riemannian manifolds through discrete approximation theorems, such as Theorem 3.2, which suggest that a net in a riemannian manifold is similar to the manifold. Furthermore we often find that problems are much easier in combinatorial category than riemannian category. So we may expect that by the aid of discrete approximation theorems we can utilize combinatorial geometry of nets to obtain results in riemannian geometry. In this section, we revisit the work of Kesten, which can be considered as a combinatorial version of the Cheeger-Buser inequalities, and from it, we will refind a weaker version of the Cheeger-Buser inequalities applying our discrete approximation theorems.

For this purpose, we should first recall the notion of isoperimetric constants for nets. Now let P be a net. For a subset S of P, its boundary is defined by

$$\partial S \;=\; \{p \in S \,:\, N_p \not\subset S\},$$

and, for each $1 < m \leq \infty$, the isoperimetric constant $I_m(P)$ of P is introduced by

$$I_m(P) \;= \inf_S \frac{\#\partial S}{(\#S)^{(m-1)/m}},$$

where S runs over all finite subsets of P. Then we have the following discrete approximation theorem which was the most essential in the proof of Theorem 1.1 (see [**K1**]):

Theorem 4.1. *Suppose that X is a complete riemannian manifold satisfying the condition $(*)$ and P is an ϵ-net in X with arbitrary $\epsilon > 0$. Then, for any $\dim X \leq m \leq \infty$, $I_m(P) > 0$ is equivalent to $I_m(X) > 0$.*

We will utilize this theorem later.

Now we refer to the work of Kesten [**Ks2**], which is stated in the following form in our language. (This work of Kesten was motivated by the study of random walks on discrete groups: For the probabilistic aspects, see Kesten's original papers [**Ks1**] and [**Ks2**].)

Proposition 4.2. *For a net P, we always have*

$$S_{2,2}(P) \geq \frac{1}{2} I_\infty(P). \tag{4.1}$$

Moreover if P is uniform, then

$$S_{2,2}(P)^2 \leq c \cdot I_\infty(P), \tag{4.2}$$

where c is a constant depending only on $\sup_{p\in P} \# N_p < \infty$.

Proof. First we prove the second inequality (4.2). Let S be an arbitrary non-empty finite subset of a uniform net P, and u the characteristic function of S. Then it is immediate to see that $\sum_{p\in P} u^2(p) = \#S$ and $\sum_{p\in P} |Du|^2(p) \leq c \cdot \#\partial S$, and consequently we have $c \cdot (\#\partial S/\#S) \geq \sum_{p\in P} |Du|^2(p) / \sum_{p\in P} u^2(p) \geq S_{2,2}(P)^2$, which implies (4.2).

Next we prove the first inequality (4.1). Let u be any non-negative finitely supported function on P. We will prove the inequality

$$\frac{\left\{\sum_{p\in P} |Du|^2(p)\right\}^{1/2}}{\left\{\sum_{p\in P} u^2(p)\right\}^{1/2}} \geq \frac{1}{2} I_\infty(P), \tag{4.3}$$

which implies (4.1) because $|Dv| \geq |D|v||$ for any function v on P. Put $S_t = \{p \in P : u^2(p) \geq t\}$ for $t > 0$. Then, by the definition of $I_\infty(P)$, we get

$$\int_0^\infty \#\partial S_t \, dt \geq I_\infty(P) \cdot \int_0^\infty \#S_t \, dt = I_\infty(P) \cdot \sum_{p\in P} u^2(p). \tag{4.4}$$

On the other hand, put $P_0 = \{p \in P : \min_{q\in N_p} u(q) < u(p)\}$. Then $p \in \partial S_t$ if and only if $p \in P_0$ and $\min_{q\in N_p} u^2(q) < t \leq u^2(p)$, and therefore we have

$$\int_0^\infty \#\partial S_t \, dt = \sum_{p\in P_0} \{u^2(p) - \min_{q\in N_p} u^2(q)\}.$$

Now, for each $p \in P_0$, take $q_0 \in N_p$ which minimizes $u(q)$ among $q \in N_p$. Then

$$\begin{aligned} u^2(p) - u^2(q_0) &= \{u(p) + u(q_0)\}\{u(p) - u(q_0)\} \\ &\leq 2u(p)\{u(p) - u(q_0)\} \\ &\leq 2u(p)|Du|(p), \end{aligned}$$

and this implies, by Schwarz's inequality, that

$$(4.5)\quad \int_0^\infty \#\partial S_t\, dt \le 2\sum_{p\in P} u(p)|Du|(p) \le 2\left\{\sum_{p\in P} u^2(p)\right\}^{1/2}\left\{\sum_{p\in P}|Du|^2(p)\right\}^{1/2}.$$

Now combining (4.4) and (4.5) we conclude (4.3). ▯

Note that, as was seen in the proof given above, the combinatorial version of Buser inequality, (4.2), is almost trivial, while Buser's inequality (1.5) needs more works, and using the combinatorial inequalities (4.1) and (4.2) proved just now, we can obtain the following assertion which is just (1.6) under a superfluous condition on the injectivity radius.

Corollary 4.3. *Let X be a complete riemannian manifold satisfying the condition $(*)$. Then $I_\infty(X) > 0$ if and only if $S_{2,2}(X) > 0$.*

This follows immediately from our discrete approximation theorems and Kesten's inequalities. In fact, for X in the corollary, take an ϵ-net P in it: Then $I_\infty(X) > 0$ iff $I_\infty(P) > 0$ by Theorem 4.1, $I_\infty(P) > 0$ iff $S_{2,2}(P) > 0$ by Proposition 4.2, and $S_{2,2}(P) > 0$ iff $S_{2,2}(X) > 0$ by Theorem 3.2.

So Corollary 4.3 can be considered as a typical example of applications of combinatorial geometry of nets to riemannian geometry.

REFERENCES

[**B**] P. Buser, *A note on the isoperimetric constant*, Ann. Sci. École Norm. Sup. **15** (1982), 213–230.

[**Ch**] J. Cheeger, *A lower bound for the smallest eigenvalue of the laplacian*, in "Rroblems in Analysis (A symposium in Honor of S. Bochner)", Princeton Univ. Press, Princeton, 1970, pp. 195–199.

[**Cr**] C. B. Croke, *Some isoperimetric inequalities and eigenvalue estimates*, Ann. Sci. École Norm. Sup. **13** (1980), 419–435.

[**FF**] H. Federer and W. H. Fleming, *Normal and integral currents*, Ann. of Math. **72** (1960), 458–520.

[**K1**] M. Kanai, *Rough isometries, and combinatorial approximations of geometries of non- compact riemannian manifolds*, J. Math. Soc. Japan **37** (1985), 391–413.

[**K2**] ———, *Rough isometries and the parabolicity of riemannian manifolds.*

[**K3**] ———, *Rough isometries and isoperimetic inequalities for non-compact riemannian manifolds*, to appear in the proceedings of "The 6th Symposium on Differential Equations and Differential Geometry", held at Fudan Univ., Shanghai, 1985.

[**Ks1**] H. Kesten, *Symmetric random walks on groups*, Trans. Amer. Math. Soc. **92** (1959), 336–354.

[**Ks2**] ———, *Full Banach mean values on countable groups*, Math. Scand. **7** (1959), 146–156.

[**M**] V. G. Maz'ya, *Classes of domains and imbedding theorems for function spaces*, Dokl. Akad. Nauk SSSR **133** (1960), 527–530; English transl., Soviet Math. Dokl. **1** (1960), 882–885.

[**O**] R. Osserman, *The isoperimetric inequality*, Bull. Amer. Math. Soc. **84** (1978), 1182–1238.

[**Y**] S.-T. Yau, *Isoperimetric constants and the first eigenvalue of a compact riemannian manifold*, Ann. Sci. École Norm. Sup. **8** (1975), 487–507.

GAP THEOREMS FOR CERTAIN SUBMANIFOLDS OF EUCLIDEAN SPACE AND HYPERBOLIC SPACE FORM II

Atsushi Kasue and Kunio Sugahara*

Osaka University
Toyonaka, Osaka 560
Japan

Osaka Kyoiku University
Tennoji, Osaka 543
Japan

Simons [17] studied minimal submanifolds of spheres and showed, among other things, that a compact minimal submanifold M of the unit n-sphere must be totally geodesic if the square length of the second fundamental form is less than $n/(2-p^{-1})$ (p = codim M). Later, Ogiue [15], [16] and Tanno [19] considered complex submanifolds in the complex projective space and obtained similar results to the Simons' theorem. On the other hand, Greene and Wu [8] have proved a gap theorem for (noncompact) Riemannian manifolds with a pole (cf. [6] [10] [13]). Roughly speaking, their theorem says that a Riemannian manifold with a pole whose sectional curvature goes to zero in faster than quadratic decay is isometric to Euclidean space if its dimension is greater than two and the curvature does not change its sign. These gap theorems suggest that one could expect similar results for certain open submanifolds of Euclidean space, the hyperbolic space form, the complex hyperbolic space form, etc.. Actually, in [12], we have proved the following results :

Theorem A.

(I) Let M be a connected, minimal submanifold of dimension m properly immersed into Euclidean space $\mathbb{R}^n$. Then M is totally geodesic if one of the following conditions holds :

(A-i) $m \geqq 3$, M has one end and the second fundamental form α_M of the immersion $M \to \mathbb{R}^n$ satisfies

$$\bar{\rho}(x)|\alpha_M|(x) \to 0$$

as $x \in M$ goes to infinity.

(A-ii) $m = 2$, M has one end and

*The second named author was partly supported by Grant-in-Aid for Scientific Research (No. 60740039), Ministry of Education.

$$\sup \bar{\rho}^2(x)|\alpha_M|(x) < +\infty .$$

(A-iii) $2m > n$, M is imbedded and

$$\sup \bar{\rho}^\varepsilon(x)|\alpha_M|(x) < +\infty$$

for some constant $\varepsilon > m$. Here $\bar{\rho}(x)$ denotes the distance in $\mathbb{R}^n$ to a fixed point of $\mathbb{R}^n$.

(II) Let M be a connected, minimal submanifold of dimension m properly immersed into the hyperbolic space form $\mathbb{H}^n(-1)$ of constant curvature -1 . Then M is totally geodesic if one of the following conditions holds :

(A-iv) $m \geqq 3$, M has one end and the second fundamental form α_M of the immersion $M \to \mathbb{H}^n(-1)$ satisfies

$$\sup \bar{\rho}^\varepsilon(x)e^{\bar{\rho}(x)}|\alpha_M|(x) < +\infty$$

for some constant $\varepsilon > 1$.

(A-v) $m = 2$, M has one end and

$$e^{2\bar{\rho}(x)}|\alpha_M|(x) \longrightarrow 0$$

as $x \in M$ goes to infinity.

(A-vi) $m = n - 1$, M is imbedded and

$$e^{m\bar{\rho}(x)}|\alpha_M|(x) \longrightarrow 0$$

as $x \in M$ goes to infinity. Here $\bar{\rho}(x)$ stands for the distance in $\mathbb{H}^n(-1)$ between x and a fixed point of $\mathbb{H}^n(-1)$.

Theorem B.

(I) Let M be a connected, noncompact Riemannian submanifold of dimension m properly immersed into $\mathbb{R}^n$. Suppose that M has one end and the second fundamental form α_M of the immersion $M \to \mathbb{R}^n$ satisfies

$$\sup \bar{\rho}^\varepsilon(x)|\alpha_M|(x) < +\infty$$

for a constant $\varepsilon > 2$. Then M is totally geodesic if $2m > n$ and the sectional curvature is nonpositive everywhere on M , or if $m = n - 1$ and the scalar curvature is nonpositive everywhere on M .

(II) Let M be a connected, noncompact Riemannian submanifold of dimension m properly immersed into $\mathbb{H}^n(-1)$. Suppose that M has one end and

$$e^{2\bar{\rho}(x)}|\alpha_M|(x) \longrightarrow 0$$

as $x \in M$ goes to infinity. Then M is totally geodesic if $2m > n$ and the sectional curvature is everywhere less than or equal to -1 or if $m = n - 1$ and the scalar curvature is everywhere less than or equal to $-m(m - 1)$.

(III) Let M be a connected hypersurface of $\mathbb{H}^n(-1)$ which bounds a totally convex domain of $\mathbb{H}^n(-1)$. Then M is totally geodesic if

$$e^{\bar{\rho}(x)}|\alpha_M|(x) \longrightarrow 0$$

as $x \in M$ goes to infinity.

Theorem C. Let M be a connected, complex submanifold properly immersed into the complex hyperbolic space form $\mathbb{CH}(-1)$ of constant holomorphic sectional curvature -1. Then M is totally geodesic if the second fundamental form α_M of M satisfies

$$e^{3\bar{\rho}(x)/2}|\alpha_M|(x) \longrightarrow 0$$

as $x \in M$ goes to infinity.

In this note, a few supplementary results to Theorem B and the gap theorem due to Greene and Wu [8] will be given. To begin with, we shall prove the following

Theorem 1. Let M be a complete, connected, noncompact Riemannian manifold of nonnegative Ricci curvature. Suppose there is an isometric immersion from M into Euclidean space such that the second fundamental form α_M of the immersion satisfies

$$|\alpha_M|(x) \operatorname{dis}_M(o,x) \longrightarrow 0 \tag{1}$$

as $x \in M$ goes to infinity, where o is a fixed point of M and $\operatorname{dis}_M(o,x)$ denotes the distance in M between o and x. Then M is isometric to Euclidean space $\mathbb{R}^m$ ($m = \dim M$).

We shall prove the above theorem after recalling the following two facts :

Lemma 1. Let $\iota : N \to \mathbb{R}^n$ be an isometric immersion from a complete Riemannian manifold N into $\mathbb{R}^n$. Given a point x of N, denote by $B_N(x;r)$ the metric ball in N around x with radius r and by $B_T(\iota(x);r)$ the metric ball in $\iota_*(T_xN)$ around $\iota(x)$ with radius r. Suppose that the second fundamental form α_N of the immersion satisfies

$$|\alpha_N| < \eta$$

on $B_N(x;r)$, for a positive constant η. Then there are positive constants $C_1(\eta)$, $C_2(\eta)$, $\delta(\eta)$ depending on η, and a smooth map F from $B_T(\iota(x);\ C_1(\eta))$ into $\iota_*(T_xN)^\perp$ such that

(i) $C_2(\eta) \leqq C_1(\eta) \leqq r$ and $C_2(\eta) \to r$ as η goes to 0,

(ii) $\delta(\eta) \to 0$ as η goes to 0,

(iii) $B_n(x;C_1(\eta))$ is imbedded into the graph of F over $B_T(\iota(x);C_1(\eta))$ and contains the graph of F over $B_T(\iota(x);C_2(\eta))$,

(iv) the C^2-norm of F is bounded by $\delta(\eta)$.

Proof. For a point $y \in B_N(x;r)$, let γ be a minimizing geodesic from x to y. If η is small, the rotation of $\iota_*(T_{\gamma(t)}N)$ along γ is also small. Then $\iota_*(T_xN)$ and $\iota_*(T_yN)$ are not orthogonal, which implies that the orthogonal projection $\pi : \iota(B_N(x;r)) \to \iota_*(T_xN)$ is of maximal rank. For two vectors $v,w \in T_xN$ $(|v|,|w| < r)$, put $y = \exp_x v$ and $z = \exp_x w$. Two geodesics $\iota(\exp_x tv)$ and $\iota(\exp_x tw)$ $(0 \leqq t \leqq 1)$ in $\iota(B_N(x;r))$ converge to straight lines of dirctions $\iota_*(v)$ and $\iota_*(w)$ in $\mathbb{R}^n$ as η tends to zero. Since π is locally diffeomorphic, $\pi(\iota(y)) = \pi(\iota(z))$ if and only if $v = w$ for small η. Hence $\pi : \iota(B_N(x;r)) \to \iota_*(T_xN)$ is an imbedding. Moreover $\pi(\iota(B_N(x;r)))$ converges to $B_T(\iota(x);r)$ as η tends to zero. So F can be defined to be the map π^{-1} composed with the orthogonal projection $\mathbb{R}^n \to \iota_*(T_xN)^\perp$.

Lemma 2 (cf. [3: p.253] and [9: 5.3. bis Lemma]). Let N be a complete Riemannian manifold of nonnegative Ricci curvature.

(I) Given a point $x \in N$, $\mathrm{Vol}(B_N(x;r))/\omega_m r^m$ ($m = \dim N$, ω_m = the volume of the unit ball in $\mathbb{R}^m$) is monotone non-increasing in r. Moreover N is isometric to $\mathbb{R}^m$ if and only if $\lim_{r\to\infty} \mathrm{Vol}(B_N(x;r))/\omega_m r^m = 1$.

(II) Given a point $x \in N$ and two positive constants $\delta \in (0.1)$,

$\varepsilon \in (0,\delta/8)$, there are finite points $x_1,\dots,x_p$ of $B_N(x;r) - B_N(x;\delta r)$ such that

(i) $B_N(x;r) - B_N(x;2\delta r) \subset \bigcup_{k=1}^{p} B_N(x_k;2\varepsilon r)$,

(ii) $\bigcup_{k=1}^{p} B_N(x_k;2\varepsilon r) \cap B_N(x;\delta r) = \phi$,

(iii) $p \leqq 2^m/\varepsilon^m$.

Proof of Theorem 1. For the sake of simplicity, let us identify a point of M with its image in $\mathbb{R}^n$ and M contain the origin o . Given a positive number r , set $M_r = \frac{1}{r} M$ and $B^r(t) = \{x \in M_r : \mathrm{dis}_{M_r}(o,x) < t\} (\subset B_{\mathbb{R}^n}(t) = \{v \in \mathbb{R}^n : |v| < t\})$. Let $\{r_i\}$ be a divergent sequence and δ be any positive number less than 1 . Then by Lemma 1 and Lemma 2 (II), we can find a subsequence of $\{r_i\}$, denoted again by $\{r_i\}$, and an m-dimensional linear subspace T of $\mathbb{R}^n$ such that $B^{r_i}(1) - B^{r_i}(\delta)$ converges (in C^2-topology) to $\{v \in T ; \delta \leqq |v| < 1\}$ as i tends to infinity, and hence $\lim \mathrm{Vol}(B^{r_i}(1) - B^{r_i}(\delta)) = \omega_m(1-\delta^m)$. This observation shows that $\liminf \mathrm{Vol}(B^r(1)) \geqq \omega_m$, so that by Lemma 2 (I), we see that M is isometric to $\mathbb{R}^m$.

In what follows, we consider a Riemannian manifold H with a pole $o \in H$ (i.e., $\exp_o : T_oH \to H$ induces a diffeomorphism between T_oH and H) whose dimension n is greater than or equal to 3 . Let us define four functions $\bar{k}_r$, $\underline{k}_r$, $\bar{k}^\perp$, $\underline{k}^\perp$ on $[0,\infty)$ as follows

$\bar{k}_r(t)$ = the maximum of the radial curvature on $S(t)$,

$\underline{k}_r(t)$ = the minimum of the radial curvature on $S(t)$,

$\bar{k}^\perp(t)$ = the maximum of the sectional curvature for planes tangent to $S(t)$,

$\underline{k}^\perp(t)$ = the minimum of the sectional curvature for planes tangent to $S(t)$,

where $S(t)$ denotes the metric sphere around o of radius t .

Theorem 2 ([8], [12]). Let H , $\bar{k}_r$, $\underline{k}_r$, $\bar{k}^\perp$ and $\underline{k}^\perp$ be as above. Then H is isometric to the n-dimensional simply connected space form $\mathbb{H}^n(-a^2)$ of constant curvature $-a^2$ $(a \geqq 0)$ if (and only if) either of the following two conditions holds:

(i) $\bar{k}_r(t) \leqq -a^2$, $\limsup\,(a^{-1}\sinh at)^2(\underline{k}^{\perp}(t) + a^2) = 0$,

(ii) $\underline{k}_r(t) \geqq -a^2$, $\liminf\,(a^{-1}\sinh at)^2(\bar{k}^{\perp}(t) + a^2) = 0$.

Here we understand $\mathbb{H}^n(-a^2) = \mathbb{R}^n$ and $a^{-1}\sinh at = t$ when $a = 0$.

Proof. For the sake of simplicity, we shall prove the theorem in case of $a = 0$. Define a metric g_t on $S(t)$ by $g_t(X,Y) = t^{-2}\langle X,Y\rangle$. Then the sectional curvature $K_t(\Pi)$ of g_t for a plane Π in $TS(t)$ is given by

$$K_t(\Pi) = t^2K_H(\Pi) + 1 + t^2(E_t(X,X)E_t(Y,Y) - E_t(X,Y)^2)$$
$$+ t(E_t(X,X) + E_t(Y,Y)) ,$$

where $\{X,Y\}$ is an orthonormal basis of Π with respect to the induced metric $\langle\ ,\ \rangle$ on $S(t)$ and we have set $E_t(X,Y) = \nabla^2\rho(X,Y) - t^{-1}\langle X,Y\rangle$ ($\rho = \mathrm{dis}_H(o,*)$) . Suppose first that the condition (i) holds. Then E_t is positive semi-definite (cf. [7]) and hence we have

$$K_t(\Pi) \geqq t^2\underline{k}^{\perp}(t) + 1 .$$

Moreover we can take a sequence $\{t_i\}$ such that $t_i^2\underline{k}^{\perp}(t_i)$ goes to 0 as $t_i \to +\infty$. This implies that

$$\limsup \mathrm{Vol}((S(t_i),g_{t_i})) \leqq \omega_{n-1} .$$

On the other hand, we know that for any $t > 0$, $\mathrm{Vol}(S(t),g_t)) \geqq \omega_{n-1}$ and further if $\limsup \mathrm{Vol}(S(t),g_t)) = \omega_{n-1}$, then H is isometric to $\mathbb{R}^n$ (cf. [8: Lemma 2]). Thus the theorem has been proved under the condition (i). Suppose next the second condition (ii) holds. Since $\nabla^2\rho$ is positive semi-definite (cf. [8:Lemma 5]), we see that

$$K_t(\Pi) \leqq t^2K_H(\Pi) + 1$$
$$\leqq t^2\bar{k}^{\perp}(t) + 1 .$$

By the assumption, we can take a sequence $\{t_i\}$ such that $t_i^2\bar{k}^{\perp}(t_i)$ goes to 0 as $t_i \to \infty$. Moreover it follows from Rauch comparison theorem that a diffeomorphism ϕ_t from the unit sphere $S_o(1)$ in T_oH onto $S(t)$ defined by $\phi_t(v) = \exp_o tv$ satisfies : $\phi_t{}^*g_t \leqq g_o$,where

g_o is the metric on $S_o(1)$ of constant curvature 1 . Then the next lemma and the above argument show that H is isometric to $\mathbb{R}^n$.

Lemma 3 ([12: Lemma 12]). Let (S^n, g_o) be the standard sphere of constant curvature 1 and g a Riemannian metric on S^n . Suppose the curvature of (S^n, g) is bounded above by $1 + \varepsilon$ for a constant ε : $0 \leqq \varepsilon < 3$, and suppose there is a diffeomorphism $\phi : S^n \to S^n$ such that $\phi^* g \leqq g_o$. Then the injectivity radius of (S^n, g) is greater than or equal to $2\pi/\sqrt{1+\varepsilon} - \pi$.

Proposition 1. Let H , $\bar{k}_r$, $\underline{k}_r$, $\bar{k}^{\perp}$ and $\underline{k}^{\perp}$ be as in Theorem 2 and denote by B(t) the metric ball around o of radius t . Then :

(i) $\sup \frac{\mathrm{Vol}(B(t))}{t^n} < +\infty$ if $\bar{k}_r(t) \leqq 0$ and $\liminf t^2 \underline{k}^{\perp}(t) > -1$.

(ii) $\inf \frac{\mathrm{Vol}(B(t))}{t^n} > 0$ if n is odd, $\underline{k}_r(t) \geqq 0$ for all t and $\limsup t^2 \bar{k}^{\perp}(t) < +\infty$.

(iii) $\inf \frac{\mathrm{Vol}(B(t))}{t^n} > 0$ if $\underline{k}_r(t) \geqq 0$ and $\limsup t^2 \bar{k}^{\perp}(t) < 3$.

Proof. The assertions (i) and (iii) are immediate consequences of the above proof for Theorem 2. In order to prove the second assertion we need the well known Klingenberg's theorem (cf. e.g., [4:Theorem 5.9]) in place of Lemma 3.

Corollary. Let H , $\bar{k}_r$, $\underline{k}_r$, $\bar{k}^{\perp}$ and $\underline{k}^{\perp}$ be as in Theorem 2. Then :

(i) $\limsup t^2 \bar{k}_r(t) = 0$ if $\bar{k}_r(t) \leqq 0$ and $\liminf t^2 \underline{k}^{\perp}(t) > -1$.

(ii) $\liminf t^2 \underline{k}_r(t) = 0$ if n is odd, $\underline{k}_r(t) \geqq 0$ for all t and $\limsup t^2 \bar{k}^{\perp}(t) < +\infty$.

(iii) $\liminf t^2 \underline{k}_r(t) = 0$ if $\underline{k}_r(t) \geqq 0$ and $\limsup t^2 \bar{k}^{\perp}(t) < 3$.

Let us now give a few examples of Riemannian manifolds with a pole, in relation with the above results.

Example 1. Let k_1 be a nonpositive smooth function on $[0,\infty)$ such that $k_1(t) = (\frac{1}{4} - c^2)t^{-2}$ for large t , where $c > \frac{1}{2}$. Let f_1 be the solution of equation : $f_1'' + k_1 f_1 = 0$, subject to the initial conditions : $f_1(0) = 0$ and $f_1'(0) = 1$. Note that $f_1(t) = at^{\frac{1}{2}+c} + bt^{\frac{1}{2}-c}$ for large t , where a and b are some positive constants. Then a Riemannian manifold $H_1 = (\mathbb{R}^n, dr^2 + f_1{}^2(r) g_0)$ $(n \geqq 3)$ with a pole $o \in \mathbb{R}^n$ has the properties : the radial curvature $k_1(r) \leqq 0$; lim inf $t^2\underline{k}^{\perp}(t) = \lim t^2(1 - (f_1')^2) f_1{}^{-2} = -(\frac{1}{2} + c) < -1$; and $\sup t^{-n}\mathrm{Vol}(B(t)) = \sup \omega_n t^{-n} f_1{}^n = +\infty$.

Example 2. Let k_2 be a nonnegative smooth function on $[0,\infty)$ such that $k_2(t) = (\frac{1}{4} - c^2)t^{-2}$ for large t , where $0 \leqq c < \frac{1}{2}$. Let f_2 be the solution of equation : $f_2'' + k_2 f_2 = 0$, with $f_2(0) = 0$ and $f_2'(0) = 1$. Note that $f_2(t) = at^{\frac{1}{2}+c} + bt^{\frac{1}{2}-c}$ for large t , where a and b are some positive constants. Then a Riemannian manifold $H_2 = (\mathbb{R}^n, dr^2 + f_2{}^2(r) g_0)$ $(n \geqq 3)$ with a pole $o \in \mathbb{R}^n$ has the properties: the radial curvature $k_2(r) \geqq 0$ everywhere and $k_2(r) > 0$ outside a compact set ; and $\inf t^{-n}\mathrm{Vol}(B(t)) = \inf \omega_n t^{-n} f_2{}^n = 0$.

Example 3. Let us consider the unit sphere $S^3(1)$ of dimension 3 in $\mathbb{H}$ (Quaternion field) as a Lie group with the multiplication in $\mathbb{H}$. Let $\{z_1, z_2, z_3\}$ be a left invariant, orthogonal frame field on $S^3(1)$ such that $[z_1,z_2] = 2z_3$, $[z_2,z_3] = 2z_1$, $[z_3,z_1] = 2z_2$ (cf. e.g., [4: 3.35]). We denote by θ_i $(i = 1,2,3)$ the dual forms of z_i and consider a Riemannian metric g_f on $\mathbb{R}^4$ of the form

$$g_f = dr^2 + r^2\theta_1{}^2 + r^2\theta_2{}^2 + f^2(r)\theta_3{}^2 ,$$

where f is a smooth function on $[0,\infty)$. Then it follows from direct computation that for any plane Π tangent to $S^3(r)$, the sectional curvature $K(\Pi)$ satisfies

$$(2) \quad \min\left\{ \frac{3r^2-3f^2}{r^4} , \frac{f^3-r^3f'}{r^4 f} \right\} \leqq K(\Pi) \leqq \max\left\{ \frac{3r^2-3f^2}{r^4} , \frac{f^3-r^3f'}{r^4 f} \right\}$$

and for a plane Π_r spanned by ∇r and a unit vector $a_1z_1+a_2z_2+a_3z_3$, the sectional curvature $k_r(\Pi_r)$ is given by

$$(3) \qquad k_r(\Pi_r) = -|a_3|^2 ff'' .$$

Let us now choose a suitable f . We first take a smooth function h on $[0,\infty)$ such that $h(r) = 1$ on $[0, a]$ $(a > 0)$, $0 < h(r) \leqq 1$ on $[a,\infty)$, $h'(r) \leqq 0$ on $[0,\infty)$ and $h(r)$ goes to zero as $r \to +\infty$. Let f be the solution of equation : $f'(r) = (f(r)/r)^3 h(r)$ with $f(a) = a$.

Then it is easy to check that f satisfies : $f(r) \leqq r$, $f' \geqq 0$, $f'' \leqq 0$, $\lim f(r)/r = 0$ and $r^{-4}f^{-1}(f^3 - r^3 f') \geqq 0$. Thus by (2), (3) and these properties of f , we have obtained a Riemannian manifold $H_3 = (\mathbb{R}^4, g_f)$ with a pole such that

(4) the radial curvature $\geqq 0$, $\limsup t^2 \bar{k}^{\perp}(t) = 3$, $\inf \frac{\mathrm{Vol}(B(t))}{r^4} = 0$.

Moreover, making use of Hopf fibration $S^{2m-1}(1)$ ($\subset \mathbb{C}^m$) $\to \mathbb{C}P^{m-1}$ ($m \geqq 3$) and the above example ($m = 2$), we can construct a Riemannian manifold with a pole whose dimension is $2m$ and which satisfies (4) with $2m$ in place of 4 .

Before concluding this note, let us give several remarks and propose an open question.

(a) The first part (I) of Theorem A with a stronger condition instead of (A-i) has been proved in [11] and a few examples are given there to illustrate the roles of several hypothesis in Theorem A. We should mention here the recent paper of Anderson [2] in which he has investigated complete minimal submanifolds in $\mathbb{R}^n$ of finite total scalar curvature. Especially as a consequence derived from his main theorem, which is a generalization of the well known Chern-Osserman theorem on minimal surfaces in $\mathbb{R}^n$ of finite total curvature, he showed that a complete minimal submanifold M immersed into $\mathbb{R}^n$ is an affine m-space if $m = \dim M \geqq 3$, M has one end and the total scalar curvature $\int_M |\alpha_M|^m$ is finite, where α_M denotes as before the second fundamental form of M . Moreover the proof of his main theorem suggests that for a complete minimal submanifold M of dimension $m \geqq 3$ immersed into $\mathbb{R}^n$, the immersion is proper and $|\alpha_M| \leqq c/|x|^m$ for some positive constant c if the total scalar curvature is finite. It is easy to see that the total scalar curvature is finite if the immersion is proper and $|\alpha_M| \leqq c/|x|^{\varepsilon}$ for some positive constants c and $\varepsilon > 1$ (cf. [11]).

(b) Recall that $\mathbb{H}^n(-1)$ has a natural smooth compactification $\overline{\mathbb{H}}^n(-1) = \mathbb{H}^n(-1) \cup S(\infty)$, where $S(\infty)$ can be identified with asymptotic classes of geodesic rays in $\mathbb{H}^n(-1)$. In [1], Anderson has proved that any closed (m-1)-dimensional submanifold $M(\infty)$ of $S(\infty)$ is the asymptotic boundary of a complete, absolutely area-minimizing locally integral m-current M in $\mathbb{H}^n(-1)$. As is noted in [1], M is smooth in case of $m = n-1 \leqq 6$. It would be interesting to investigate the curvature behavior of his solution M in relation with the 'regularity' of $\bar{M} =$

$M \cup M(\infty)$. One should also consult the recent paper of do Calmo and Lawson [5] for a related resúlt to Theorem A (II).

(c) In [14], Mori constructed a family of complete minimal surfaces $I_\lambda : S^1 \times \mathbb{R} \to \mathbb{H}^3(-1)$. The second fundamental forms α_λ of these imbedded surfaces $M_\lambda = I_\lambda(S^1 \times \mathbb{R})$ have the property that $|\alpha_\lambda|(x) \sim \exp -2\bar{\rho}(x)$, where $\bar{\rho}$ is the distance in $\mathbb{H}^3(-1)$ to a fixed point.

(d) The last part (III) of Theorem B is concerning the boundary of a totally convex domain in $\mathbb{H}^n(-1)$. When we replace $\mathbb{R}^n$ for $\mathbb{H}^n(-1)$, we have a similar result (cf. [18]).

(e) Let $\bar{\Sigma}$ be a compact smooth surface with genus > 1 and set $\Sigma = \bar{\Sigma}\setminus\{p\}$ $(p \in \bar{\Sigma})$. In Section 4 of [12], we have constructed a proper imbedding of Σ into $\mathbb{R}^3$ such that the second fundamental form α_Σ satisfies : $|\alpha_\Sigma|(x) \leqq c/|x|^2$ and further the Gaussian curvature with respect to the induced metric is everywhere nonpositive (cf. the first part of Theorem B).

(f) In the first part (I) (resp. the second part (II)) of Theorem B, the conditions on the dimension and the curvature of M can be replaced with the following weaker condition : for every $x \in M$, there is a subspace T of T_xM such that $\dim T > n - m$ and the sectional curvature for any plane in T is nonpositive (resp. less than or equal to -1) (cf. [12]).

(g) Let M be a complete, connected, noncompact Riemannian manifold of nonnegative Ricci curvature. Suppose there is an isometric and proper immersion ι from M into $\mathbb{R}^n$ such that for some positive constant ε , the second fundamental form α_M satisfies :

(5) $$|\alpha_M|(x)|\iota(x)|^{1+\varepsilon} \longrightarrow 0$$

as $x \in M$ goes to infinity. Then M is isometric to $\mathbb{R}^m$, since (5) implies (1) (cf. [11: Section 1]).

(h) Let M be a complete, connected, noncompact Riemannian manifold of nonnegative Ricci curvature. Suppose there is an isometric immersion ι from M into $\mathbb{R}^n$ whose second fundamental form α_M satisfies :

(6) $$\sup_{x \in M} |\alpha_M|(x)\ \mathrm{dis}_M(x,o) < +\infty$$

Then $\lim \mathrm{Vol}(B_M(o,r))/\omega_m r^m > 0$ (cf. the proof of Theorem 1). Hence the Riemannian manifolds as in Examples 2 and 3 have no isometric immersions into $\mathbb{R}^n$ satisfying the above condition (6).

(i)(cf. Theorem C) Let $H = (H,g,J)$ be a complete, connected, simply connected Kaehler manifold of nonpositive sectional curvature. Suppose that the sectional curvature $K_H(\Pi)$ for a tangent plane Π at a point $x \in H$ satisfies:

$$|K_H(\Pi) + \frac{1}{4}(1 + 3g(X,JY)^2)| \leqq c \exp -\varepsilon\rho(x) \ ,$$

where c , ε are positive constants, $\{X,Y\}$ is an orthonormal basis of Π , and $\rho(x)$ denotes the distance to a fixed point of H . Then (after choosing an appropriate number ε), can we assert that H must be $\mathbb{C}H^n(-1)$, if the holomorphic sectional curvature of $H \geqq -1$ or $\leqq -1$?

References

1. M.T. Anderson, Complete minimal varieties in Hyperbolic space, Inventiones math. 69 (1982), 477-494.
2. M.T. Anderson, The compactification of a minimal submanifold in Euclidean space by the Gauss map, to appear.
3. R. Bishop and R. Crittenden, Geometry of Manifolds, Academic Press, New York, 1964.
4. J. Cheeger and D.G. Ebin, Comparison Theorems in Riemannian Geometry, North-Holland Math. Library 9, North-Holland Publ. Amsterdam-Oxford-New York, 1975.
5. M. do Calmo and H.B. Lawson, Jr., On Alexander-Bernstein theorems in Hyperbolic space, Duke Math. J. 50 (1983), 995-1003.
6. D. Elerath, Open nonnegatively curved 3-manifold with a point of positive curvature, Proc. Amer. Math. Soc. 75 (1979), 92-94.
7. R.E. Greene and H. Wu, Function Theory on Manifolds Which Possess a Pole, Lecture Notes in Math. 699, Springer-Verlag, Berlin-Heidelberg-New York, 1979.
8. R.E. Greene and H. Wu, Gap theorems for noncompact Riemannian manifolds, Duke Math. J. 49 (1982), 731-756.
9. M. Gromov, Structures métriques pour les variétés riemanniennes, Cedic-Fernand Nathan, 1981.
10. Th. Hasanis and D. Koutroufiotis, Flatness of Riemannian metrics outside compact sets, Indiana Univ. Math. J. 32 (1983), 119-128.
11. A. Kasue, Gap theorems for minimal submanifolds of Euclidean space, to appear in J. Math. Soc. of Japan 38 (1986).
12. A. Kasue and K. Sugahara, Gap theorems for certain submanifolds of Euclidean space and Hyperbolic space form, to appear.

13. N. Mok, Y.-T. Siu and S.T. Yau, The Poincaré-Lelong equation on complete Kähler manifolds, Compositio Math. 44 (1981), 183-218.

14. H. Mori, Minimal surfaces of revolutions in H^3 and their global stability, Indiana Math. J. 30 (1981), 787-794.

15. K. Ogiue, Complex submanifolds of complex projective space with second fundamental form of constant length, Kodai Math. Sem. Rep., 21 (1969), 252-254.

16. K. Ogiue, Differential geometry of Kaehler subamnifolds, Advances in Math. 13 (1974), 73-114.

17. J. Simons, Minimal varieties in riemannian manifolds, Ann. of Math. 88 (1968), 62-105.

18. K. Sugahara, Gap theorems for hypersurfaces in R^N, Hokkaido Math. J. 14 (1985), 137-142.

19. S. Tanno, Compact complex submanifolds immersed in complex projective spaces, J. Differential Geometry 8 (1973), 629-641.

After the symposium, M. Gromov informed us that he had proved the following

Theorem(M. Gromov). Let M be a connected, complete, noncompact Riemannian manifold of nonnegative sectional curvature K_M. Suppose that M is simply connected at infinity and K_M satisfies:

$$K_M(x)\ \mathrm{dis}_M(x,o)^2 \longrightarrow 0$$

as $x \in M$ goes to infinity, where o is a fixed point of M. Then M is isometric to Euclidean space.

A PINCHING PROBLEM FOR LOCALLY HOMOGENEOUS SPACES

Atsushi Katsuda
Department of Mathematics
Nagoya University
Nagoya, 464 Japan

1. Introduction and earlier results

In this paper, we concern the following problem.

"If a complete connected Riemannian manifold M^n of dimension n is geometrically similar to some standard Riemannian manifold N - we call it Model space -, then is M diffeomorphic to N ?"

Up to now, there are several answers to this problem.
For a Riemannian manifold M, we denote by $K = K(M)$ the sectional curvature, $Ric(M)$ the Ricci curvature, $V(M)$ the volume, $D(M)$ the diameter of M and $\lambda_1(M)$ the first eigenvalue of Laplacian on M. The constants $\delta = \delta(a,b,c,\cdots)$ depend only on a, b, c, $\cdots$.

1.1 (Gromoll, Calabi, Shikata, Sugimoto, Shiohama, Ruh, Im Hof, $\cdots$, see the reference of [17])

There is a constant $\delta > 0$ such that if M^n satisfies $1 \geqq K(M) \geqq 1 - \delta$, then M^n is diffeomorphic to the space form of positive curvature S^n/Γ.

1.2 (Brittain [3], Croke [4], Kasue [10], Katsuda [11])

There are constants $\delta_1 = \delta_1(n,\Lambda) > 0$ and $\delta_2 = \delta_2(n,\Lambda,v) > 0$ such that if M^n satisfies $|K_M| \leqq \Lambda^2$ and one of the following conditions, then M^n is diffeomorphic to the standard sphere S^n.

(i) $V(M) \geqq V(S^n) - \delta_1$.

(ii) $V(M) \geqq v$ and $D(M) \geqq \pi - \delta_2$

(iii) $V(M) \geqq v$ and $\lambda_1(M) \leqq n + \delta_2$.

1.3 (Berger [2])

There is a constant $\delta = \delta(2n) > 0$ such that a simply connected

Riemannian manifold M^{2n} of even dimension satisfies $1 \geqq K(M) \geqq (1/4) - \delta$, then M^{2n} is homeomorphic to S^{2n} or diffeomorphic to the compact symmetric space of rank one.

1.4 (Gromov [7])

There is a constant $\delta(n,v,D) > 0$ such that if M^n satisfies $|K(M)| \leqq \delta$, $V(M) \geqq v$ and $D(M) \leqq D$, then M^n is diffeomorphic to a flat manifold.

1.5 (Gromov [7], Ruh [16])

There is a constant $\delta(n) > 0$ such that if M^n satisfies $|K(M)|D(M)^2 \leqq \delta$, then M^n is diffeomorphic to a infranilmanifold.

There are examples which have no flat metric and have a metric satisfying the assumption of 1.5 for any $\delta > 0$

1.6 (Gromov [7] c.f. [5])

There is a constant $\delta(n,D) > 0$ such that if M^n satisfies $-1 \leqq K(M) \leqq -1 + \delta$ and $D(M) \leqq D$, then M^n is diffeomorphic to a hyperbolic manifold.

1.7 (Min-Oo and Ruh [13])

Let $N = G/K$ denote an irreducible simply connected Riemannian symmetric space of compact type. There is a constant $\delta > 0$ depending only on N such that if the norm of the curvature of a Cartan connection on M with respect to the Model N is smaller than δ, then M is diffeomorphic to N/Γ, where Γ is a finite subgroup of G.

They also obtained the result for symmetric space of noncompact type [14].

2. Results

We consider the problem for more general Model space. Firstly we give a result for locally symmetric space under the condition different to 1.7. It is well known that M is locally symmetric if and only if the cuvature tensor $R = R(M)$ of M is parallel. So it is natural to relax this condition. Let ∇ denote the Levi-Civita connection on M and $|\cdot|$ denote the norm of the tensor.

<u>Theorem A</u>. There is a constant $\delta = \delta(n,\Lambda,v,D) > 0$ such that if M^n satisfies $|K(M)| \leqq \Lambda^2$, $V(M) \geqq v$, $D(M) \leqq D$ and $|\nabla R| \leqq \delta$, then M^n is diffeomorphic to a locally symmetric space.

Moreover we can say something about locally homogeneous space.

<u>Theorem B</u>. There is a constant $\delta = \delta(n,\Lambda,v,D) > 0$ such that if M^n satisfies $|K(M)| \leqq \Lambda^2$, $V(M) \geqq v$, $D(M) \leqq D$ and has a tensor field T of type (1,2) satisfying the following condition (*), then M^n is diffeomorphic to a locally homogeneous space.

$$(*)\quad \begin{array}{ll} \text{(i)} & g(T_X Y, T_X Y) \leqq \Lambda^2 \\ \text{(ii)} & |g(T_X Y, Z) + g(Y, T_X Z)| \leqq \delta \\ \text{(iii)} & |(\nabla_X R)_{YZ} - [T_X, R_{YZ}] + R_{T_X Y Z} + R_{Y T_X Z}| \leqq \delta \\ \text{(iv)} & |(\nabla_X T)_Y - [T_X, T_Y] + T_{T_X Y}| \leqq \delta \end{array}$$

for unit vectors X, Y, Z.

<u>Remarks</u>.

1. In the case when $\delta = 0$, the condition (*) is a local characterization of locally homogeneous space due to Ambrose and Singer [1]. Then, the tensor field T is written by the difference between the Levi Civita connection and the canonical connection. (cf. [18])
2. Theorem A is the special case $T = 0$ of theorem B.
3. Comparing 1.4 and 1.5, the dependence of the constants in δ is essential. So, investigating this in δ of theorems A and B is an interesting problem. Ruh told the author that the manifold in 1.5 is an example such that the dependence of the volume in δ can not remove. Gromov knows another example of this, which has positive Ricci curvature.

3. Preliminaries.

Let $M(n,\Lambda,v,D)$ (resp. $M(n,\Lambda,i_0)$) be the category of all complete n-dimensional Riemannian manifolds with $|K(M)| \leqq \Lambda^2$, $V(M)$

$\geqq v$, $D(M) \leqq D$ (resp. the injectivity radius $i(M) \geqq i_0$). For $M, M' \in \mathcal{M}(n,\Lambda,i_0)$, the following distances are defined.

Definition. (cf. [9])

1. Lipschiz distance : $d_L(M,M')$.

$$d_L(M,M') = \inf(|\log(\mathrm{dil}\ f)| + |\log(\mathrm{dil}\ f^{-1})|)$$

where f ranges over bi-Lipschiz homeomorphisms of M to M' and

$$\mathrm{dil}\ f = \sup\{d(f(x),f(y))/d(x,y) \mid x,y \in M,\ x \neq y\}.$$

2. Hausdorff distance : $d_H(M,M')$.

$$d_H(M,M') = \inf(d_H^Z(f(M),f'(M')))$$

where Z ranges over metric spaces with a distance d and f and f' range over isometric embeddings from M and M' to Z respectively and

$$d_H^Z(X,X') = \max\{\sup_{x\in X}(\inf_{y\in X'}(d(x,y))),\ \sup_{x\in X'}(\inf_{y\in X}(d(x,y)))\}.$$

Following theorems play essential roles in our proof.

Theorem I (Gromov [9] 5.2). $\mathcal{M}(n,\Lambda,V,D)$ is precompact with respect to the Hausdorff distance.

Theorem II (Gromov [9] 8.28, Greene and Wu [6]). If $\{M_i\}_{i=1}^{\infty} \subset \mathcal{M}(n,\Lambda,V,D)$ is a Cauchy sequence with respect to the Hausdorff distance, there exists C^∞ manifold M with $C^{1,1}$ Riemannian metric such that M_i is diffeomorphic to M for large $i > 0$ and

$$\lim_{i\to\infty} d_L(M_i,M) = \lim_{i\to\infty} d_H(M_i,M) = 0.$$

Theorem III (cf. [12] 3.10) If $\{M_i\}_{i=1}^{\infty} \subset \mathcal{M}(n,\Lambda,V,D)$ is a Cauchy sequence with respect to the Hausdorff distance and there exists a constant $\Lambda_1 > 0$ such that $|\nabla R_{M_i}| \leqq \Lambda_1$, then the limit manifold M satisfies the following property.

" Any continuous isometry on M is automatically C^1 isometry."

Here, we would like to commemt about the assumption of this theorem. To prove "...", it needs that exponential map of M is C^1. From the assumption of theorem II, we only conclude that the exponential map is lipschitz continuous. This is the reason why we need condition $|\nabla R_{M_i}| \leqq \Lambda_1$.

<u>Theorem IV</u> ([15] p. 208, Theorem 2, Theorem 3). Let G be a locally compact effective transformation group of a connected C^1 manifold M and let each transformation of G is C^1. Then G is a Lie group and the map $G \times M \to M$ is C^1.

4. Outline of the proof.

Assume that the conclusion of theorem B does not hold. Then there exists a sequence $\{M_i\} \subset M(n,\Lambda,V,D)$ such that

(i) M_i satisfies the condition (*) for $\delta = 1/i$,

(ii) M_i is not diffeomorphic to a locally homogeneous space.

By theorem I, taking a subsequence if necessary, we may assume that $\{M_i\}$ is a Cauchy sequence with respect to the Hausdorff distance. Then, by theorem II, there exists a limit M. Assume for a moment the following condition (T) is satisfied.

(T) The identity component of the group of isometries G_0 of the universal covering space $\tilde{M}$ of M acts transitively on $\tilde{M}$.

Then, by theorems III and IV, G_0 is a Lie group and $\tilde{M}$ is C^1 diffeomorphic to G_0/K, where K is the isotropy subgroup of G_0. Since $\tilde{M}$ is C^∞ and G_0/K is so, these are C^∞ diffeomorphic. The induced metric g_0 of G_0/K from the C^0 metric tensor g of $\tilde{M}$ is G_0-invariant and so, C^∞. Using theorem II, we see M_i is diffeomorphic to a locally homogeneous space for large i. This is a contradiction. Hence the conclusion holds.

To prove (T), we firstly see that if a Riemannian manifold N satisfies the condition (*) for small δ, then there exists a local quasi isometry Φ between the neighborhoods of any two points. Nextly, extend this map Φ to larger domain. Finally, for each M_i, we define such map Φ_i on $\tilde{M}_i$ and see that Φ_i converges to an isometry of $\tilde{M}$.

Detailed proof will appear elsewhere.

<u>References</u>

[1] W. Ambrose and I. M. Singer, On homogeneous Riemannian manifolds, Duke Math. J. 25 (1958) 647-669.

[2] M. Berger, Sur les variétés riemanniennes pincées juste audes-sous de 1/4, Ann. Inst. Fourier, 33 (1983) 135-150.

[3] D. L. Brittain, A diameter pinching theorem for positive Ricci curvature, preprint.

[4] C. B. Croke, An eigenvalue pinching theorem, Inv. Math. 68 (1982) 253-256.

[5] K. Fukaya, Theory of convergence for Riemannian orbifolds, pre-print (1984).

[6] R. E. Greene and H. Wu, Lipschitz convergence of Riemannian manifolds, preprint.

[7] M. Gromov, Almost flat manifolds, J. Diff. Geom. 13 (1978) 231-241.

[8] ---------, Manifolds of negative curvature, J. Diff. Geom., 13 (1978), 223-230.

[9] ---------, _Structures métriques pour les variétés riemann-iennes_, rédigé par J. Lafontaine et P. Pansu, Cedic/Fernand Nathan 1981.

[10] A. Kasue, Applications of Laplacian and Hessian comparison theorems, Advanced Studies in Pure Math. 3, Geometry of Geodesics, 333- 386.

[11] A. Katsuda, Gromov's convergence theorems and its application, to appear in Nagoya Math. J. 100 (1985).

[12] S. Kobayashi and K. Nomizu, _Foundations of Differential Geome-try_, John Wilery, New York I 1963, II 1969.

[13] Min-Oo and E. Ruh, Comparison theorems for compact symmetric spaces, Ann. Sci. École Norm. Sup., 12 (1979) 335-353.

[14] -----------------, Vanishing theorems and almost symmetric spaces of noncompact type, Math. Ann., 257 (1981), 419-433.

[15] D. Montgomery and L. Zippin, _Topological transformation groups_, Interscience, 1955.

[16] E. Ruh, Almost flat manifolds, J. Diff. Geom., 17 (1982) 1-14.

[17] T. Sakai, Comparison and finiteness theorems in Riemannian geo-metry, Advanced Studies in Pure Math. 3, Geometry of Geodesics, 183-192.

[18] F. Tricerri and L. Vanhecke, _Homogeneous structures on Riemann-ian manifolds_, London Math. Soc., Lect. Note Ser. 83, Cambridge Univ. Press 1983.

REMARKS ON THE INJECTIVITY RADIUS ESTIMATE FOR ALMOST 1/4-PINCHED MANIFOLDS

W.Klingenberg and T.Sakai [1)]

Mathematisches Inst.
Univesitat Bonn
Wegeler strasse 10
5300 Bonn, West Germany

Department of Math.
Okayama University
700 Okayama, Japan

For a smooth Riemannian manifold (M, g) we denote by i_g the injectivity radius of (M, g). Here we discuss about the estimate for i_g when (M, g) is almost 1/4-pinched. Firstly we put

$\mathfrak{M}_\delta^n := \{ (M, g)$; compact simply connected smooth Riemannian manifold of dimension n with $\delta \leqslant K_g \leqslant 1 \}$,

where K_g denotes the sectional curvature of g. Then the following are known:

(1)(n:even) For any $\delta > 0$ and $(M, g) \in \mathfrak{M}_\delta^n$ we have $i_g \geqslant \pi$. (W.Klingenberg (15)).

If $n = 2$, then for any compact surface M we have $i_g \geqslant 2\pi$ - diameter M (C.Bavard(2)) and $i_g^2 + 1/32\,(\text{Area } S^2 - 4\pi)^2 \geqslant \pi^2$ (C.Bavard and P.Pansu(3)).

(2)(n:odd) .

If $\delta \geqslant 1/4$, then we have $i_g \geqslant \pi$ for any $(M, g) \in \mathfrak{M}_\delta^n$ (W.Klingenberg-T.Sakai(17) and J.Cheeger-D.Gromoll(7)).

If $\delta < 1/9$, then there exist $(S^n, g) \in \mathfrak{M}_\delta^n$ with $i_g < \pi$. (Berger spheres; see e.g., J.Cheeger-D.G.Ebin(6)).

There exists an $0 < H < 1$ with the following property: if $\delta < H$, then for any $\varepsilon > 0$ there exists $(M, g) \in \mathfrak{M}_\delta^7$ with $i_g < \varepsilon$. Namely we have $\inf\{ i_g;\ g \in \mathfrak{M}_\delta^7\} = 0$ (H.M.Huang (13) and J.H.Eschenburg (9)).

In case $n = 3$, Bavard has shown for any compact 3-dimensional manifold M that $\inf\{ i_g;(M^3,g) \text{ with } |K_g| \leqslant 1 \} = 0$ ((1)).

Injectivity radius estimate plays important roles in global Riemannian geometry. For instance in sphere theorem for $\delta(> 1/4)$-pinched manifold

, the estimate $i_g \geq \pi$ was crucial. On the other hand very recently J.H.Eschenburg proved the sphere theorem via convexity argument due to Gromov and got the injectivity radius estimate as a consequence ((21)).

Now we may ask the following question for odd $n \geq 3$.

(a) What is $\delta_0 := \inf\{ \delta > 0 ;\ i_g \geq \pi$ for every $(M, g) \in \mathfrak{M}^n_\delta \}$?
(b) What is $\delta_1 := \inf\{ \delta > 0 ;\ \inf\{ i_g ;\ (M, g) \in \mathfrak{M}^n_\delta \} > 0 \}$?

We take interest in (a) because π is the best possible value which we may expect and (b) is related to the finitness of the diffeomorphism types of $\mathfrak{M}^n_\delta$. We suspect that $\delta_0 < 1/4$, namely for almost 1/4-pinched manifolds we have $i_g \geq \pi$.

Similarly we fix a compact simply connected manifold M of odd dimension n and put

$\mathfrak{M}_\delta(M) := \{ g$: smooth Riemannian structure on M with $\delta \leq K_g \leq 1 \}$.

and ask the similar questions:

(a)' What is $\delta_0(M) := \inf\{ \delta > 0 ;\ i_g \geq \pi$ for every $g \in \mathfrak{M}_\delta(M) \}$?
(b)' What is $\delta_1(M) := \inf\{ \delta > 0 ;\ \inf\{ i_g ;\ g \in \mathfrak{M}_\delta(M) \} > 0 \}$?

Hereafter we assume that M is always simply connected compact manifold of odd dimension $n \geq 3$, unless otherwise stated.

Firstly we consider the very restricted case. Namely we fix a manifold M which admits a Riemannian metric g_0 with $1/4 \leq K_{g_0} \leq 1$. Then we have

Proposition. There exists a positive constant $\delta < 1/4$ and $U(g_0)$, which is a neighborhood of g_0 in the space of Riemannian metrics on M with respect to the C^2-topology, such that

$i_g \geq \pi$ for all $g \in \mathfrak{M}_\delta(M) \cap U(g_0)$.

Proof. Suppose the contrary. Then there exists a sequence g_n with $1/4a_n^2 \leq K_{g_n} \leq 1$ ($a_n \uparrow 1$) and $i_{g_n} < \pi$ which converges to g_0 with respect to the C^2-topology. We have closed g_n-geodesics c_n' of length $2\, i_{g_n} < 2\pi$ and we consider homotopies H^n_s $(0 \leq s \leq 1)$ from a fixed point curve in M to c_n'. We fix g_n for a while. We use Morse theory for the energy integral E_{g_n} acting on the space of closed curves on M. Since we may approximate g_n by non-degenerate (with respect to the Morse theory) bumpy metrics, we may assume that g_n are non-degenerate from the first. Now we want to deform H_s^n ($0 \leq s \leq 1$) to a homotopy consisting of shorter curves via the flow generated by -grad E_{g_n} and standard retracting technique for critical points of E_{g_n} of index ≥ 2. Recall that closed g_n-geodesic of length l is nothing but a critical point of E_{g_n} of energy $l^2/2$. Then we assert that we have a closed g_n-geodesic c_n of index ≤ 1 and $2\pi \leq \text{length}_{g_n} c_n \leq 2a_n\pi$. To see this firstly note that any closed g_n-geodesic of length greater than $2a_n\pi$ is of index ≥ 2 via the index comparison theorem. Thus if our assertion does not hold , all closed g_n -geodesics of length $\geq 2\pi$ are of index ≥ 2. Then we may deform H_s^n by the above Morse theory technique to a homotopy $\bar{H}_s^n$ from a point curve to c_n^1 so that $\text{length}_{g_n}\bar{H}_s^n$ is less than 2π for all $0 \leq s \leq 1$. Secondly recall that the map $\Phi: TM \to M \times M$ given by $\Phi(v) := (pv, \exp_{g_n} v)$, where $p: TM \to M$ denotes the tangent bundle, is a local diffeomorphism on $W := \{v \in TM;\ \|v\|_{g_n} < \pi\}$. We lift the curves $\bar{H}_s^n$ of the homotopy to the closed curves $\tilde{H}_s^n$ in the tangent bundle so that $\tilde{H}_s^n \subset T_{H_s^n(0)}M \cap W$ and $\exp_{g_n}\tilde{H}_s^n = \bar{H}_s^n$ by starting from a point curve and standard lifting arguement. Then we may lift the whole curves of the homotopy because $\text{length}_{g_n}\bar{H}_s^n$ is less than 2π. But this contradicts the fact that the final curve $\bar{H}_1^n = H_1^n = c_n^1$ of the homotopy is a closed geodesic. Thus we have for any n a closed g_n-geodesic c_n of index ≤ 1 such that $2\pi \leq \text{length}_{g_n} c_n \leq 2a_n\pi$.

Now taking a subsequence if necessary we may assume that $c_n \to c_0$ with respect to the C^1-topology as is easily seen by considering the geodesic flows of g_n, where c_0 is a closed g_0 -geodesic of length 2π. Then by an elementary continuity arguement for the index form of closed geodesics we have index $c_0 \leq 1$, because $g_n \longrightarrow g_0$ with respect to the C^2-topology. Then our assertion follows from the following lemma.

Lemma. Let (M, g_0) be as above. Then every closed g_0-geodesic c of length 2π is of index ≥ 2.

This is in fact given in (17) or (7). We give a brief outline of proof of this lemma. The difficulty is that in this case we can not assume that g_0 is a non-degenerate metric. Let K' be the set of all closed geodesics of length 2π (i.e., of energy $2\pi^2$) and of index ≤ 1. Then what we should show is that $K' = \emptyset$. Suppose the contrary and take an element $c_0 \in K'$. We consider the family $\mathcal{H}$ of all homotopies from a fixed point curve to c_0. Then $\mathcal{H}$ is a so called ϕ-family (namely it is invariant under the flow generated by -grad E_{g_0}) and we define its critical value κ as $\inf_{H \in \mathcal{H}} \max_{0 \leq s \leq 1} E_{g_0}(H_s)$. Let K'' be the set of closed geodesics of length $(2\kappa^2)^{1/2}$ and of index ≤ 1. Then our main tool is the following modified Lyusternik-Schnirelmann argument:

(*) For any neighborhood $U(K'')$ of K'' in the space $\Lambda(M)$ of closed curves, there exists an $H \in \mathcal{H}$ such that $\{H_s\} \subset \Lambda^{\kappa-}(M) \cup U(K'')$, where $\Lambda^{\kappa-}(M) := \{c \in \Lambda(M) \ ; \ E_{g_0}(c) < \kappa\}$ (see (17)).

From this we see that $\kappa = 2\pi^2$ and consequently $K' = K''$.Then by taking a sequence of neighborhoods $U_k(K') \longrightarrow K'$ $(k = 1,2,\ldots)$ in $\Lambda(M)$, we have a sequence of closed curves c_k $(= H^k_{s_k}$ for some $H^k \in \mathcal{H})$ which converges to a closed geodesic c in K' and may be lifted via the above map Φ to the tangent space to M at $H^k_{s_k}(0)$ using the homotopy $H^k|_{[0,\ s_k]}$ from the given point curve. Then we can show the following:

1°. $C(\pi)$ (resp. $c(2\pi) = c(0)$) is conjugate to $c(0)$ (resp. $c(\pi)$) along $c|_{[0,\ \pi]}$ (resp. $c|_{[\pi,\ 2\pi]}$).

2°. Comparing with the round sphere of radius 1 we have Jacobi field ξ (resp. η) along $c|_{[0,\ \pi]}$ (resp. $c|_{[\pi,\ 2\pi]}$) vanishing at the end points which should take the form $\xi(t) = \sin t\ X_1(t)$ (resp. $\eta(t) = \sin t\ X_2(t)$)) with parallel vector fields X_1, X_2.

3°. We extend ξ (resp. η) to a vector field along c by putting 0 outside $c|_{[0,\pi]}$ (resp. $c|_{[\pi,\ 2\pi]}$) and we get vector fields ξ_1 (resp. η_1) along c.

Then from the assumption that index $c < 1$ we may show that $\dim(\{\xi_1,\ \eta_1\}_{\mathbf{R}} \cap \text{Null space of } D^2E_{g_0}(c)) = 1$ and we get a parallel periodic vector field $X(t)$ along c. Then Synge's trick gives the second parallel periodic vector field $Y(t)$ by dimensionality assumption. Finally for the index form we have $D^2E_{g_0}|\{X,\ Y\}_{\mathbf{R}}$ is negative definite , which means that index c is greater than or equal to 2, a contradiction.

Remark. In the above we only need that g_0 is of class C^2 and the energy integral is of class C^2 on $\Lambda(M)$. Also note that on a fixed manifold M the injectivity radius function $g \longrightarrow i_g$ is continuos with respect to the C^2-topology ((8),(20)).

Now we return to $\mathfrak{M}^n_\delta$ and ask whether there exists $\delta < 1/4$ such that $i_g \geq \pi$ for all $(M, g) \in \mathfrak{M}^n_\delta$. We want to use the Gromov's convergence theorem and the collapsing theorem to this problem. We set

$\mathfrak{M}^n_{d,\varepsilon} := \{ (M, g)$; smooth Riemannian manifold of dimension n with $|K_g| \leq 1$, diameter of $M \leq d$ and $i_g \geq \varepsilon \}$.

Then for $(M_k, g_k) \in \mathfrak{M}^n_{d,\varepsilon}$ there exists riemannian manifold (M_0, g_0) of class $C^{1,a}$ $(0 < a < 1)$ such that (M_k, g_k) converges to (M_0, g_0) with respect to the Lipschitz distance if we take a subsequence (M.Gromov-P.Pansu-J.Lafontaine(12), A.Katsuda(14), R.Greene-H.Wu (11)). Especially M_k is diffeomorphic to M for sufficiently large k. So first assuming that $(M_k, g_k) \in \mathfrak{M}^n_{d,\varepsilon}$ and $1/4a_k^2 \leq K_{g_k} \leq 1$ $(a_k \uparrow 1)$ and if we could show that $i_{g_0} \geq \pi$ then we suspect whether some continuity arguement as above would imply that $i_{g_k} \geq \pi$ for sufficiently large k. But to check the above lemma for $C^{1,a}$ case we have some difficulties because we have no curvature in (M_0, g_0) and index form for the energy integral at the closed geodesics of (M_0, g_0). On the other hand if we assume the uniform upper bound on the norm of derivative of curvature tensors R, then we have more regularity of the limit manifold (M_0, g_0) ((4)). Namely we consider the following class

$$\mathfrak{M}^n_{d,\varepsilon,c} := \{ (M, g) \in \mathfrak{M}^n_{d,\varepsilon};\ \|\nabla R_g\| < c \},$$

where ∇R_g denotes the covariant derivative of the curvature tensor. Then for a sequence $(M_k, g_k) \in \mathfrak{M}^n_{d,\varepsilon,c}$ there exists a Riemannian manifold of class $C^{2,a}$ $(0 < a < 1)$ such that $(M_k, g_k) \longrightarrow (M_0, g_0)$ taking a subsequence if necessary. Since M_k are diffeomorphic to M_0 and g_k may be considered as metrics on M_0 we may assume that $g_k \longrightarrow g_0$ with respect to $C^{2,b}$ $(0 < b < a)$-topology. Then our previous argument implies the following:

Theorem. Let c be any fixed positive number and set

$$\mathfrak{M}^n_{\delta,c} := \{ (M, g) \in \mathfrak{M}^n_\delta ;\ \|\nabla R_g\| < c \} .$$

Then for any $\varepsilon > 0$ there exists $\delta := \delta(\varepsilon,c) < 1/4$ such that we have either $i_g < \varepsilon$ or $i_g \geq \pi$ for any $g \in \mathfrak{M}^n_{\delta,c}$.

In fact otherwise there exist $\varepsilon > 0$ and (M_k,g_k) with $1/4a_k^2 \leq K_{g_k} \leq 1$ ($a_k \uparrow 1$) and $\varepsilon \leq i_{g_k} < \pi$. Because of curvature assumption we see that $(M_k,g_k) \in \mathfrak{M}^n_{\varepsilon,d,c}$ for some d and there exists an (M_0,g_0), which is a Riemannian manifold of class $C^{2,a}$ with $1/4 \leq K_{g_0} \leq 1$ and $g_k \longrightarrow g_0$ (with respect to the C^2-topology). Then the situation is the same as in the Proposition. To be precise we should show that the energy integral is of class $C^{2,a}$ on $\Lambda(M_0)$. To see this take a harmonic coordinate system with respect to which metric tensor $g_{0_{ij}}$ is of class $C^{2,a}$.

Corollary. In case of $n = 3$ we have $i_g \geq \pi$ for any $g \in \mathfrak{M}^3_{\delta,c}$.

Proof. Use a theorem of Burago-Toponogov((5),(19)).

Now how we may exclude the assumption on the uniform boundedness of derivative of curvature tensor and collapsing (i.e.,$i_g < \varepsilon$)? Here we give an observation due to the first named author which seems to be useful for our purpose.

Let (M, g) be an oriented Riemannian manifold of odd dimension $n \geq 3$ with $\delta \leq K_g \leq 1$ (we don't assume that M is simply connected). Let c be a closed geodesic of length 2π and index ≤ 1, and $P: \dot{c}(0)^{\perp} \rightarrow \dot{c}(0)^{\perp}$ be the parallel translation around c. If P has an eigenvalue 1, namely if we have a periodic parallel vector field around c, then as before Synge's trick implies a cotradiction to the fact that index $c \leq 1$. Then we have the following:

Proposition. Let $\alpha \in (0, \pi]$ be one of the rotational angles of P. Then we have $\alpha^2 \geq (\text{length } c)^2\delta \geq 4\pi^2\delta$.

Proof. By definition we have parallel orthonormal vector fields ξ_1, ξ_2 along c such that $\xi_i \perp \dot{c}$ $(i = 1,2)$ and

$$\xi_1(l) = P\xi_1(0) = \cos\alpha\, \xi_1(0) + \sin\alpha\, \xi_2(0)$$
$$\xi_2(l) = P\xi_2(0) = -\sin\alpha\, \xi_1(0) + \cos\alpha\, \xi_2(0),$$

where l denotes the length of c. Next we define the unit periodic vector fields η_1, η_2 as

$$\eta_1(t) := \cos \alpha t/l\ \xi_1(t) - \sin \alpha t/l\ \xi_2(t)$$
$$\eta_2(t) := \sin \alpha t/l\ \xi_1(t) + \cos \alpha t/l\ \xi_2(t).$$

Note that $\eta_1(t)$ and $\eta_2(t)$ are pointwise orthogonal and we get $\nabla\eta_i = (-1)^i \alpha/l\ \eta_{i+1}$ $(i = 1,2 \bmod 2)$.
Now we consider the family of periodic unit vector fields

$$x_\theta := \cos\theta\ \eta_1 + \sin\theta\ \eta_2 \quad (0 \leqslant \theta \leqslant 2\pi).$$

Then we have for the index form

$$I(x_\theta, x_\theta) = \int_0^\ell \{\langle \nabla x_\theta(t), \nabla x_\theta(t)\rangle - K(x_\theta(t), \dot{c}(t))\} dt$$
$$< \int_0^\ell \{(\alpha/l)^2 - \delta\} dt.$$

Since index $c \leqslant 1$ we have for some θ $I(x_\theta, x_\theta) \geqslant 0$, namely we get $(\alpha/l)^2 \geqslant \delta$.

Remark. What happens in the limit case in the above proposition? Namely let M be an orientable Riemannian manifold of odd dimension $n \geqslant 3$ with $1/4 \leqslant K \leqslant 1$ and c a closed geodesic of length 2π and of index $\leqslant 1$. Then the parallel translation P around c is equal to -identity and the space $\mathcal{N}$ of all Jacobi fields perpendicular to c is given by $\{\cos t/2\ \xi(t) + \sin t/2\ \eta(t);\ \xi, \eta$ are parallel around c with $\xi(0), \eta(0) \in \dot{c}(0)^\perp\}$. Thus $\mathcal{N}$ consists of periodic Jacobi fields and we see that index $c = 0$ by the index theorem (see e.g. (17)).This happens in the case of real projective space of constant curvature 1/4. Now if we assume that M is simply connected we see from 1° in the proof of lemma that index $c \geqslant 1$. Thus we may replace 2° and 3° in the proof of lemma with the above argument.

We apply the above proposition to the earlier situation:Suppose that we have a sequence (M_n, g_n) of compact simply connected Riemanniann manifolds of odd dimension $n \geqslant 3$ with $1/4a_n^2 \leqslant K_{g_n} \leqslant 1$ ($a_n \uparrow 1$) and $\pi > i_{g_n} \geqslant \varepsilon (> 0)$. Considering homotopy as before we get closed g_n-geodesics c_n of index $\leqslant 1$ and $2\pi \leqslant$ length $c_n \leqslant 2a_n\pi$. Then the rotational angles α_n of parallel translations P_{g_n} around c_n satisfy $\alpha_n \geqslant \pi/a_n$. Namely eigenvalues of parallel translations go to -1 and curvatures containig $c_n(t)$ go to 1/4 as $n \longrightarrow +\infty$. By Gromov's compactness theorem, taking a subsequence if necessary, we have a limit Riemannian manifold of class $C^{1,a}$ and a limit closed geodesic c.

Thus if we could show that there is the eigenvalue 1 for the parallel translation around c in the limit manifold, we may prove the following conjecture:

(*) For any $\varepsilon > 0$ there exists $0 < \delta := \delta(\varepsilon) < 1/4$ such that $i_g < \varepsilon$ or $i_g \geq \pi$ holds for every $g \in \mathfrak{M}^n_\delta$.

We suspect that the above parallel translation argument is also useful to show that there does not occur collapsing for positively curved Riemannian structure. Namely we fix a compact simply connected manifold M of odd dimension ≥ 3. Then we conjecture

(**) For any $\delta > 0$ there exists $i_0 := i_0(M, \delta) > 0$ such that $i_g \geq i_0$ for any $g \in \mathfrak{M}_\delta(M)$.

Especially this would imply

(***) There exists $\delta := \delta(M) < 1/4$ such that $i_g \geq \pi$ for any $g \in \mathfrak{M}_\delta(M)$.

Here we may only mention some ideas how to see (**). Otherwise we have for some $0 < \delta < 1/4$ a family of Riemannian metrics $g_n \in \mathfrak{M}_\delta(M)$ with $i_{g_n} \to 0$. Assume firstly that (M, g_n) collapses toward a Riemannian manifold B. Then for large n $(M, g_n) \longrightarrow B$ may be viewed as an almost Riemannian submersion with almost flat fibers (K.Fukaya (10)). In our positively curved case such fibers seem to be circles. On the other hand for (M, g_n) we have a short closed geodesic c_n of length $2i_{g_n}$ and considering the homotopy from a point curve we get a closed geodesic c_n of length 2π and of index ≤ 1. Then the above Proposition show that for rotationally angle α_n of the parallel translation around c_n, we have

$$\alpha_n^2 \geq (\text{length}_{g_n} c_n)^2 \delta \geq 4\pi^2\delta,$$

namely rotational angles bounded away from zero. On the other hand c_n may be viewed as an almost horizontal closed geodesic. As $n \to +\infty$ the vertical subspaces along an almost horizontal geodesic are left almost invariant under the parallel translation, which contradicts the above. Even in the case that collapsed manifolds admits singularities, as long as generically the fibers are circles, one may argue that for most part of the geodesic c_n vertical subspaces are almost parallel. Of course we need explicit computation for the turning angle of the vertical subspaces under the parallel translation along almost horizontal geodesic, on which Peng Xian-Wei is now carring.

REFERENCES

(1) Bavard,C., La borne supérieure du rayon d'injectivité en dimension 2 et 3 ,These(Orsay),1984.

(2) ________ , Le rayon d'injectivité des surfaces a courbure majorée,J.Diff.Geo., **20** (1984), 137-142.

(3) ________ et Pansu,P., Sur le volume minimal de R^2,Preprit(1984).

(4) Bemelmans,J.,Min-Oo and Ruh,E.A., Smoothing Riemannian metric,Math Z., **188**(1984),69-74.

(5) Burago,Yu.D. and Toponogov,V.A., On 3-dimensional Riemannian spaces with curvature bounded above, Math Zametic, **13**(1973),881-887.

(6) Cheeger,J. and Ebin,D.G., Comparison Theorems in Riemannian Geometry, American Elsevier, New-York, 1982.

(7) Cheeger,J. and Gromoll,D., On the lower bound for the injectivity radius of 1/4-pinched manifolds, J.Diff.Geo., **15**(1980), 437-442.

(8) Ehrlich,P., Continuity properties of the injectivity radius function, Compositio Math., **29**(1974), 151-178.

(9) Eschenburg,J.H., New examples of manifolds of strictly positive curvature, Invent.Math., **29**(1982), 469-480.

(10) Fukaya,K., Collapsing Riemannian manifolds to lower dimensional one, Preprint(1985).

(11) Greene,R.E. and Wu,H., Lipschitz convergence of Riemannian manifolds, Preprint(1985).

(12) Gromov,M., Structures métriques pour les variét's riemanniennes, rédigé par Lafontaine,J. et Pansu,P.,Textes Math.n **1**, Cedic-Nathan, Paris(1980).

(13) Huang,H.M., Some Remarks on the pinching problems, Bull.Inst.Math.Acad.Sinica, **9**(1981),321-340.

(14) Katsuda,A., Gromov's convergence theorem and its application,to appear in Nagoya Math.J.

(15) Klingenberg,W.,Contribution to Riemannian geometry in the large, Ann. Math., **69**(1959),654-666.

(16) ______________ ,Riemannian Geometry, de Gruyter Studies in Math. **1**, Walter de Gruyter, Berlin-New York,1982.

(17) ______________ and Sakai,T.,Injectivity radius estimate for $\frac{1}{4}$-pinched manifolds, Arch. Math., **34**(1980),371-376.

(18) Pansu,P., Degenerescence des varietes riemanniennes d'apres J.Cheeger et M.Gromov.,Seminare Bourbaki **36** année 1983/84, n 618.

(19) Sakai,T., On a theorem of Burago-Toponogov, Indiana Univ.Math.J., **32**(1983), 165-175.

(20) _________ , On continuity of injectivity radius function, Math J. Okayama Univ., **25**(1983), 91-97.

(21) Eschenburg,J.H., Local convexity and non-negative curvature - Gromov's proof of the sphere theorem, Preprint (1985).

1) The research by this author is partially supported by Grand-in-Aid for Scientific Research (NO.59940042),Ministry of Education.

NON-HOMOGENEOUS KÄHLER-EINSTEIN METRICS ON COMPACT COMPLEX MANIFOLDS

Norihito KOISO and Yusuke SAKANE

College of General Education, Osaka University
Faculty of Science, Osaka University
Toyonaka, Osaka, 560 JAPAN

0. Introduction

On the existence of Kähler-Einstein metric on a compact complex manifold there are a well-known theorem due to T. Aubin [1] for negative first Chern class case and that due to S. T. Yau [11] for vanishing first Chern class case. However, for positive first Chern class case, all known examples were homogeneous ones and some obstruction theorems are known.

On the other hand, in the real category, D. Page and L. Bérard Bergery [4] constructed non-homogeneous Einstein metric. Recently the second author applied their method to $P^1(\mathbf{C})$-bundles over hermitian symmetric spaces and got many examples of non-homogeneous compact Kähler-Einstein manifolds [10].

The first purpose of the paper is to generalize his result and to get more examples of non-homogeneous Kähler-Einstein manifolds.

<u>Theorem</u> (Example 5.8). Let (N, g_N) be a compact Kähler-Einstein manifold with positive first Chern class. Let L be a holomorphic line bundle over N such that $C_1(L) = a \cdot C_1(N)$ for some real number $a \in (0,1)$. Then the canonical compactification of the line bundle $L \otimes L^{-1}$ over $N \times N$, that is the $P^1(\mathbf{C})$-bundle $P(1 \oplus L \otimes L^{-1})$, admits a Kähler-Einstein metric.

The second purpose is to clarify the relation between the above examples and obstruction theorems, precisely Futaki's obstruction theorem [6]. Futaki's theorem states that if an integral related with a $\mathbf{C}^*$-action does not vanish, then there are no Kähler-Einstein metrics. Therefore we may conjecture that there is some relationship between our construction and Futaki's obstruction, which is in fact true.

Theorem (Theorem 5.2). Let X be a compact almost homogeneous space with $C_1(X) > 0$ and with a disconnected exceptional set. Then X admits a (unique) Kähler-Einstein metric if and only if Futaki's obstruction vanishes.

1. Kähler metric and Ricci tensor on a $\mathbf{C}^*$-bundle

Let $\pi : L \to M$ be a hermitian line bundle over a compact Kähler manifold. Denote by $\mathring{L}$ the open set $L \setminus \{0\text{-section}\}$ of L. Then the manifold $\mathring{L}$ is a $\mathbf{C}^*$-bundle over M. Let t be a function on $\mathring{L}$ which depends only on the norm and increses for the norm. Then the horizontal lift $X^\sim$ of a vector field of M to $\mathring{L}$ is characterized by

$$(1.0.1) \qquad \pi_* X^\sim = X, \quad X^\sim[t] = (\tilde{J}X^\sim)[t] = 0,$$

where $\tilde{J}$ is the almost complex structure of L. Remark that the horizontal lift is invariant under the $\mathbf{C}^*$-action. We decompose the group $\mathbf{C}^*$ into $S^1 \times \mathbf{R}^+$ and define holomorphic vector field S (resp. H) on $\mathring{L}$ generated by S^1-action (resp. $\mathbf{R}^+$-action) so that

$$(1.0.2) \qquad \exp 2\pi S = \mathrm{id}, \quad H = -\tilde{J}S, \quad H[t] > 0.$$

If we denote by ρ_L the Ricci form of L, then we have

$$(1.0.3) \qquad [X^\sim, Y^\sim] - [X,Y]^\sim = -\rho_L(X,Y)S.$$

Define a hermitian 2-form B on M by

$$(1.0.4) \qquad \rho_L(X,Y) = B(X,JY).$$

Then we have

$$(1.0.5) \qquad [X^\sim, Y^\sim] - [X,Y]^\sim = -B(X,JY)S.$$

Now we consider a riemannian metric on $\mathring{L}$ of the form

$$(1.0.6) \qquad \tilde{g} = dt^2 + (dt \circ \tilde{J})^2 + \pi^* g_t,$$

where $\{g_t\}$ is a one-parameter family of riemannian metrics on M. Define a function u on $\mathring{L}$ depending only on t by

$$(1.0.7) \qquad u(t)^2 = \tilde{g}(H,H).$$

Lemma 1.1. The metric $\tilde{g}$ on $\mathring{L}$ is a Kähler metric if and only if each g_t is a Kähler metric on M and $\frac{d}{dt} g_t = -u(t)\, B$. In particular, if we assume that the range of t contains 0 and set

$$(1.1.1) \qquad U = \int_0^t u(t)dt,$$

then

$$(1.1.2) \qquad g_t = g_0 - UB.$$

Proof. The metric $\tilde{g}$ is a hermitian metric if and only if each g_t is hermitian. Denote by $\tilde{\omega}$ (resp. ω_t) the Kähler form of $\tilde{g}$ (resp. g_t). Then we easily see, by the invariantness of the horizontal lift under the $\mathbb{C}^*$-action, that

$$(d\tilde{\omega})(H,S,X^{\sim}) = 0,$$

$$(d\tilde{\omega})(S,X^{\sim},Y^{\sim}) = 0,$$

$$(d\tilde{\omega})(X^{\sim},Y^{\sim},Z^{\sim}) = (d\omega_t)(X,Y,Z).$$

Moreover, by (1.0.5), we see that

$$(d\tilde{\omega})(H,X^{\sim},Y^{\sim}) = H[g_t(X,JY)] + B(X,JY)\cdot g_t(S,\tilde{J}H)$$

$$= u\cdot\frac{d}{dt}g_t(X,JY) + u^2B(X,JY). \qquad \text{Q.E.D.}$$

From now on, we assume that 1) $\tilde{g}$ is a Kähler metric, 2) the range of t contains 0 and 3) the eigenvalues of B with respect to g_0 are constant on M. Remark that condition 3) does not depend on the choice of the origin of t (see (1.1.2)).

Let $z^1,\cdots,z^n$ be a local coordinate system of M. Using a local trivialization of $\mathring{L}$ we take a local coordinate system $z^0,\cdots,z^n$ of $\mathring{L}$ so that $\frac{\partial}{\partial z^0} = H - \sqrt{-1}S$. Denote by $\hat{\partial}_\alpha$ $(0 \le \alpha \le n)$ the partial derivation on $\mathring{L}$ and by ∂_α $(1 \le \alpha \le n)$ on M.

<u>Lemma 1.2</u>. $\tilde{g}_{\bar{0}0} = 2u^2,$

$$\tilde{g}_{\bar{0}\beta} = 2u\hat{\partial}_\beta t,$$

$$\tilde{g}_{\bar{\alpha}\beta} = g_{t\bar{\alpha}\beta} + 2\hat{\partial}_{\bar{\alpha}}t\cdot\hat{\partial}_\beta t,$$

where $1 \le \alpha, \beta \le n$. In particular,

$$\det(\tilde{g}_{\bar{\alpha}\beta}) = 2u^2\det(g_{t\bar{\alpha}\beta}).$$

Proof. We see the following equalities, from which we easily derive the above equalities.

$$(dt)(H) = u,$$

$$(dt\circ\tilde{J})(\hat{\partial}_\beta) = \sqrt{-1}(dt)(\hat{\partial}_\beta) = \sqrt{-1}\hat{\partial}_\beta t. \qquad \text{Q.E.D.}$$

We define a new function p on $\mathring{L}$ by

$$p = \det(g_0^{-1}\cdot g_t). \tag{1.2.1}$$

Then, by (1.1.2), p depends only on t, and

$$\det(\tilde{g}) = 2u^2\cdot p\cdot \det(g_0). \tag{1.2.2}$$

Now, we may assume that

$$\hat{\partial}_{\bar{\alpha}}t = \hat{\partial}_\alpha t = 0 \quad (1 \le \alpha \le n) \tag{1.2.3}$$

on a fiber. Then we have the following on the fiber.

<u>Lemma 1.3</u>. Let f be a function on $\mathring{L}$ depending only on t. Then, under assumption (1.2.3),

$$\hat{\partial}_{\bar{0}}\hat{\partial}_0 f = u\cdot\frac{d}{dt}(u\cdot\frac{df}{dt}),$$

$$\hat{\partial}_{\bar{0}}\hat{\partial}_\beta f = 0,$$

$$\hat{\partial}_{\bar{\alpha}}\hat{\partial}_\beta f = -\frac{1}{2}u\cdot\frac{df}{dt}\cdot B_{\bar{\alpha}\beta}.$$

Proof. The first two equalities are obvious. Since

$$\hat{\partial}_{\bar{\alpha}}\hat{\partial}_\beta f = \hat{\partial}_{\bar{\alpha}}(\hat{\partial}_\beta t\cdot\frac{df}{dt}) = \hat{\partial}_{\bar{\alpha}}\hat{\partial}_\beta t\cdot\frac{df}{dt},$$

we have to compute the term $\hat{\partial}_{\bar{\alpha}}\hat{\partial}_\beta t$. Since $\tilde{g}$ is a Kähler metric, we know that

$$\hat{\partial}_0\tilde{g}_{\bar{\alpha}\beta} = \hat{\partial}_\beta\tilde{g}_{\bar{\alpha}0}.$$

The right hand side

$$= \hat{\partial}_\beta(2u\cdot\hat{\partial}_{\bar{\alpha}}t) = 2u\cdot\hat{\partial}_\beta\hat{\partial}_{\bar{\alpha}}t,$$

and the left hand side

$$= \hat{\partial}_0(g_{t\bar{\alpha}\beta} + 2\hat{\partial}_{\bar{\alpha}}t\cdot\hat{\partial}_\beta t) = H[g_{t\bar{\alpha}\beta}]$$

$$= u\cdot\frac{d}{dt}g_{t\bar{\alpha}\beta} = -u^2\cdot B_{\bar{\alpha}\beta},$$

from which the third equality follows. Q.E.D.

Lemma 1.4. Under assumption (1.2.3), the Ricci tensor $\tilde{r}$ of $\tilde{g}$ becomes

$$\tilde{r}_{\bar{0}0} = -u\cdot\frac{d}{dt}(u\cdot\frac{d}{dt}(\log(u^2\cdot p)),$$

$$\tilde{r}_{\bar{0}\beta} = 0,$$

$$\tilde{r}_{\bar{\alpha}\beta} = r_{0\bar{\alpha}\beta} + \frac{1}{2}u\cdot\frac{d}{dt}\log(u^2 p)\cdot B_{\bar{\alpha}\beta},$$

where r_0 is the Ricci tensor of g_0.

Proof. For $0 \le \alpha,\beta \le n$, we have a general formula

$$\tilde{r}_{\bar{\alpha}\beta} = -\hat{\partial}_{\bar{\alpha}}\hat{\partial}_{\beta}(\log(\det(\tilde{g})),$$

and from (1.2.2) the right hand side is given by

$$-\hat{\partial}_{\bar{\alpha}}\hat{\partial}_{\beta}(\log(u^2\cdot p) + \log(\det(g_0))). \qquad \text{Q.E.D.}$$

2. Chern class of a compactification of a $\mathbf{C}^*$-bundle

From now on, we assume that a compactification $\hat{L}$ of $\mathring{L}$ with following properties is given.

(Assumption A) Let (mint, maxt) be the range of t. The function t extends to a continuous function on $\hat{L}$ with range [mint,maxt], and the subset M_{min} (resp. M_{max}) of $\hat{L}$ defined by t = mint (resp. t = maxt) is a complex submanifold of $\hat{L}$ with codimension D_{min} (resp. D_{max}). Moreover, the Kähler metric $\tilde{g}$ extends to a Kähler metric on $\hat{L}$, which is also denoted by $\tilde{g}$.

(Assumption B) (1) The Kähler form of the metric $\tilde{g}$ on $\hat{L}$ gives the first Chern class of $\hat{L}$. (2) The eigenvalues of r_0 with respect to g_0 are constant on M.

Under assumption A, we see that the function t - mint gives the distance from M_{min} in $\hat{L}$, and that the function u have Taylor expansion with first term t - mint, since a fiber $\mathring{L}_x$ is compactified to S_x^2. Therefore, by (1.1.1), the function U is extended to a C^∞-function on a neighbourhood of M_{min}. By an analogous observation for M_{max}, U may be regarded as a C^∞-function on $\hat{L}$. Moreover, if we pay attention to a fiber S_x^2, then we easily see the following

Lemma 2.1. A function on $\mathring{L}$ depending only on t extends to a C^∞-function on $\hat{L}$ if and only if it extends C^∞-ly over minU and maxU as a function of U, where [minU, maxU] is the range of U on $\hat{L}$.

Assumption (B1) is equivalent to the existence of a C^∞-function f on $\hat{L}$ such that

$$\tilde{r}_{\bar{\alpha}\beta} - \tilde{g}_{\bar{\alpha}\beta} = \hat{\partial}_{\bar{\alpha}}\hat{\partial}_{\beta} f \qquad (0 \le \alpha, \beta \le n). \tag{2.1.1}$$

Here, $\tilde{g}$ is S^1-invariant and so is f. Therefore f may be regarded as a function on [mint, maxt] × M. Let $f(t)$ be a function on M defined by $f(t)(x) = f(t,x)$.

Consider equation (2.1.1) for $\alpha = \beta = 0$. By Lemmas 1.2 and 1.4, we see that, omitting t of $f(t)$,

$$- u\cdot\frac{d}{dt}(u\frac{d}{dt}\log(u^2p)) - 2u^2 = u\cdot\frac{d}{dt}(u\frac{d}{dt}f), \tag{2.1.2}$$

and so

$$\frac{d}{dt}(u\cdot\frac{d}{dt}\log(u^2p)) + 2u + \frac{d}{dt}(u\frac{d}{dt}f) = 0. \tag{2.1.3}$$

By integration we see that

$$u\cdot\frac{d}{dt}\log(u^2p) + 2U + u\frac{d}{dt}f \tag{2.1.4}$$

is constant on t, that is, a function on M. Now, we change the variable t to U and set

$$\varphi(U) = u^2, \quad Q(U) = p. \tag{2.1.5}$$

The function Q is a polynomial of U (see (1.1.2), (1.2.1)) and u^2 is a C^∞-function on $\hat{L}$, from which we see that, by Lemma 2.1, the functions φ and Q extend C^∞-ly over minU and maxU.

From the equality

$$\frac{d}{dt} = \frac{dU}{dt}\cdot\frac{d}{dU} = u\frac{d}{dU}, \tag{2.1.6}$$

we see that function (2.1.4) is given by

$$\begin{aligned} u^2\frac{d}{dU}\log(u^2p) + 2U + u\frac{d}{dt}f \\ = \varphi\cdot\frac{d}{dU}\log(\varphi Q) + 2U + H[f], \end{aligned} \tag{2.1.7}$$

which coincides with the function

$$\frac{d}{dU}\varphi + \frac{\varphi}{Q}\cdot\frac{d}{dU}Q + 2U + H[f]. \tag{2.1.8}$$

Let the variable U tend to minU. Then $H \to 0$ and thus $H[f] \to 0$. For the function φ, we see that $\varphi(U) \to 0$ and

$$\frac{d}{dU}\varphi = \frac{dt}{dU}\cdot\frac{d}{dt}\varphi = \frac{1}{u}\cdot 2u\frac{du}{dt} = 2\frac{du}{dt} \to 2, \tag{2.1.9}$$

and so $\varphi(U)$ has Taylor expansion with the first term $2(U-\min U)$. Since the codimension of $M_{\min}$ in $\hat{L}$ is $D_{\min}$, the hermitian metric g_t on M converges to a hermitian form on M of rank $n - D_{\min} + 1$ when t tends to $\min t$. Hence the function $Q(U)$ has Taylor expansion with the first term $a(U-\min U)^{D_{\min}-1}$ $(a \neq 0)$, and so the term $\frac{\varphi}{Q}\cdot\frac{d}{dU}Q$ has Taylor expansion with first term

$$(2.1.10) \qquad 2(U-\min U)\cdot(D_{\min}-1)(U-\min U)^{D_{\min}-2}/(U-\min U)^{D_{\min}-1}$$

$$= 2(D_{\min}-1).$$

Therefore, the function (2.1.8) on M is a constant $2 + 2(D_{\min}-1) + 2\min U = 2(\min U + D_{\min})$. By an analogous observation at $U = \max U$, this constant coincides with $2(\max U - D_{\max})$. Moving the origin of U (and the origin of t) if necessary, we may assume that

$$(2.1.11) \qquad \min U = -D_{\min}.$$

Then we see that

$$(2.1.12) \qquad \max U = D_{\max},$$

and function (2.1.8) identically vanishes. In particular, the term $\frac{d}{dt}f$ is a function depending only on t (or U). Equation (2.1.1) for $\alpha = 0$ and $\beta > 0$ now holds automatically. In fact, by Lemma 1.4, we see that

$$\tilde{r}_{\bar{0}\beta} - \tilde{g}_{\bar{0}\beta} = 0, \quad \hat{\partial}_{\beta}\hat{\partial}_{\bar{0}}f = \hat{\partial}_{\beta}(Hf) = 0.$$

Next we consider equation (2.1.1) for $\alpha, \beta > 0$. Under assumption (1.2.3), we see that

$$\tilde{r}_{\bar{\alpha}\beta} - \tilde{g}_{\bar{\alpha}\beta} = r_{0\bar{\alpha}\beta} + \frac{1}{2}u\frac{d}{dt}(\log(u^2p))\cdot B_{\bar{\alpha}\beta} - g_{0\bar{\alpha}\beta} + U\cdot B_{\bar{\alpha}\beta},$$

$$\begin{aligned}\hat{\partial}_{\bar{\alpha}}\hat{\partial}_{\beta}f &= \hat{\partial}_{\bar{\alpha}}(\hat{\partial}_{\beta}t\cdot\frac{d}{dt}f + \partial_{\beta}f)\\ &= \hat{\partial}_{\bar{\alpha}}\hat{\partial}_{\beta}t\cdot\frac{d}{dt}f + \partial_{\bar{\alpha}}\partial_{\beta}f\\ &= -\frac{1}{2}u\cdot B_{\bar{\alpha}\beta}\cdot\frac{d}{dt}f + \partial_{\bar{\alpha}}\partial_{\beta}f.\end{aligned}$$

Therefore, since (2.1.7) = 0, we have

$$\partial_{\bar{\alpha}}\partial_{\beta}f = r_{0\bar{\alpha}\beta} - g_{0\bar{\alpha}\beta}.$$

By Assumption (B2), the signature of the right hand side is constant on M, and so f must be constant on M, that is, it depends only on t. In particular, we see that $r_0 = g_0$. We resume the results of this section as

Lemma 2.2. Assume (A) and (B). We may also assume that $\min U = -D_{min}$. Then the function f is a function of t, and given by

$$\frac{d}{dU}\varphi + 2U + \frac{\varphi}{Q}\cdot\frac{d}{dU}Q + H[f] = 0.$$

Moreover we have $\max U = D_{max}$ and $r_0 = g_0$. In particular, the Kähler metric $\tilde{g}$ on $\hat{L}$ is an Einstein metric if and only if $r_0 = g_0$ and

$$\frac{d}{dU}\varphi + 2U + \frac{\varphi}{Q}\cdot\frac{d}{dU}Q = 0.$$

3. Futaki's obstruction

Let X be a compact Kähler manifold with $C_1(X) > 0$ and g a Kähler metric on X whose Kähler form belongs to $C_1(X)$. Then there is a C^∞-function f on X such that $r_{\bar{\alpha}\beta} - g_{\bar{\alpha}\beta} = \partial_{\bar{\alpha}}\partial_\beta f$. Let a holomorphic vector field H on X correspond to the integeral $\int_X H[f]v_g$.

Proposition 3.1 ([6]). If the above linear correspondance does not vanish, then X admits no Kähler-Einstein metrics.

In our case, the above integral for the holomorphic vector field H becomes

$$\begin{aligned}\int_{\hat{L}} H[f]v_{\tilde{g}} &= \int_{\min t}^{\max t} H[f]\cdot \mathrm{Vol}(M,g_t)dt \\ &= \mathrm{Vol}(M,g_0)\int_{\min t}^{\max t} H[f]\cdot u p dt \\ &= \mathrm{Vol}(M,g_0)\int_{\min U}^{\max U} H[f]\cdot Q dU \\ &= -\mathrm{Vol}(M,g_0)\int_{\min U}^{\max U}\left(\frac{d\varphi}{dU} + 2U + \frac{\varphi}{Q}\cdot\frac{dQ}{dU}\right)Q dU \\ &= -\mathrm{Vol}(M,g_0)\left\{[Q\varphi]_{\min U}^{\max U} + 2\int_{\min U}^{\max U} QU dU\right\} \\ &= -2\mathrm{Vol}(M,g_0)\int_{-D_{min}}^{D_{max}} QU dU.\end{aligned}$$

Therefore if we set

$$F(\hat{L}) = \int_{-D_{min}}^{D_{max}} QU dU, \tag{3.1.1}$$

then we have the following

Lemma 3.2. Under assumption (A) and (B), the logical implication (1) => (2) => (3) holds for the following conditions. (1) $\hat{L}$ admits a Kähler-Einstein metric. (2) The linear correspondence in Proposition 3.1 vanishes. (3) $F(\hat{L}) = 0$.

4. Existence of Kähler-Einstein metric

Theorem 4.1. Under assumption in section 2, if $F(\hat{L}) = 0$ then $\hat{L}$ admits a Kähler-Einstein metric.

Proof. Let Φ be a function on $\hat{L}$ depending only on t and we seek the condition that the metric $\tilde{h}$ on $\hat{L}$ defined by

$$\tilde{h}_{\bar{\alpha}\beta} = \tilde{g}_{\bar{\alpha}\beta} + \hat{\partial}_{\bar{\alpha}}\hat{\partial}_{\beta}\Phi \tag{4.1.1}$$

is a Kähler-Einstein metric. The metric $\tilde{h}$ satisfies condition (1.0.6) and there exists a parameter s (depending only on t) so that $\tilde{h}$ has the form

$$\tilde{h} = ds^2 + (ds\circ\tilde{J})^2 + \pi^* h_s \tag{4.1.2}$$

on $\dot{L}$, and so we can apply the results in section 1,2,3. Take parameters v, V, ψ for $\tilde{h}$ which correspond to u, U, φ for $\tilde{g}$. We will construct a solution such that $h_0 = g_0$, $\min V = -D_{min}$ and $\max V = D_{max}$. By Lemma 2.2, the condition of Kähler-Einstein reduces to the equation

$$Q(V)\frac{d}{dV}\psi(V) + 2VQ(V) + \psi(V)\frac{d}{dV}Q(V) = 0, \tag{4.1.3}$$

and it has a solution given by

$$\psi(V) = -2\{\int_{-D_{min}}^{V} VQ(V)dV\}/Q(V). \tag{4.1.4}$$

Then the function $\psi(V)$ is positive on $(-D_{min}, D_{max})$. Since the function $Q(V)$ has Taylor expansion with first term $a(V + D_{min})^{D_{min}-1}$, the function $\psi(V)$ extends C^∞-ly over $-D_{min}$ and has the same Taylor expansion with

$$-2\{\int_{-D_{min}}^{V} V(V+D_{min})^{D_{min}-1} dV\}/(V+D_{min})^{D_{min}-1},$$

whose first term is given by

$$2(V + D_{min}). \tag{4.1.5}$$

For the point $V = D_{max}$, the condition : $F(\hat{L}) = 0$ implies that

$$(4.1.6)\qquad \psi(V) = 2\left(\int_V^{D_{max}} VQ(V)dV\right)/Q(V),$$

and so $\psi(V)$ extends C^∞-ly over D_{max} with first term $-2(V-D_{max})$. Now, H coincides with $u\frac{d}{dt}$ and $v\frac{d}{ds}$, and we have

$$(4.1.7)\qquad \frac{ds}{dt} = \frac{v}{u},$$

$$(4.1.8)\qquad \frac{dV}{dU} = \frac{dt}{dU}\cdot\frac{ds}{dt}\cdot\frac{dV}{ds} = \frac{1}{u}\cdot\frac{v}{u}\cdot v = \frac{\psi(V)}{\varphi(U)},$$

$$(4.1.9)\qquad \int\frac{dV}{\psi(V)} = \int\frac{dU}{\varphi(U)}.$$

Therefore, using Taylor expansion at $-D_{min}$, we see that

$$(4.1.10)\qquad \log(V + D_{min}) + C^\infty\text{-function of } V$$

$$= \log(U + D_{min}) + C^\infty\text{-function of } U,$$

hence V extends C^∞-ly over $-D_{min}$ as a function of U. By an analogous observation at D_{max}, we see that V extends to a C^∞-function of U defined on an open set containing $[-D_{min}, D_{max}]$.

However, the C^∞-ness of the metric $\tilde{h}$ is not observed from C^∞-ness of V and ψ. We have to come back to representation (4.1.1) and show that the function ϕ is C^∞. For $\alpha = \beta = 0$, we get

$$(4.1.11)\qquad 2v^2 = 2u^2 + u\frac{d}{dt}\left(u\frac{d}{dt}\phi\right).$$

Therefore,

$$(4.1.12)\qquad \frac{d}{dt}\left(u\frac{d}{dt}\phi\right) = 2\frac{v^2}{u} - 2u,$$

$$(4.1.13)\qquad u\frac{d\phi}{dt} = 2\int\frac{v^2}{u}dt - 2\int udt = 2\int\frac{v^2}{u}\cdot\frac{u}{v}ds - 2\int udt$$

$$= 2(V - U) + C.$$

But here, if $t \to \text{mint}$ then $H[\phi] \to 0$, and so $C = 0$. Hence

$$(4.1.14)\qquad u\frac{d\phi}{dt} = 2(V - U),$$

$$(4.1.15)\qquad \frac{d}{dU}\phi = \frac{dt}{dU}\cdot\frac{d\phi}{dt} = \frac{2(V-U)}{u^2} = 2\frac{V-U}{\varphi(U)}.$$

Here $\varphi(U)$ has Taylor expansion with first term $2(U + D_{min})$, and so ϕ extends C^∞-ly over $-D_{min}$ as a function of U. By an analogous observation at the point $U = D_{max}$ and Lemma 2.1, we see that ϕ is a C^∞-function on $\hat{L}$, hence $\tilde{h}$ is a C^∞ symmetric 2-form on $\hat{L}$.

Finally, we show that $\tilde{h}$ is positive definite. On $\mathring{L}$,

(4.1.16) $$ds = \frac{ds}{dt}dt = \frac{v}{u}dt$$

holds and $\psi > 0$ implies that $ds \neq 0$. Combining with $Q > 0$, we see that h_s is positive definite for $s \in (\min s, \max s)$, and so $\tilde{h}$ is positive definite on $\mathring{L}$. Consider a tubular neighbourhood of M_{min} in $\hat{L}$. Since ϕ is a function depending only on U, $\hat{\partial}_{\bar{\alpha}}\hat{\partial}_{\beta}\phi$ coincides with $\hat{\partial}_{\bar{\alpha}}\hat{\partial}_{\beta}U$ up to constant factor at each point of M_{min}, hence vanishes on $T(M_{min})$ and coincides with $\tilde{g}_{\bar{\alpha}\beta}$ up to constant factor for the fiber direction of the tubular neighbourhood. Therefore, on the tubular neighbourhood, $\tilde{h}$ differs from $\tilde{g}$ only for the fiber direction, and even for the fiber direction $\tilde{h}$ coincides with $\tilde{g}$ up to constant factor. But here H belongs to the fiber direction, and

(4.1.17) $$\frac{\tilde{h}(H,H)}{\tilde{g}(H,H)} = \frac{v^2}{u^2} = \frac{\psi(V)}{\varphi(U)} = \frac{dV}{dU}$$

converges into a non-zero value by (4.1.10). Thus $\tilde{h}$ is positive definite on M_{min}. By an analogous observation on M_{max}, we see that $\tilde{h}$ is in fact a (C^{∞}, positive definite) Kähler-Einstein metric on $\hat{L}$.

Q.E.D.

We sum up our results as the following

Theorem 4.2. Let M be a compact Kähler-Einstein manifold whose Kähler form represents $C_1(M)$ and L a hermitian line bundle over M. Assume that the eigenvalues of the Ricci tensor B of L are constant on M with respect to g_0, and that there is a Kähler metric $\tilde{g}$ on a compactification $\hat{L}$ of $\mathring{L}$ of form (1.0.6) whose Kähler form represents $C_1(\hat{L})$. Then the following three conditions are equivalent.

(1) The manifold $\hat{L}$ admits a Kähler-Einstein metric.

(2) Futaki's obstruction vanishes.

(3) The integral $F(\hat{L})$ vanishes.

5. Examples

First, we apply Theorem 4.2 to almost homogeneous spaces. A complex manifold is called an almost homogeneous space if it has an open orbit for the action of the automorphism group.

Lemma 5.1 ([8, Proposition 3.1]). Let X be a compact almost homogeneous space. If $C_1(X) > 0$ and the exceptional set is not connected, then there exist a Kähler C-space M and a line bundle L over M so that X becomes $\hat{L}$ in section 2.

Theorem 5.2. Let X be a compact almost homogeneous space with $C_1(X) > 0$ and with a disconnected exceptional set. Then X admits a Kähler-Einstein metric if and only if Futaki's obstruction vanishes. Moreover, the Kähler-Einstein metric is unique up to constant factor and the action of the automorphism group.

Proof. Let K be the identity component of a maximal compact subgroup of the automorphism group. Since $C_1(X)$ is K-invariant, there is a K-invariant Kähler metric $\tilde{g}$ in $C_1(X)$, which satisfies conditions (A),(B) in section 2. Therefore the first half holds by Theorem 4.2.

On the other hand, by [9, Theorem 3], we may assume that a Kähler-Einstein metric on X is K-invariant, using transformation by an automorphism. Therefore a such metric is of the form (1.0.6). But by proof of Theorem 4.1, it is obvious that such a Kähler-Einstein metric is unique up to constant factor. Q.E.D.

Remark 5.3. For the uniqueness of Einstein metric, there is a more general result in [2]: A Kähler-Einstein metric on a compact complex manifold with positive first Chern class is unique up to constant and automorphism.

Next, we consider a line bundle L over a compact manifold M and construct a Kähler-Einstein manifold $\hat{L}$ as a $P^1(\mathbb{C})$-bundle $P(1 \oplus L)$ over M. We assume that M has a Kähler-Einstein metric g_0 with $r_0 = g_0$ and that L has a hermitian inner product such that the eigenvalues of the Ricci tensor B are constant on M and their absolute values are less than 1.

In representations (1.0.6) and (1.0.7), if we set $u(t) = \cos t$ with t in $[-\frac{\pi}{2}, \frac{\pi}{2}]$, then each fiber $P^1(\mathbb{C})$ is isometric with the standard sphere. Moreover,

$$U(t) = \int_0^t \cos t \, dt = \sin t,$$

and $\min U = -1$, $\max U = 1$. Therefore, the metric $\tilde{g}$ becomes a Kähler metric on $\hat{L}$, and we see

$$Q(U) = \det(g_0^{-1}(g_0 - UB)) = \det(\mathrm{id} - Ug_0^{-1}B)$$

and

$$\varphi(U) = u^2 = \cos^2 t = 1 - U^2.$$

Thus equation (2.1.1) has a C^∞-solution

$$f = - \log Q,$$

and we can apply Theorem 4.2.

Theorem 5.4. The above manifold $\hat{L}$ admits a Kähler-Einstein metric if and only if

$$\int_{-1}^{1} U \det(id - Ug_0^{-1}B)dU = 0.$$

Definition 5.5. Let N be a riemannian manifold. The codimension of the principal orbits for the action of the isometry group is called the cohomogeneity of N and is denoted by $\operatorname{cohom}(N)$.

Proposition 5.6. In the above situation, if B is non-trivial on any irreducible factor of M then the Kähler-Einstein manifold $\hat{L}$ is irreducible and is not riemannian homogeneous. More precisely, the inequality : $\operatorname{cohom}(\hat{L}) \geq \operatorname{cohom}(M)+1$ holds.

Proof. Assume that the Kähler manifold $\hat{L}$ admits a non-trivial decomposition : $\hat{L} = X \times Y$. Since the fixed point set for $\mathbf{C}^*$-action is the union of two complex hypersurfaces M_{-1} and M_1, $\mathbf{C}^*$ acts on X or Y trivially and the other manifold becomes a $P^1(\mathbf{C})$-bundle. Let X be the $P^1(\mathbf{C})$-bundle. Then we see that the base manifold M admits a decomposition : $M = N \times Y$, and L and B are trivial on Y, which contradicts the assumption. Next we show the inequality. Let γ be an element of the identity component $\mathrm{Isom}^0(\hat{L})$ of the isometry group. Since γ is a holomorphic automorphism, by [5], γ maps a fiber to a fiber, and defines an isometry of the base manifold M. Thus we have a homomorphism : $\mathrm{Isom}^0(\hat{L}) \to \mathrm{Isom}^0(M)$. Let γ be an element of the kernel. Then γ preserves each fiber, and we can easily check that such an isometry is an element of the S^1-action. Therefore, for the fiber direction, we gain one more cohomogeneity. Q.E.D.

Remark 5.7. Up to now, the authors have no explicit examples of Kähler-Einstein manifolds except $P^1(\mathbf{C})$-bundles (i.e., the cases that $D_{min} = D_{max} = 1$).

Example 5.8. Let N be a compact Kähler-Einstein manifold with $r^N = g^N$ and L^N a line bundle over N such that $C_1(L^N) = a \cdot C_1(N)$ for some $a \in (0,1)$. Let the fiber bundles $L_1 \to N_1$ and $L_2 \to N_2$ be copies of $L^N \to N$, and consider the fiber bundle $L = L_1 \otimes L_2^{-1} \to M = N_1 \times N_2$. We see that $C_1(M) = C_1(N) \oplus C_1(N)$ and $C_1(L) = C_1(L^N) \oplus (-C_1(L^N))$, and so for the canonical hermitian inner product h^N of L^N and a hermitian inner product h of L, we see that $h = h^N \oplus (h^N)^{-1} = a\cdot(g^N \oplus (g^N)^{-1})$. Then the integrand in Theorem 5.4 is an odd function and so the integral vanishes. Therefore the manifold $\hat{L} = P(1 \oplus L) = P(L_1 \oplus L_2)$ admits a Kähler-Einstein metric.

Example 5.9. As a special case of Example 5.8 we get Kähler-Einstein manifolds of cohomogeneity > 1. Let $N_0 = P^n(\mathbf{C})$, H the hyperplane section bundle over N_0 and $L_0 = H^m$. Then we have $C_1(L_0) = m/(n+1)\cdot C_1(N_0)$ and so we can apply Example 5.8 for $0 < m < n+1$, and get almost homogeneous Kähler-Einstein manifolds $P(L_0 \oplus L_0)$ (of cohomogeneity one). We repeat this procedure. Let $N' = P(L_0 \oplus L_0)$. As is well-known (e.g. [7]), we have

$$C_1(N') = C_1(\pi^*(T(N_0 \times N_0) \otimes (\det(L_0 \otimes L_0))^{-1}) \otimes \xi^2),$$

where π is the projection : $N' \to N_0 \times N_0$ and ξ the tautological line bundle over N'. Thus,

$$C_1(N') = (n-m+1)\pi^*(C_1(H) \oplus C_1(H)) + 2C_1(\xi).$$

If $n-m+1$ is even, then

$$C_1(N') = 2\{(n-m+1)/2\cdot\pi^*(C_1(H) \oplus C_1(H)) + C_1(\xi)\},$$

and so there exists a holomorphic line bundle L' over N' with $C_1(L') = 1/2\cdot C_1(N')$. We set $N = N' \times \cdots \times N'$ and $L = L' \oplus \cdots \oplus L'$ (ℓ copies). Then we see that $C_1(L) = 1/2\cdot C_1(N)$, and get a Kähler-Einstein manifold $P(L \oplus L)$ of cohomogeneity at least $2\ell+1$, provided that $0 < m < n$ and $n-m$ is odd.

Example 5.10. Let M be a flag manifold $SL(3,\mathbf{C})/T$, where T is a Borel subgroup of $SL(3,\mathbf{C})$. It is known that the group $H^1(M,\theta^*)$ of all isomorphism classes of holomorphic line bundles over M is $\mathbf{Z} \oplus \mathbf{Z}$ and that the first Chern class $C_1(M)$ of M is given by

$$C_1(M) = 2(h_1 + h_2),$$

where h_1 and h_2 are the Chern classes of the line bundles H_1 and H_2 which are generators of $H^1(M,\theta^*)$ (c.f. [3]). Let L be the line bundle $H_1^{-1} \otimes H_2$ over M. Since L is a homogeneous line bundle over M, we can compute the eigenvalues of the Ricci tensor B of L, and, in our case, these eigenvalues are given by -1/2, 1/2 and 0. Thus the integral in Theorem 5.4 vanishes and <u>the $P^1(C)$-bundle $\hat{L} = P(1 \oplus L)$ over M admits a Kähler-Einstein metric</u>.

References

[1] T. Aubin : Équation du type Monge-Ampère sur les variétés kähleriennes compactes, Bull. Sc. Math. 102 (1978) 63-95.

[2] S. Bando and T. Mabuchi : to appear.

[3] A. Borel and F. Hirzebruch : Characteristic classes and homogeneous spaces I, Am. J. Math. 80 (1958) 458-538.

[4] L. Bérard Bergery : Sur de nouvelles variétés riemanniennes d'Einstein, preprint 1981.

[5] M. A. Blanchard : Sur les variétés analytiques complexes, Ann. Sci. Ecole Norm Sup. 73 (1956), 157-202.

[6] A. Futaki : An obstruction to the existence of Einstein Kähler metrics, Inventiones math. 73 (1983), 437-443.

[7] P. A. Griffiths : Hermitian differential geometry, Chern classes and positive vector bundles, in Global Analysis, Papers in honor of K. Kodaira, Princeton Univ. Press, 1969, 181-251.

[8] A. T. Huckleberry and D. M. Snow : Almost-homogeneous Kähler manifolds with hypersurface orbits, Osaka J. Math., 19 (1982) 763-786.

[9] Y. Matsushima : Remarks on Kähler-Einstein manifolds, Nagoya Math. J. 46 (1972) 161-173.

[10] Y. Sakane : Examples of compact Einstein Kähler manifolds with positive Ricci tensor, to appear.

[11] S. T. Yau : On the Ricci curvature of a compact Kähler manifold and the complex Monge-Ampère equation, I, Comm. on Pure and Appl. Math. 31 (1978) 339-411.

CURVATURE DEFORMATIONS *

Maung Min-Oo, Universität Bonn and
Mc Master University, Hamilton,

and

Ernst A. Ruh, Universität Düsseldorf and
Ohio State University, Columbus.

1. Introduction.

In [H], Hamilton introduced an important evolution equation for Riemannian metrics in his study of three dimensional manifolds with positive Ricci curvature. In the present note we study this equation with emphasis not so much on the metric but more on the connections involved. The evolution equation for the connection is the gradient flow $\dot{\omega} = -\delta^{\omega}\Omega$, where $\Omega = d\omega + [\omega,\omega]$ is the curvature form and $\dot{\omega}$ indicates the infinitesimal change in the connection. The Lagrangean of this flow is the well known Yang-Mills integral $\int|\Omega|^2$. The connection to be used is a Cartan connection of hyperbolic type. The deformation is closely related to our previous work [MR] on non-compact almost symmetric spaces. Even for the study of metrics with positive curvature the hyperbolic model seems to be the appropriate one.

In the next section we give a derivation of Hamilton's equations without explicitly using the notion of Cartan connections. This section also motivates the computations for the deformation equations for Cartan connections which we derive in the last section. While the effect on the deformation of metrics is the same in both approaches the definition of the control function of the process is not. This provides additional flexibility in choosing the quantities to be estimated.

2. Deformation of the Levi-Civita connection.

In this section we reformulate Hamilton's deformation in a different set-up so that the corresponding evolution equations for the Levi-Civita

---

*) This work was done under the program "Sonderforschungsbereich Theoretische Mathematik" (SFB 40) at the University of Bonn.

connection and the Riemannian curvature can be derived in a natural and easy fashion.

Let (M^n,g) be a compact Riemannian manifold. First of all, we deform metrics not through tensors of type (0,2) directly but by using gauge transformations, i.e., tensors of type (1,1): θ: TM $\longrightarrow$ TM. A metric deformation is therefore a curve g_t of metrics defined by:

(2.1) $$g_t(X,Y) = g(\theta_t X,\theta_t Y),$$

where θ_t is a 1-parameter family of invertible maps

θ_t: TM $\longrightarrow$ TM with $\theta_o = \mathrm{id}$.

In order to be able to calculate the infinitesimal changes in the Levi-Civita connection and the curvature tensor caused by such a metric deformation in an efficient manner, we introduce the bundle $\mathrm{Aff}(M) = TM \oplus TM^* \otimes TM$.

An infinitesimal gauge transformation $\dot{\theta} = \frac{d}{dt}\theta_t\Big|_{t=o}$ can now be considered as a 1-form with values in TM or sometimes as a o-form with values in $TM^* \otimes TM$. An infinitesimal change in the connection is a 1-form with values in $TM^* \otimes TM$ and we interpret curvature as a 2-form with values in $TM^* \otimes TM$.

The Levi-Civita connection ∇ of the metric g induces a natural direct sum connection in Aff(M) and we denote the corresponding exterior covariant derivative on p-forms with values in Aff(M) by d^∇.

(2.2) $$(d^\nabla\alpha)(X_o\ldots X_p) = \sum_{i=o}^{p}(-1)^i\,\nabla_{X_i}(\alpha(\ldots\hat{X}_i\ldots)) + \sum_{i<j}(-1)^{i+j}\alpha([X_i,X_j],\ldots\hat{X}_i\ldots\hat{X}_j\ldots)\ .$$

We also introduce an algebraic operator d_2 acting on a p-form α by the formula:

(2.3) $$(d_2\alpha)(X_o\ldots X_p) = \sum_{i=o}^{p}(-1)^i[X_i,\alpha(\ldots\hat{X}_i\ldots)],$$

where the bracket [,] is defined to be:

$[X,A] = -AX \in TM$ for $X \in TM$, $A \in TM^* \otimes TM$

$[X,Y] = X \wedge Y \in TM^* \otimes TM$ for $X,Y \in TM$,

where $X \wedge Y$ is the map $Z \longmapsto g(Z,Y)X - g(Z,X)Y$.

For $A,B \in TM^* \otimes TM$, we have of course the usual definition for [A,B].

If R is the Riemannian curvature tensor of g, interpreted as an Aff(M)-valued 2-form, then the Bianchi identities can be stated as follows.

Lemma 1. $d_2R = o$ (1st Bianchi identity)

$d^{\nabla}R = o$ (2nd Bianchi identity)

We will make use of the adjoint operator of d^{∇}:

(2.4) $$(\delta^{\nabla}\alpha)(X_2 \ldots X_p) = - \sum_{k=1}^{n} (\nabla_{e_k}\alpha)(e_k, X_2 \ldots X_p),$$

where $\{e_k\}$ is an orthonormal basis for the metric g.

We also define an algebraic operator δ_2 by the formula:

(2.5) $$(\delta_2\alpha)(X_2 \ldots X_p) = \sum_{k=1}^{n} [e_k, \alpha(e_k, X_2 \ldots X_p)]$$

The Ricci tensor, interpreted as a TM-valued 1-form can now be defined as δ_2R. Using this notation, we derive a consequence of the Bianchi identities which is fundamental for our deformation equations.

Lemma 2. $d^{\nabla}\delta_2R + d_2\delta^{\nabla}R = 0.$

Proof. $$\begin{aligned} d^{\nabla}\delta_2R(X,Y) &= (\nabla_X\delta_2R)(Y) - (\nabla_Y\delta_2R)(X) \\ &= (\nabla_XR)(Y,e_k,e_k) - (\nabla_YR)(X,e_k,e_k) \\ &= (\nabla_{e_k}R)(Y,X,e_k), \end{aligned}$$

where we sum over k from 1 to n and used the 2nd Bianchi identity. On the other hand,

$$\begin{aligned} d_2\delta^{\nabla}R(X,Y) &= - (\nabla_{e_k}R)(e_k,X,Y) + (\nabla_{e_k}R)(e_k,Y,X) \\ &= (\nabla_{e_k}R)(X,e_k,Y) + (\nabla_{e_k}R)(e_k,Y,X) \\ &= - (\nabla_{e_k}R)(Y,X,e_k), \end{aligned}$$

where the 1st Bianchi identity is used. The sum of the two terms is zero.

Let $\dot{\gamma}$ be the infinitesimal change in the Levi-Civita connection caused by an infinitesimal gauge transformation $\dot{\theta}$: TM $\longrightarrow$ TM, which we assume from now on, without loss of generality, to be symmetric with respect to g. $\dot{\gamma}$ is a 1-form with values in TM* ⊗ TM and can be decomposed as:

(2.6) $$\dot{\gamma} = \dot{\eta} + \dot{\sigma},$$

where $\dot{\eta}$ has values in the skew-symmetric, and $\dot{\sigma}$ in the symmetric component of $TM^* \otimes TM$.

Differentiating the condition $\nabla g = 0$, we get $\dot{\gamma} \cdot g + \nabla \dot{g} = 0$. It is easy to see that $\dot{\gamma} \cdot g = -2\dot{\sigma}$ and that $\nabla \dot{g} = 2\nabla\theta$, where θ is considered as a o-form with values in the symmetric part of $TM^* \otimes TM$. Hence $\dot{\sigma}$ is given by

(2.7) $\dot{\sigma} = \nabla\dot{\theta}$.

Differentiating the condition that the Levi-Civita connection is torsion-free, we obtain:

$\dot{\gamma}(X)Y - \dot{\gamma}(Y)X = 0$, which we write as

$$d_2\dot{\gamma} = d_2\dot{\eta} + d_2\dot{\sigma} = d_2\dot{\eta} + d_2\nabla\dot{\theta} = 0.$$

Now, $d_2\nabla\dot{\theta} = d^{\nabla}\dot{\theta}$, where $\dot{\theta}$ on the left is viewed as a section in $TM^* \otimes TM$ and on the right as a 1-form with values in TM. Moreover, d_2 resticted to 1-forms with values in the skew symmetric endomorphisms is well know to be an isomorphism onto the 2-forms with values in TM. Incidentally, this fact is responsible for the uniqueness of the Levi-Civita connection among metric connections. Hence $\dot{\eta}$ is determined uniquely by the equation:

(2.8) $d_2\dot{\eta} + d^{\nabla}\dot{\theta} = 0$, and the change in the Levi-Civita connection is given by

(2.9) $\dot{\gamma} = \dot{\eta} + \nabla\dot{\theta}$.

By Lemma 2, the relation (2.8) is satisfied if we set

(2.10) $\dot{\theta} = -\delta_2 R$, $\dot{\eta} = -\delta^{\nabla} R$, and we have proved the following result.

Lemma 3. The infinitesimal change in the Levi-Civita connection caused by the infinitesimal gauge transformation $\dot{\theta} = -\delta_2 R$ is given by

(2.11) $\dot{\gamma} = -\delta^{\nabla} R - \nabla\delta_2 R$.

The corresponding metric deformation is computed to be $\dot{g}(X,Y) = g(\dot{\theta}X,Y) + g(X,\dot{\theta}Y) = -2\,\mathrm{Ric}(X,Y)$, which is exactly Hamilton's deformation without the normalizing term. Hamilton [H] proved that the deformation exists at least for a short time and the above infinitesimal computations are not just formal. Introduction of a normalization

(2.12) $\dot{\theta} = -\delta_2 R + c\ \mathrm{id}$, c any constant, does not alter the formula $\dot{\gamma} = -\delta^{\nabla} R - \nabla\delta_2 R$.

The infinitesimal change of the curvature tensor, considered as a 2-form with values in $TM^* \otimes TM$ is then given by (compare also [M]):

Lemma 4.

(2.13) $\dot{R} = - d^\nabla \delta^\nabla R - R.\delta_2 R = - \Delta^\nabla R - R.\delta_2 R$, where $\Delta^\nabla = d^\nabla \delta^\nabla + \delta^\nabla d^\nabla$ is the Laplacian and $(R.\delta_2 R)(X,Y) = [R(X,Y), \delta_2 R] \in TM^* \otimes TM$.

Proof. $\dot{R} = d^\nabla \dot{\gamma} = - d^\nabla \delta^\nabla R - d^\nabla \nabla \delta_2 R$
$= - \Delta^\nabla R - d^\nabla \nabla \delta_2 R$, ($2^{nd}$ Bianchi id.)

and $d^\nabla \nabla$ applied to a zero-form is by definition of curvature, the curvature applied to this zero-form. In our notation, $(d^\nabla \nabla \delta_2 R)(X,Y) =$ $= [R(X,Y), \delta_2 R]$.

In order to compare (2.11) to Hamilton's formula [H, Theorem 7.1] for the evolution of the curvature, we need to recall first the following Weitzenböck formula for the Laplacian $\Delta^\nabla R$. (compare eg. [B])

$$(2.14)\quad (\Delta^\nabla R)(X,Y) = (\bar{\Delta} R)(X,Y) + R(\mathrm{Ric}\,X,Y) + R(X,\mathrm{Ric}\,Y) - R(R(X,Y)) + 2\sum_{p=1}^{n} [R(e_p,X), R(e_p,Y)],$$

where $\bar{\Delta}$ is the rough Laplacian and $\{e_p\}$ is an orthonormal basis.

Using index notation, we can write the last two terms of the algebraic expression on the right hand side of (2.14) as:

(2.15) $R_{ij}{}^{pq} R_{pqk}{}^{l} + 2 R_{piq}{}^{l} R_{pjk}{}^{q} - 2 R_{pjq}{}^{l} R_{pik}{}^{q}$, where $R_{ijkl} =$ $= g(R(e_i,e_j)e_k,e_l)$ and $R_i{}^j$ denotes the Ricci tensor.

As in [H] we define $B_{ijkl} = R_{piqj} R_{pkql}$, where an orthonormal frame is used. Now,

$$\begin{aligned} B_{ijkl} - B_{ijlk} &= R_{piqj}(R_{pkql} - R_{plqk}) \\ &= R_{piqj} R_{pqkl} = (- R_{pqji} - R_{pjiq}) R_{pqkl} \\ &= R_{ijpq} R_{pqkl} - R_{piqj} R_{pqkl}, \text{ and hence} \end{aligned}$$

$$B_{ijkl} - B_{ijlk} = R_{piqj} R_{pqkl} = \frac{1}{2} R_{ijpq} R_{pqkl} .$$

By definition, we have

$$\begin{aligned} B_{ijkl} - B_{iljk} &= R_{piqk} R_{pjql} - R_{piql} R_{pjqk} \\ &= -R_{pjql} R_{pikq} + R_{piql} R_{pjkq} , \end{aligned}$$

and we accounted for all the terms involving the whole curvature tensor in Hamilton's formula and in our expression (2.15).

The four terms containing the Ricci curvature in Hamilton's formula can be written as follows:

$$R(\mathrm{Ric}\,X,Y)Z + R(X,\mathrm{Ric}\,Y)Z + R(X,Y)\mathrm{Ric}\,Z + \mathrm{Ric}\,(R(X,Y)Z) .$$

The first two terms coincide with the corresponding terms in (2.14), but instead of the last two terms our formula (2.13) gives us $R(X,Y)\,\mathrm{Ric}\,Z - \mathrm{Ric}\,(R(X,Y)Z)$. The difference $2\,\mathrm{Ric}\,(R(X,Y)Z)$ is due to the fact that Hamilton's equation is for the curvature of type (0,4) and Lemma 4 treats the curvature as a tensor of type (1,3). We have

$$\dot{R}_{ijkl} = g_{ml}\,\dot{R}_{ijk}{}^{m} + \dot{g}_{ml}\,R_{ijk}{}^{m}\,, \quad \text{and}$$
$$\dot{g} = -\,2\,\mathrm{Ric}\,.$$

This proves Hamilton's Theorem 7.1:

$$\dot{R}_{ijkl} + (\bar{\Delta}R)_{ijkl} = 2(B_{ijkl} - B_{ijlk} + B_{ikjl} - B_{iljk})$$
$$- R_{pi}\,R_{pjkl} - R_{pj}\,R_{ipkl} - R_{pk}\,R_{ijpl} - R_{pl}\,R_{ijkp}\,.$$

3. Deformation of Cartan connections.

Let (M,g) as before, denote a compact Riemannian manifold. In this section we identify $X \wedge Y \in TM \wedge TM$ with the skew-symmetric endomorphism $Z \longrightarrow g(Z,Y)X - g(Z,X)Y$, and define $E = TM \oplus TM \wedge TM$, a vector bundle over M with fibre metric $\langle\,,\rangle$ defined by g. For skew-symmetric maps $A,B\colon TM \longrightarrow TM$ we have $\langle A, X\wedge Y\rangle = -\langle AX,Y\rangle$ and $\langle A,B\rangle = -\frac{1}{2}\,\mathrm{tr}\,AB$.

Let ∇ be a metric connection for TM, not necessarily torsion-free. A gauge transformation θ is simply an invertible map $\theta\colon TM \longrightarrow TM$. The image, im θ, should be viewed as the subspace $TM \subset E$. We define a Cartan connection on E by:

$$D_X Y = \nabla_X Y + \theta X \wedge Y$$
$$D_X A = -\,A\theta X + \nabla_X Y, \qquad X,Y \in TM, \quad A \in TM \wedge TM\,.$$

D defines a Cartan connection of <u>hyperbolic type</u> for M. Note that the metric is not invariant under D. To simplify the above formula we define the structure of a Lie algebra, isomorphic to the Lie algebra of the isometry group of hyperbolic space o(n,1), on the fibres by

$$[(X,A),(Y,B)] = (AY - BX, [A,B] + X \wedge Y)\,.$$

D leaves this bracket invariant and is expressed as follows:

(3.1) $D_X s = \nabla_X s + [\theta X, s]$, s a section in E.

Let R^D denote the curvature tensor of the connection D, i.e., $R^D(X,Y)s = (D_X D_Y - D_Y D_X - D_{[X,Y]})s$. We define the Cartan curvature form Ω by the formula

(3.2) $[\Omega(X,Y), s] = R^D(X,Y)s$.

It will be convenient to split the curvature tensor R^D, and accordingly Ω, into the components $TM \wedge TM$ and TM of E. We write

$$R^D(X,Y) = R_1(X,Y) + R_2(X,Y),$$
$$\Omega(X,Y) = \Omega_1(X,Y) + \Omega_2(X,Y),$$

where $R_1(X,Y) = R^\nabla(X,Y) + \theta X \wedge \theta Y$ with R^∇ the curvature of the connection ∇, and $R_2 = T^{(\nabla,\theta)}$, with $T^{(\nabla,\theta)}(X,Y) = \nabla_X(\theta Y) - \nabla_Y(\theta X) - \theta[X,Y]$ the Cartan torsion.

The connection D defines an exterior covariant derivative for E-valued p-forms on M.

(3.3) $$(d^D\alpha)(X_0 \dots X_p) = \sum_{i=0}^{p} (-1)^i D_{X_i}(\alpha(\dots \hat{X}_i \dots)) + \sum_{i<j} (-1)^{i+j} \alpha([X_i,X_j], \dots \hat{X}_i \dots \hat{X}_j \dots),$$

which we also write as $d^D\alpha = d_1\alpha + d_2\alpha$, where

$$(d_2\alpha)(X_0 \dots X_p) = \sum_{i=0}^{p} (-1)^i [\theta X_i, \alpha(\dots \hat{X}_i \dots)].$$

In this notation the Bianchi identities take the form $d^D\Omega = 0$. Corresponding to Lemma 1 of the previous section we have

Lemma 5. If $\Omega_2 = 0$, then $d_1\Omega = 0$, and $d_2\Omega = 0$.

The purpose of this section is to study deformations of metrics on M via deformations of Cartan connections. We will start with ∇^0 equal to the Levi-Civita connection of the initial metric $g_0 = g$. In the course of the deformation the metric g on M of course will change but we will keep the fixed metric <,> as well as the fixed Lie algebra bracket defined by g_0 in the fibres. A 1-parameter family $D = D^t$ of Cartan connections determines a family (∇^t, θ_t) of connections and gauge transformations on TM. D_t and (∇^t, θ_t) are related by the formula $D_X^t s = \nabla_X^t s + [\theta_t X, s]$ of (3.1). The changing metric g_t on M is related to the fixed metric <,> on E by the formula $g_t = \theta_t^* g_0$, where $(\theta_t^* g_0)(X,Y) = g_0(\theta_t X, \theta_t Y)$ as in (2.1).

We will mainly be interested in deformations of Cartan connections with the property $\nabla^t g_o = 0$, and $T^{(\nabla^t,\theta_t)} = 0$ at all times t. The reason is the following

Lemma 6. If $\nabla g_o = 0$, and $T^{(\nabla,\theta)} = 0$, then θ-gauge transform $\theta^{-1}\nabla\theta$ is the Levi-Civita connection of the metric $g = \theta^* g_o$.

So, a metric deformation is equivalent to a deformation of Cartan connections with vanishing Cartan torsion on the vector bundle $E = TM \oplus TM \wedge TM$ with fiexed metric g_o and fixed Lie algebra structure $o(n,1)$ on the fibres. The curvature R of the Levi-Civita connection of Lemma 6, of course, is $R(X,Y)Z = \theta^{-1}(R^\nabla(X,Y)\theta Z)$.

In order to define a suitable deformation of Cartan connections we introduce the adjoint δ^D of the exterior derivative d^D, defined by $D = D_t$, with respect to the variable metric $g = g_t$ on the base, and the fixed metric <,> (defined by g_o) on the fibres of E, compare (2.4).

$$(3.4) \qquad (\delta^D\alpha)(X_2 \ldots X_p) = -\sum_{k=1}^{n} (\nabla_{e_k}\alpha)(e_k, X_2 \ldots X_p) + \sum_{k=1}^{n} [\theta e_k, \alpha(e_k, X_2 \ldots X_p)],$$

where $\{e_k\}$ is an orthonormal basis with respect to the variable metric $g = g_t$. We also write $\delta^D\alpha = \delta_1\alpha + \delta_2\alpha$, where

$$(3.5) \quad (\delta_2\alpha)(X_2 \ldots X_p) = \sum_{k=1}^{n} [\theta e_k, \alpha(e_k, X_2 \ldots X_p)].$$

This explains the definition (2.5). Note that the positive sign occurs because we are using the non-compact Lie algebra $o(n,1)$ as typical fibre in E. For spherical Cartan connections the sign would be negative.

We consider the evolution equation

$$(3.6) \qquad \dot{\omega} = -\delta^D\Omega ,$$

where $\dot{\omega}$ is an E-valued 1-form on M and defines an infinitesimal deformation of the Cartan connection D by $\dot{D}_X s = [\dot{\omega}(X), s]$.

The equation (3.6) is not just formal. The integrability condition $\delta^D(\delta^D\Omega) = 0$ makes it parabolic and by [H, Theorem 5.1] the equation has a solution for $o \leqq t < \varepsilon$ and some $\varepsilon > o$. The equation (3.6) yields the following evolution equation for the Riemannian metric $g = g_t$ on M.

(3.7) $\dot{g} = -2\,\mathrm{Ric}(g) - 2(n-1)g$, which coincides with (2.12) and is Hamilton's equation except for a normalization. The purpose of section 2 was to study this evolution emphasizing the evolution of the Levi-Civita connection. In the present section we stay with Cartan connections.

The evolution equation for the Cartan curvature form Ω is

(3.8) $\dot{\Omega} = -\Delta^D\Omega.$

Next we prove that the evolution (3.6) is tangent to the space of Cartan connections with vanishing Cartan torsion, see definition (3.2). The following result corresponds to (2.8).

Lemma 7. Let $\dot{\omega} = \dot{\eta} + \dot{\theta}$ denote the splitting of the infinitesimal connection form $\dot{\omega}$ of (3.6) into $TM \wedge TM$ and TM components respectively. If the Cartan torsion vanishes, then $d_2\dot{\eta} + d_1\dot{\theta} = 0$.

Proof. Vanishing Cartan torsion means $\Omega_2 = 0$ and Lemma 5 applies. Lemma 6 applies also, i.e., the Levi-Civita connection is the gauge transform by θ of the connection defined by D. Since we are working with the Levi-Civita connection on the base M we can choose vector fields X,Y,Z,e_k in TM which at a given point have vanishing covariant derivative and vanishing (vector fields) bracket. Since the connection ∇ in the fibre is related by the gauge transformation θ to the Levi-Civita connection on the base M, the sections θX, θY, θZ, θe_k in $TM \subset E$ have vanishing covariant derivative with respect to ∇ as well. This simplifies the following computation. Now, $d_1\dot{\theta} = -d_1\delta_2\Omega$, and $d_2\dot{\eta} = -d_2\delta_1\Omega$, and

$$\begin{aligned} -(d_1\delta_2\Omega)(X,Y) &= -\nabla_X[\theta e_k,\Omega(e_k,Y)] + \nabla_Y[\theta e_k,\Omega(e_k,X)] \\ &= -[\theta e_k,\nabla_X\Omega(e_k,Y)] + [\theta e_k,\nabla_Y\Omega(e_k,X)] \\ &= -[\theta e_k,\nabla_{e_k}\Omega(X,Y)], \text{ since } d_1\Omega = 0. \end{aligned}$$

$$\begin{aligned} -(d_2\delta_1\Omega)(X,Y) &= [\theta X,\nabla_{e_k}\Omega(e_k,Y)] - [\theta Y,\nabla_{e_k}\Omega)(e_k,X)] \\ &= [\theta e_k,\nabla_{e_k}\Omega(X,Y)], \quad \text{since } d_2\Omega = 0. \end{aligned}$$

The two terms add up to zero.

The following Lemma states that the evolution defined by (3.6) is tangent to the space of Cartan connections with vanishing Cartan torsion.

Lemma 8. Let ω evolve according to (3.6). If $\Omega_2 = 0$ for a given time t, then $\dot{\Omega}_2 = 0$ at that time.

Proof. $\Omega_2(X,Y) = \nabla_X(\theta Y) - \nabla_Y(\theta X) - \theta[X,Y]$.

$$\dot{\Omega}_2(X,Y) = \dot{\eta}(X)\theta Y + \nabla_X(\dot{\theta}Y) - \dot{\eta}(Y)\theta X - \nabla_Y(\dot{\theta}X) - \dot{\theta}[X,Y]$$
$$= d_2\dot{\eta}(X,Y) + d_1\dot{\theta}(X,Y) = 0 \quad \text{by Lemma 7.}$$

Since the original Cartan connection $D = D^o$ is constructed from the Levi-Civita connection of the original metric g_o, the Cartan torsion remains zero at all times. For this reason, the Laplacian $\Delta^D = d^D\delta^D + \delta^D d^D$ can be conveniently written as the sum of two non-negative operators as the following Lemma states.

Lemma 9. If the Cartan torsion Ω_2 vanishes, then $\Delta^D = \Delta_1 + \Delta_2$, where $\Delta_1 = d_1\delta_1 + \delta_1 d_1$, and $\Delta_2 = d_2\delta_2 + \delta_2 d_2$.

Proof. To simplify the computation we choose vector fields as in the proof of Lemma 7. We prove it for a $TM \wedge TM$-valued 2-form α only to keep things simple. This is the case we need anyway. We have to prove: $(d_1\delta_2 + d_2\delta_1 + \delta_1 d_2 + \delta_2 d_1)\alpha = 0$.

$$(d_1\delta_2\alpha)(X,Y) = \nabla_X[\theta e_k, \alpha(e_k,Y)] - \nabla_Y[\theta e_k, \alpha(e_k,X)]$$

$$(d_2\delta_1\alpha)(X,y) = -[\theta X, \nabla_{e_k}\alpha(e_k,Y)] + [\theta Y, \nabla_{e_k}\alpha(e_k,X)]$$

$$(\delta_1 d_2\alpha)(X,Y) = -\nabla_{e_k}[\theta e_k, \alpha(X,Y)] - \nabla_{e_k}[\theta X, \alpha(Y,e_k)] - \nabla_{e_k}[\theta Y, \alpha(e_k,X)]$$

$$(\delta_2 d_1\alpha)(X,Y) = [\theta e_k, \nabla_{e_k}\alpha(X,Y)] + [\theta e_k, \nabla_X\alpha(Y,e_k)] + [\theta e_k, \nabla_Y\alpha(e_k,X)]$$

Each summand occurs twice with opposite signs if we take the choice of the vector fields into consideration.

The main result of this section in the following simple form of the evolution equation for the Cartan curvature.

Theorem. Assume that the family of Cartan connections $D = D^t$ evolves according to $\dot{\omega} = -\delta^D\Omega$. Then, the Cartan curvature satisfies the parabolic equation

$$\dot{\Omega} = -\Delta^D\Omega.$$

If in addition the initial Cartan connection is torsion free, i.e., $\Omega_2 = 0$ initially, then the torsion remains zero at all times and we have $\dot{\Omega}_1 = -\Delta_1\Omega_1 - \Delta_2\Omega_1$.

REFERENCES

[B] J.P. BOURGUIGNON : Les variêtês de dimension 4 à signature non nulle dont la courbure est harmonique sont d'Einstein; Invent. Math. 63(1981), p. 263-268.

[H] R.S. HAMILTON: Three-manifolds with positive Ricci curvature; J. Diff. Geom., 17(1982), p. 255-306.

[M] C. MARGERIN: Pointwise pinched manifolds are space forms; Preprint, 1984.

[MR] MIN-OO and E.A. RUH: Vanishing theorems and almost symmetric spaces of non-compact type; Math. Ann., 257(1981), p. 419-433.

The first eigenvalue of the Laplacian of an isoparametric minimal hypersurface in a unit sphere

Hideo Muto
Department of Mathematics,
Tokyo Institute of Technology
Oh-okayama 2-12-1
Meguroku, Tokyo 152, Japan

1. Introduction

We announce a result on the first eigenvalues of the Laplacian of some embedded minimal hypersurfaces in a unit sphere [13] . Let M^{n-1} be an (n-1)-dimensional compact connected Riemannian manifold without boundary. We denote the Laplacian acting on smooth functions on M by Δ and its eigenvalues by $\{0=\lambda_0 \le \lambda_1 \le \lambda_2 \le \dots\}$. Let $f : M^{n-1} \to S^N \subset \mathbf{R}^{N+1}$ be an isometric immersion and $(x_1, x_2, \dots, x_{N+1})$ the canonical coordinate system on $\mathbf{R}^{N+1}$. Then, by Takahashi's theorem [18] , M^{n-1} is minimal in S^N if and only if (N+1)-functions $x^i \circ f$ satisfy $\Delta(x_i \circ f) = -(n-1)(x_i \circ f)$.

Our problem is, posed by Ogiue [14] : What kind of embedded closed minimal hypersurfaces in $S^n(1)$ have n-1 as their first eigenvalue? Similar problem is posed by Yau [19] . Choi and Wang [4] showed that the first eigenvalue of every minimally embedded hypersurface in $S^n(1)$ is not less than $(n-1)/2$.

Let M^{n-1} be an isoparametric hypersurface in $S^n(1)$, that is, M^{n-1} has constant principal curvatures in $S^n(1)$. For these hypersurfaces, Münzner obtained beautiful results. Let ν be a unit normal vector field along M in $S^n(1)$, g the number of distinct principal curvatures, $\cot\theta_\alpha$ $(\alpha=0,\dots,g-1,\ 0<\theta_0<\dots<\theta_{g-1}<\pi)$ the principal curvatures with respect to ν and m_α the multiplicity of $\cot\theta_\alpha$. Then Münzner showed that (1) $m_\alpha = m_{\alpha+2}$ (indices mod g) (2) $\theta_\alpha = \theta_0 + \alpha\pi/g$ (3) $g \in \{1,2,3,4,6\}$

Our purpose is to show that some of nonhomogeneous closed minimal hypersurfaces in $S^n(1)$ are embedded by their first eigenfunctions.

Theorem A. Let M^{n-1} be a closed minimal isoparametric hypersurface in $S^n(1)$ satisfying the following condition (C). Then

$$\lambda_1(M^{n-1}) = n-1 .$$

Condition (C) A closed isoparametric hypersurface M in $S^n(1)$ satisfies one of the following conditions for some integer $k \geq 1$.

(1) $g=1$

(2) $g=2$

(3) $g=3$: $(m_0, m_1) = (4,4)$, $(8,8)$.

(4) $g=4$: $(m_0, m_1) = (3,4), (3,8), \ldots, (3,4k), \ldots$

$(4,5)$

$(4,7), (4,11), \ldots, (4,4k+3), \ldots$

$(5,9), (5,10), (5,18), (5,26), (5,34)$

$(6,9), (6,17), (6,25), (6,33)$

$(7,8), (7,16), \ldots, (7,8k), \ldots$

$(8,15), (8,23), (8,31), (8,39)$

$(9,22), (9,38)$

$(10,21), (10,53)$

Remarks (1) It is known [1] that if $g \leq 3$, M is homogeneous, and in particular, that when $g = 1$, $M = S^{n-1}(1)$ and when $g = 2$, M is a generalized Clifford torus $S^p(\sqrt{p/(n-1)}) \times S^q(\sqrt{q/(n-1)})$ $(p+q=n-1)$. But for $g = 4$, Ozeki and Takeuchi [16] and Ferus, Karcher and Münzner [5] found infinite nonhomogeneous minimal isoparametric hypersurfaces. ($(m_0, m_1) = (3,4k), (4,4k-1), (7,8k), \ldots\ldots$)

(2) Closed homogeneous minimal hypersurfaces in $S^n(1)$ are classified in [6] , [17] , [16] . And the pairs (m_0, m_1) of these hypersurfaces are (1,1), (2,2), (4,4), (8,8) when $g = 3$, $(1,k)$, $(2,2k-1)$, $(4,4k-1)$, (2,2), (4,5), (6,9) $(k \geq 1)$ when $g = 4$, (1,1), (2,2) when $g = 6$. By this classification and the representation theory of groups, Muto, Ohnita and Urakawa [12] and Kotani [7] showed that every homogeneous

minimal hypersurface in a unit sphere with $g = 3$ or 6 is embedded by the first eigenfunctions.

We summarize our results. We review Münzner's results [8] , [9] in section two. He showed that every closed isoparametric hypersurface M in a unit sphere $S^n(1)$ has two smooth closed focal embedded submanifolds M_+ and M_- whose codimensions in the sphere are greater than one and M is a normal sphere bundle over M_+ and M_-. Hence M is embedded. In section three, we extend the k-th eigenfunction on M to a suitable function satisfying the Dirichlet boundary condition on a domain in $S^n(1)$ obtained by excluding ε-neighborhoods of two focal submanifolds M_+ and M_- from $S^n(1)$. Using these extended functions, we give a relation between $\lambda_k(M^{n-1})$ and the (k+1)-th eigenvalue $\lambda_{k+1}(\varepsilon)$ of the Laplacian on the domain under the Dirichlet boundary condition ($k=0,1,2,\ldots$). And this implies Theorem B in setion three which shows that there is a constant G depending on m_0, m_1, g and θ_0 satisfying $\lambda_k(M^{n-1}) \geq G\,\lambda_k(S^n(1))$. In section four, we show that for every hypersurface in Theorem A, $\lambda_{n+2}(M^{n-1}) > n-1 = \dim M$ using Theorem B. Since every minimal hypersurface fully immersed in $S^n(1)$ which is not isometric to $S^{n-1}(1)$ has its dimension as an eigenvalue with multiplicity $\geq (n+1)$, the first eigenvalue of M^{n-1} must be equal to its dimension $n-1$ with multiplicity $n+1$.

2. An isoparametric hypersurface in a unit sphere.

In this section, we recall Münzner's results ([8] , [9]).

Let $f : M^{n-1} \dashrightarrow S^n(1)$ $(\subset \mathbf{R}^{n+1})$ be an isoparametric hypersurface of $S^n(1)$. Let ν be a unit normal vector field along M in $S^n(1)$. And let E^α ($\alpha=0,\ldots,g-1$) be the eigenspace of the shape operator with eigenvalues $\cot\theta_\alpha$ $(0<\theta_0<\ldots<\theta_{g-1}<\pi)$ with respect to ν. We define $f_\theta : M \dashrightarrow S^n(1)$ $(-\pi<\theta<\pi)$ by the following: for any $p \in M$,

$$f_\theta(p) = \exp_{f(p)}\theta\nu ,$$

$$// \cos\theta\, f(p) + \sin\theta\, \nu .$$

Here $x // y$ means that x and y are parallel as vectors in $\mathbf{R}^{n+1}$.

Theorem 1. (Münzner [8] , [9]) Let M be a closed isoparametric hypersurface in $S^n(1)$. Then (1)-(5) hold:

(1) $\theta_\alpha=\theta_0+\alpha\pi/g$ $(\alpha=0,1,\dots,g-1)$,

(2) $m_\alpha=m_{\alpha+2}$ (indices mod g),

(3) $g\in\{1,2,3,4,6\}$ and $2g=\dim_{\mathbf{R}}H^*(M:\mathbf{R})$,

(4) Set $M_+=f_{\theta_0}(M)$ (resp. $M_-=f_{-\pi+\theta_{g-1}}(M)=f_{-\frac{\pi}{g}+\theta_0}(M)$). Then M_+ (resp. M_-) is a smooth embedded closed submanifold of dimension $(n-m_0-1)$ (resp. $(n-m_1-1)$). Moreover, set $\widetilde{M}_+=\cup_{\theta\in[0,\theta_0]}f_\theta(M)$ (resp. $\widetilde{M}_-=\cup_{\theta\in[\theta_0-\frac{\pi}{g},0]}f_\theta(M)$). Then $\widetilde{M}_+$ (resp. $\widetilde{M}_-$) is a normal disk bundle over M_+ (resp. M_-) induced by f_θ and $\widetilde{M}_+$ and $\widetilde{M}_-$ satisfy $\widetilde{M}_+\cup\widetilde{M}_-=S^n(1)$ and $\widetilde{M}_+\cap\widetilde{M}_-=M$.

(5) $f_\theta(M)$ $(\theta\in(-\frac{\pi}{g}+\theta_0,\theta_0))$ is an isoparametric hypersurface which is diffeomorphic to M.

We prepare some formula (see Münzner [8]).

(2.1) For $X\in E^\alpha$,

$$f_{\theta_*}X=\frac{\sin(\theta_\alpha-\theta)}{\sin(\theta_\alpha)}\widetilde{X}\ .$$

Here $\widetilde{X}$ // X .

(2.2) Set $t(p)=\mathrm{dist}(p,M_+)$ for any $p\in S^n(1)$. If $p\in f_\theta(M)$ $(\theta\in[-\frac{\pi}{g}+\theta_0])$, then $t(p)=\theta_0-\theta$.

(2.3) Let h be the mean curvature of M^{n-1} in $S^n(1)$ with respect to ν. Then

$$(n-1)\,h=\begin{cases} m_0 g\cot(gt)\ , & (g:\text{odd or } m_0=m_1),\\ \frac{m_0 g}{2}\cot\frac{gt}{2}-\frac{m_1 g}{2}\tan\frac{gt}{2}\ , & (g:\text{even or } m_0\neq m_1). \end{cases}$$

3. Volume elements and eigenvalues

Let M^{n-1} be a closed isoparametric hypersurface in $S^n(1)$ and D a domain in M with a smooth boundary H. We use the same notations as in section two. Set $D_\theta = f_\theta(D)$, $H_\theta = f_\theta(H)$ and

$$\widetilde{D}_+ = \cup_{\theta\in[0,\theta_0]} D_\theta , \quad \widetilde{D}_- = \cup_{\theta\in[-\frac{\pi}{g}+\theta_0, 0]} D_\theta ,$$

$$\widetilde{H}_+ = \cup_{\theta\in[0,\theta_0]} D_\theta , \quad \widetilde{H}_- = \cup_{\theta\in[-\frac{\pi}{g}+\theta_0, 0]} H_\theta ,$$

$$\widetilde{D} = \widetilde{D}_+ \cup \widetilde{D}_- , \quad \widetilde{H} = \widetilde{H}_+ \cup \widetilde{H}_- .$$

Then, by the construction of $\widetilde{D}$ and $\widetilde{H}$, we easily have the following Lemma 1.

Lemma 1. We have

$$\partial\widetilde{D} = \widetilde{H} , \quad \widetilde{H}_+ \cap \widetilde{H}_- = H , \text{ and } \widetilde{D}_+ \cap \widetilde{D}_- = D .$$

Lemma 2. (see [13]). Let M be a closed isoparametric hypersurface in $S^n(1)$ with $g\geq 2$. For $\theta \in (-\pi/g+\theta_0, \theta_0)$, let g_θ be the Riemannian metric of $M_\theta = f_\theta(M)$ induced from $S^n(1)$ and dM_θ the volume element of M_θ. Set $\ell = g/2$. Then we have

$$f_\theta^* \, dD_\theta = F(\theta)\, dD_0 . \tag{3.1}$$

Here

$$F(\theta) = \frac{\sin^{m_0}\ell(\theta_0-\theta)\cos^{m_1}\ell(\theta_0-\theta)}{\sin^{m_0}\ell\theta_0 \cos^{m_1}\ell\theta_0} .$$

Remark 1. Lemma 2 implies that $\mathrm{vol}(D)/\mathrm{vol}(M) = \mathrm{vol}(\widetilde{D})/\mathrm{vol}(S^n(1))$. Therefore, we can estimate Cheeger's isoperimetric constant of M from below in terms of such a constant of $S^n(1)$ (see [13]).

Remark 2. We can show that every closed isoparametric hypersurface M in a unit sphere $S^n(1)$ is tight, that is, every non-degenerate function $\xi_p(x)=(f(x),p)$ $(x\in M,\ p\in S^n(1))$ has the minimum number of critical points required by the Morse inequalities where (,) denotes the canonical Euclidean inner product. Because we can calculate the absolute total curvature and show that it coincides with $\dim H^*(M;Z) = 2g$ (see Theorem 1 (3)). This fact was first proved by Cecil and Ryan [2] through the other method.

Theorem B. Let M be a closed isoparametric hypersurface satisfying $g \geqq 2$ and $\min(m_0, m_1) \geqq 2$. Then for any $k=0,1,2,\ldots$, we have

$$\lambda_k(M^{n-1}) \geqq G(g:m_0,m_1,\theta_0)\, \lambda_k(S^n(1)) .$$

Here $\ell = g/2$ and

$$G(g:m_0,m_1,\theta_0) = \frac{G_1(m_0,m_1)}{G_2(g:m_0,m_1,\theta_0) + G_3(g:m_0,m_1,\theta_0)} ,$$

$$G_1(m_0,m_1) = \int_0^{\frac{\pi}{2}} \sin^{m_0} x \cos^{m_1} x \, dx ,$$

$$G_2(g:m_0,m_1,\theta_0) = \sin^2\theta_0 \int_0^{\ell\theta_0} \frac{\sin^{m_0} x}{\sin^2\frac{x}{\ell}} \cos^{m_1} x \, dx ,$$

$$G_3(g:m_0,m_1,\theta_0) = \sin^2\left(\frac{\pi}{2\ell} - \theta_0\right)\int_0^{\frac{\pi}{2}-\ell\theta_0} \frac{\sin^{m_1} x}{\sin^2\frac{x}{\ell}} \cos^{m_0} x \, dx .$$

Moreover, when M is minimal, then $\ell\theta_0 = \operatorname{arc\,cot} \sqrt{m_1/m_0}$.

Proof. For sufficiently small $\varepsilon > 0$, set

$$M(\varepsilon) = \bigcup_{\theta\in[-\frac{\pi}{g}+\theta_0+\varepsilon,\,\theta_0-\varepsilon]} f_\theta(M) .$$

Then, by Münzner's theorem (Theorem 1), $M(\varepsilon)$ is a domain of $S^n(1)$ obtained by excluding ε-neighborhood of M_+ and M_- from $S^n(1)$ and is diffeomorphic to $M \times [-\pi/g + \theta_0+\varepsilon, \theta_0-\varepsilon]$ under $f_\theta(p)$. By the mini-max principle, we have $\lambda_k^D(M(\varepsilon)) \geqq \lambda_{k-1}(S^n(1))$ $(k=1,2,\ldots)$ (see, in detail, [3] [15],). Therefore, we may estimate $\lambda_k^D(M(\varepsilon))$ from above in terms of $\lambda_{k-1}(M^{n-1})$. Let $\{X_{\alpha,i} : i=1,\ldots,m_\alpha,\ \alpha=0,\ldots,g-1,\ X_{\alpha,i}\in E^\alpha\}$ be a local orthonormal frame field on M. Then $\{\partial/\partial\theta$, $\sin\theta_\alpha/\sin(\theta_\alpha-\theta)\, X_{\alpha,i} : i=1,\ldots,m_\alpha$, $\alpha=0,\ldots,g-1$, $X_{\alpha,i}\in E^\alpha$, $\theta\in[-\frac{\pi}{g}+\theta_0+\varepsilon, \theta_0-\varepsilon]\ \}$ is a local orthonormal frame field on $M(\varepsilon)$ by the diffeomorphisms f_θ. (3.1) implies that the volume element $dM(\varepsilon)$ of $M(\varepsilon)$ is represented by the following:

$$dM(\varepsilon) = \frac{\sin^{m_0}\ell(\theta_0-\theta)\cos^{m_1}\ell(\theta_0-\theta)}{\sin^{m_0}\ell\theta_0 \cos^{m_1}\ell\theta_0} \, d\theta \, dM . \tag{3.2}$$

Let $\{f_k\}$ $(k=0,1,\ldots)$ be the eigenfunctions on M which are orthogonal to each other with respect to the square integral inner product on M. Let Ψ be a nonnegative, non-decreasing smooth function on $[0,\infty)$ satisfying $\Psi=1$ on $[2,\infty)$ and $\Psi=0$ on $[0,1]$. For sufficiently small $\eta > 0$, let ψ_η be a nonnegative smooth function on $[\eta,\frac{\pi}{2}-\eta]$ such that (1) $\psi_\eta(\eta)=\psi_\eta(\frac{\pi}{2}-\eta)=0$ (2) ψ_η is symmetric with respect to $x=\frac{\pi}{4}$ and (3) $\psi_\eta(x)=\Psi(\frac{x}{\eta})$ on $[\eta,\pi/4]$. Let L_k be the space of functions spanned by $\{f_0, f_1, \ldots, f_k\}$ $(k \geq 0)$. For any $\varphi \in L_k$, define a function Φ_ε on $M(\varepsilon)$ by

$$\Phi_\varepsilon(x,\theta) = \psi_{\ell\varepsilon}(\ell(\theta_0-\theta))\varphi(x) .$$

Then Φ_ε is a smooth function on $M(\varepsilon)$ satisfying the Dirichlet boundary condition and is square integrable on $M(\varepsilon)$. By (2.1), (3.2), we see that

$$\begin{aligned}
&\frac{\|d\Phi_\varepsilon\|_2^2}{\|\Phi_\varepsilon\|_2^2} \\
&\leq \ell^2 \frac{\int_{\ell\varepsilon}^{\frac{\pi}{2}-\ell\varepsilon} \psi_{\ell\varepsilon}'(x)^2 \sin^{m_0}x \cos^{m_1}x\,dx}{\int_{\ell\varepsilon}^{\frac{\pi}{2}-\ell\varepsilon} \psi_{\ell\varepsilon}(x)^2 \sin^{m_0}x \cos^{m_1}x\,dx} \\
&\quad + \frac{1}{\int_{\ell\varepsilon}^{\frac{\pi}{2}-\ell\varepsilon} \psi_{\ell\varepsilon}(x)^2 \sin^{m_0}x \cos^{m_1}x\,dx} \times \frac{\|d\varphi\|_2^2}{\|\varphi\|_2^2} \\
&\quad \times \Big(\sin^2\theta_0 \int_{\ell\varepsilon}^{\ell\theta_0} \psi_{\ell\varepsilon}^2(x) \frac{\sin^{m_0}x}{\sin^2\frac{x}{\ell}} \cos^{m_1}x\,dx \\
&\qquad + \sin^2\Big(\frac{\pi}{2\ell}-\theta_0\Big) \int_{\ell\theta_0}^{\frac{\pi}{2}-\ell\varepsilon} \psi_{\ell\varepsilon}^2(x) \sin^{m_0}x \frac{\sin^{m_1}x}{\sin^2\frac{1}{\ell}(\frac{\pi}{2}-x)}\,dx \Big).
\end{aligned}$$

By the condition $\min(m_0,m_1) \geq 2$, we see that the first term in the right hand side tends to zero as ε tends to zero. Therefore, by the mini-max principle, we see that

$$\lambda_k(S^n(1)) \leq \lim_{\varepsilon\to 0} \lambda^D_{k+1}(M(\varepsilon)) ,$$

$$\leq \lim_{\varepsilon\to 0} \sup_{\varphi\in L_k} \frac{\|d\Phi_\varepsilon\|^2}{\|\Phi_\varepsilon\|^2}$$

$$\leq \lambda_k(M^{n-1}) \frac{G_2(g:m_0,m_1,\theta_0)+G_3(g:m_0,m_1,\theta_0)}{G_1(m_0,m_1)} .$$

We have the required inequality. Moreover, when M is minimal, $\ell\theta_0$ must be equal to arc tan $\sqrt{m_0/m_1}$ by (2.3).

3. Proof of Theorem A.

When g = 1 or 2, it is known (see [1]) that $M^{n-1} = S^{n-1}(1)$ or $S^p(\sqrt{p/(n-1)}) \times S^q(\sqrt{q/(n-1)})$ (p+q=n-1) and the first eigenvalue of M must be equal to its dimension. When g = 3 and $(m_0,m_1) = (4,4)$ or (8,8), Kotani (see [7]) first showed that $\lambda_1(M^{n-1}) = n-1$. We also prove this fact by our method. To prove Theorem A, we may assume $m_0 \leq m_1$ and show that $\lambda_{n+2}(M^{n-1}) > n-1 = \dim M$ for each closed isoparametric minimal hypersurface M^{n-1} of $S^n(1)$ satisfying condition (C). Because the multiplicity of the eigenvalue n-1 of every minimal submanifold M^{n-1} fully immersed in $S^N(1)$ which is not isometric to the unit sphere is not less than N+1 (Takahashi's theorem, see [18]). From Theorem B, we may show that $G(g:m_0,m_1,\theta_0)\lambda_{n+2}(S^n(1)) > n-1$ in each case of condition (C). We notice here that $\ell(n-1) = m_0 + m_1$ because $m_\alpha = m_{\alpha+2}$ (indices mod g) by Theorem 1 (2).

Let M^{n-1} be a closed isoparametric minimal hypersurface of $S^n(1)$ with g = 4,that is, $2\theta_0$ = arc cot $\sqrt{m_1/m_0}$, and satisfies one of the following : (m_0,m_1) = (4,7), (4,11),..., (4,4k-1),..., (k ≧ 2) . We first show that $G(4:4,m,\theta_0) > 1/2$ for any m ≧ 34 and, by Theorem B, that $G(4:4,m,\theta_0)\lambda_{n+2}(S^n(1)) > 1/2\times 2(n+1) > n-1$ for m ≧ 34 . And for the other cases (g = 4, m_0 = 4), we can verify the inequality $G(4:4,m,\theta_0)\lambda_{n+2}(S^n(1)) > n-1$ by using a computer. We assume here that m ≧ 34. Since M is minimal, (2.3) implies that $2\theta_0$= arc cot $\sqrt{4/m}$. Set $\eta = 2\theta_0$, $A = \sin\eta = 2/\sqrt{m+4}$ and $B = \cos\eta = \sqrt{m}/\sqrt{m+4}$. Then we have :

$$\sin^2\theta_0 = \frac{m}{2(m+4)} \frac{1}{1+B} , \quad \sin^2(\frac{\pi}{4} - \theta_0) = \frac{4}{2(m+4)} \frac{1}{1+A} .$$

Set $J(m) = \int_0^{\pi/2} \sin^m dx$. Then we see that

$$G(4:4,m,\theta_0)^{-1}\ 3\ J(m)$$

$$\leq \frac{4}{1+B}\Big[\int_0^{2\theta_0}\cos^m x dx + \frac{m+2}{m+3}\int_0^{2\theta_0}\cos^{m+1} x dx$$

$$- AB^{m+1} - \frac{m+2}{m+3}AB^{m+2}\Big]$$

$$+ \frac{1}{1+B}\Big[\frac{8(m+2)(5m^2+14m+6)}{(m^2-1)(m+3)(m+4)}B^{m+1} + \frac{7m+8}{m-1}AB^{m+1}$$

$$+ \frac{3m}{m-1}\int_0^{\pi/2-2\theta_0}\sin^m x dx\Big] .$$

Thererfore we have

$$3\ G(4:4,m,\theta_0)^{-1}$$

$$< \frac{8}{1+m} + \Big[\frac{2(7m+8)}{m^2-1} - \frac{8}{1+B}\frac{(2m+5)m}{(m+3)(m+4)}\Big]\sqrt{2/\pi}B^{m+1}$$

$$+\frac{8}{1+A}\frac{(m+2)(5m^2+14m+6)}{(m^2-1)(m+3)(m+4)}\frac{m}{\sqrt{m-1}}\sqrt{2/\pi}B^{m+1}$$

Since $G(4:4,m,\theta_0)^{-1} > 1/2$ when $m \geq 34$, we have $\lambda_1(M^{n-1}) > n-1$. For $m \leq 33$, we can directly show that $G(4:4,m,\theta_0) > 1/2$ by using a computer. We use the double exponential formula (see [11]) and the language of the program is FORTRAN. A subroutine program using the double exponential formula is written in an appendix of a book [10]. This is a subroutine program to integrate an analytic function on $(-1, 1)$ or $(0, \infty)$ and has an absolute error 10^{-16} . But it is easy to make a partial revision of this program so that we have relatively very small errors which depend on this program and our machine. For example, $G(4:4,7,\theta_0) = 0.5424236654321133 > 1/2$, $G(4:4,33,\theta_0) = 0.6146834883261047 > 1/2$. By these computations, we obtain the required inequality for $m_0 = 4$.

Set $D(g:m_0,m_1) = G(g:m_0,m_1,\theta_0) - (n-1)$. Then, by the similar estimate, we have that $G(4:m,3,\theta_0) > 1/2$ for any $m \geq 46$ and $G(4:m,7,\theta_0) > 1/2$ for any $m \geq 36$ and by a computer, we have, for example, $G(4:4,3,\theta_0) = 0.441152996992993$, $D(4:4,3) = 0.1168863903775783 > 0$, $G(4:5,4,\theta_0) = 0.5110829726081493 > 1/2$, $G(4:8,7,\theta_0) = 0.6258833686366021 > 1/2$, and $G(4:35,7,\theta_0) = 0.7061682378135796 > 1/2$.

For the other cases, we directly compute G and D , for example, $G(4:6,9,\theta_0) = 0.6185633191383751 > 1/2$, $G(4:8,15,\theta_0) = 0.6856918246775244 > 1/2$, $G(3:4,4,\theta_0) = 0.495059684\ldots$, $D(3:4,4) =$

1.86167115... , $G(3:8,8,\theta_0) = 0.648727497... > 1/2$. Therefore we have theorem A.

Remark. We have the limits $G(4:m_1)$ of $\lim_{m_0\to\infty} G(4:m_0,m_1)$ ($m_1=3,4,7$) as follows: $G(4:3) = 1/(1.5+e^{-3/2}) = 0.5803392124 > 1/2$, $G(4:4) \geq 3/(4+6e^{-2}\sqrt{27/\pi}) = 0.64545394898 > 1/2$ and $G(4:7) = 0.7524581288 > 1/2$.

References

[1] E. Cartan, Sur des familles remarquables d'hypersurfaces isoparamétriques dans les espaces sphériques, Math. Z., **45** (1939), 335-367.

[2] T. Cecil and R. Ryan, Tight spherical embeddings, Lect. Notes in Math., No.838, Springer-Verlag, Berlin, Heidelberg, New York, 1981, 94-103.

[3] I. Chavel and E. A. Feldman, Spectra of domains in compact manifolds, J. Funct. Analysis, **30** (1978), 198-222.

[4] H. I. Choi and A. N. Wang, A first eigenvalue estimate for minimal hypersurfaces, J. Differential Geom., **18** (1983), 559-562.

[5] D. Ferus, H. Karcher, and H. F. Münzner, Cliffordalgebren und neue isoparametrische Hyperflächen, Math. Z., **177** (1981), 479-502.

[6] W. Y. Hsiang and H. B. Lawson Jr., Minimal submanifolds of low cohomogeneity, J. Differential Geom., **5** (1971), 1-36.

[7] M. Kotani, The first eigenvalue of homogeneous minimal hypersurfaces in a unit sphere $S^{n+1}(1)$. preprint

[8] H. F. Münzner, Isoparametrische Hyperflächen in Sphären, Math. Ann., **256** (1981), 57-71.

[9] H. F. Münzner, Isoparametrische Hyperflächen in Sphären, II, Math. Ann., **256** (1981), 215-232.

[10] M. Mori, Curves and Surfaces, Kyoiku-shuppan, Tokyo, 1984. (in Japanese)

[11] M. Mori and H. Takahashi, Double exponential formulas for numerical integration, Publ. RIMS. Kyoto Univ., **9** (1974), 721-741.

[12] H. Muto, Y. Ohnita, and H. Urakawa, Homogeneous minimal hypersurfaces in the unit sphere and the first eigenvalue of the Laplacian, Tôhoku Math. J., **36** (1984), 253-267.

[13] H. Muto, The first eigenvalue of the Laplacian of an isoparametric minimal hypersurface in a unit sphere. preprint

[14] K. Ogiue, Open Problems, Geometry of the Laplace Operator, ed. T. Ochiai, 1980/1981. (in Japanese)

[15] S. Ozawa, Singular variation of domains and eigenvalues of the Laplacian, Duke Math. J., **48** (1981), 767-778.

[16] H. Ozeki and M. Takeuchi, On some types of isoparametric hypersurfaces in spheres, I, Tôhoku Math. J., **27** (1975), 515-559.

[17] R. Takagi and T. Takahashi, On the principal curvatures of homogeneous hypersurfaces in a unit sphere, Differential Geometry, Kinokuniya, Tokyo, 1972, 469-481.

[18] T. Takahashi, Minimal immersions of Riemannian manifolds, J. Math. Soc. Japan, **18** (1966), 380-385.

[19] S. T. Yau, Problem Section, Seminar on Differential Geometry, Annals of Math. Studies, No.102 , Princeton Univ. Press, Princeton, 1982.

ON DEFORMATION OF RIEMANNIAN METRICS AND MANIFOLDS WITH POSITIVE CURVATURE OPERATOR

Seiki Nishikawa
Department of Mathematics
Kyushu University 33
Fukuoka 812, Japan

§1. Introduction

1.1. Let (M, g) be a Riemannian n-manifold with Riemannian metric $g = (g_{ij})$. By $Rm = (R_{ijk\ell})$, $Rc = (R_{ij})$ and R, we denote, respectively, the Riemannian curvature tensor, the Ricci curvature tensor and the scalar curvature of (M, g).

Let $x \in M$ and V be the tangent space of M at x. We identify V and its dual V^* via the metric g. Let $\Lambda^2 V$ denote the space of skewsymmetric 2-tensors over V. Then the Riemannian curvature tensor Rm defines a linear endomorphism $\hat{R}$, which we call the *curvature operator of first kind*, of $\Lambda^2 V$ by

$$\hat{R} : \omega = (\omega_{ij}) \mapsto \hat{R}(\omega) = (-R_{ijk\ell}\omega^{k\ell}).$$

From the symmetries of Rm it follows that $\hat{R}$ is a selfadjoint endomorphism of $\Lambda^2 V$, with respect to the inner product naturally defined on $\Lambda^2 V$ by the metric g. Hence the eigenvalues of $\hat{R}$ are all real. We note that if the eigenvalues of $\hat{R}$ are $\geqq \lambda$ then the sectional curvatures at x are $\geqq \lambda/2$. $\hat{R}$ is said to be *positive* (resp. *nonnegative*), or simply $\hat{R} > 0$ (resp. $\hat{R} \geqq 0$), if all the eigenvalues of $\hat{R}$ are positive (resp. nonnegative).

1.2. Let $S^2 V$ denote the space of symmetric 2-tensors over V. Then the Riemannian curvature tensor Rm defines also a linear endomorphism $\mathring{R}$, which we call the *curvature operator of second kind*, of $S^2 V$ by

$$\mathring{R} : \zeta = (\zeta_{ij}) \mapsto \mathring{R}(\zeta) = (R_{ikj\ell}\zeta^{k\ell}).$$

Again the symmetries of Rm imply that $\mathring{R}$ is a selfadjoint endomorphism of $S^2 V$, with respect to the natural inner product on $S^2 V$. However, contrary to the case of $\hat{R}$, we see that if the eigenvalues of

$\mathring{R}$ are $\geqq 0$ then the sectional curvatures at x must be zero, that is, $Rm \equiv 0$.

The new feature here is that the space S^2V is not irreducible under the action of the orthogonal group $O(V)$ of V. Let S_0^2V denote the space of *traceless* symmetric 2-tensors over V. Then S^2V splits into $O(V)$-irreducible subspaces as $S^2V = S_0^2V \oplus \mathbb{R}\cdot g$. We note that if the eigenvalues of $\mathring{R}$ restricted to S_0^2V are $\geqq \lambda$ then the sectional curvatures at x are $\geqq \lambda$. Thus, we say $\mathring{R}$ is *positive* (resp. *nonnegative*), or simply $\mathring{R} > 0$ (resp. $\mathring{R} \geqq 0$), if all the eigenvalues of $\mathring{R}$ *restricted* to S_0^2V are positive (resp. nonnegative). It should be noted that $\mathring{R}$ does not preserve the subspace S_0^2V in general, but it does, for instance, when g is an Einstein metric.

1.3. Let (S^n, g_0) be a Euclidean n-sphere with standard metric g_0. Then, at each $x \in M$, $\hat{R}$ is identical with $2R/n(n-1) \times$ (the identity map) on Λ^2V, and $\mathring{R}$ with $R/n(n-1) \times$ (the identity map) on S_0^2V. Thus, on (S^n, g_0), the curvature operators $\hat{R}$ and $\mathring{R}$ are both positive everywhere.

Next, let $(\mathbb{C}P^n, g_0)$ be a complex projective n-space with Fubini-Study metric g_0. Then, at each $x \in M$, $\hat{R}$ has two positive eigenvalues and one zero eigenvalue on Λ^2V, and $\mathring{R}$ has one positive eigenvalue and one negative eigenvalue on S_0^2V (cf.[4]). Hence, on $(\mathbb{C}P^n, g_0)$, the curvature operator $\hat{R}$ is nonnegative (but not positve) everywhere, while the curvature operator $\mathring{R}$ is neither positive nor nonnegative everywhere.

1.4. Let (M, g) be a *compact* Riemannian n-manifold. Then there have been the following two conjectures.

CONJECTURE I. *If* $\hat{R} > 0$ *everywhere, then* M *is diffeomorphic to a spherical space form* S^n/Γ. *(More generally, if* $\hat{R} \geqq 0$ *everywhere, then* M *is diffeomorphic to a Riemannian locally symmetric space.)*

CONJECTURE II. *If* $\mathring{R} > 0$ *everywhere, then* M *is diffeomorphic to a spherical space form* S^n/Γ. *(More generally, if* $\mathring{R} \geqq 0$ *everywhere, then* M *is diffeomorphic to a Riemannian locally symmetric space.)*

The aim of this note is to summarize what is known so far about these conjectures and what needs to be done.

§2. Manifolds with positive curvature operator

2.1. Since the positivity (resp. nonnegativity) of each curvature operator implies the positivity (resp. nonnegativity) of the sectional curvatures, it is obvious that Conjectures I and II hold in dimension 2.

Also, in dimension 3, we can see from the classification of compact 3-manifolds with positive Ricci curvature due to Hamilton[7] together with a remark in Bando[2] that Conjectures I and II are both true.

Recently, Hamilton[8] has succeeded in proving Conjecture I affirmatively in dimension 4.

2.2. However, in the case of higher dimensions, we know only a few general results supporting the conjectures. Among them we first note the following

THEOREM 1(Berger[3], Meyer[13]). *Let* (M, g) *be a compact Riemannian n-manifold for which* $\hat{R} > 0$ *everywhere. Then* M *is a real homology n-sphere.*

Indeed, using harmonic theory, Berger proved $H^2(M, \mathbb{R}) = 0$ and then Meyer improved it to $H^k(M, \mathbb{R}) = 0$ for all $k \neq 0, n$.

The counterpart for the curvature operator $\mathring{R}$ is also known.

THEOREM 2(Ogiue-Tachibana[16]). *Let* (M, g) *be a compact Riemannian n-manifold for which* $\mathring{R} > 0$ *everywhere. Then* M *is a real homology n-sphere.*

2.3. On the other hand, for Einstein manifolds both conjectures are known to be true. In fact, we have the following

THEOREM 3(Tachibana[19]). *Let* (M, g) *be a compact Einstein n-manifold* $(n \geqq 3)$. *If* $\hat{R} > 0$ *everywhere, then* (M, g) *is a space of constant curvature. More generally, if* $\hat{R} \geqq 0$ *everywhere, then* (M, g) *is a Riemannian locally symmetric space.*

THEOREM 4(Kashiwada[10]). *Let* (M, g) *be a compact Einstein n-manifold* $(n \geqq 3)$. *If* $\mathring{R} > 0$ *everywhere, then* (M, g) *is a space of constant curvature. More generally, if* $\mathring{R} \geqq 0$ *everywhere, then* (M, g) *is a Riemannian locally symmetric space.*

2.4. A sketch of the proof of Theorem 3 goes as follows. Let $Z = (Z_{ijk\ell})$ denote the concircular curvature tensor of (M, g), that is

$$Z_{ijk\ell} = R_{ijk\ell} - \frac{R}{n(n-1)}(g_{i\ell}g_{jk} - g_{ik}g_{j\ell}) ,$$

which is the scalar curvature free part of the Riemannian curvature tensor. Note that g is a metric of constant curvature if and only if $Z \equiv 0$. We then calculate the Laplacian of the squared norm $|Z|^2 = Z_{ijk\ell}Z^{ijk\ell}$ of Z. Since $\nabla_p R = 0$ and $\nabla^p R_{pjk\ell} = 0$ on an Einstein manifold, it follows from the second Bianchi identity and the Ricci formula that

$$\frac{1}{2}\Delta|Z|^2 = \Delta R_{ijk\ell}\cdot Z^{ijk\ell} + |\nabla Rm|^2$$
$$= P + |\nabla Rm|^2 ,$$

where

$$P = 2R_{pq}R^{pjk\ell}R^{q}{}_{jk\ell} + R^{ijk\ell}R_{ij}{}^{pq}R_{pqk\ell} + 4R^{ijk\ell}R^{p}{}_{i}{}^{q}{}_{k}R_{pjq\ell} .$$

Now, for each fixed p,q,r,s, we define a local skewsymmetric 2-tensor field $\omega^{(pqrs)} = (\omega_{ij}^{(pqrs)})$ by

$$\omega_{ij}^{(pqrs)} = R_{iqrs}g_{jp} + R_{pirs}g_{jq} + R_{pqis}g_{jr} + R_{pqri}g_{js}$$
$$- R_{jqrs}g_{ip} - R_{pjrs}g_{iq} - R_{pqjs}g_{ir} - R_{pqrj}g_{is} .$$

A straightforward computation then gives

$$\sum_{p,q,r,s} < \hat{R}(\omega^{(pqrs)}), \omega^{(pqrs)} > = 8P ,$$
$$\sum_{p,q,r,s} < \omega^{(pqrs)}, \omega^{(pqrs)} > = 8(n-1)|Z|^2 .$$

Since M is compact and $\hat{R} > 0$ (resp. $\geqq 0$) everywhere, there is a constant $\varepsilon > 0$ (resp. $\geqq 0$) such that

$$< \hat{R}(\omega), \omega > \geqq \varepsilon|\omega|^2$$

holds for any skewsymmetric 2-tensor field ω on M. Hence, using the relations above, we finally obtain

$$\frac{1}{2}\Delta|Z|^2 \geqq (n-1)\varepsilon|Z|^2 + |\nabla Rm|^2 ,$$

which shows that $|Z|^2$ is a subharmonic function on M and hence is constant. In consequence, we see $Z \equiv 0$ (resp. $\nabla Rm \equiv 0$) if $\hat{R} > 0$ (resp. $\geqq 0$).

Theorem 4 can be proved in a similar fashion.

§3. Deformation of Riemannian metrics

3.1. Theorems 3 and 4 provide a recipe for the proof of Conjectures I and II. Indeed, it is suggested to try to deform a given Riemannian metric to an Einstein metric preserving the positivity of the curvature operators. To achieve this, it is also suggested to use the following parabolic Einstein equation

$$\frac{\partial}{\partial t} g_{ij} = -2R_{ij} + \frac{2}{n} r\, g_{ij} \tag{1}$$

where $r = \int R\, d\mu / \int d\mu$ is the average of the scalar curvature R. This equation was first studied by Hamilton[7] and has stemmed from the variational problem on the total scalar curvature functional in the following fashion.

3.2. Let M be a compact n-manifold and $\mathcal{M}(M)$ denote the set of Riemannian metrics g on M. We consider the total scalar curvature of (M, g),

$$I(g) = \int_M R(g)\, d\mu(g)$$

as a functional defined on $\mathcal{M}(M)$.

Let $\mathcal{H}(M)$ denote the subset of $\mathcal{M}(M)$ consisting of those metrics with a fixed total volume. A classical theorem due to Hilbert then says that $g \in \mathcal{H}(M)$ is an Einstein metric if and only if g is a critical point of the functional $I|\mathcal{H}(M)$, the restriction of I to $\mathcal{H}(M)$, though in general it is not a local maximum nor minimum (cf.[11, 14]).

Let us now calculate the gradient field of the functional $I|\mathcal{H}(M)$ formally. Projecting the vector field obtained from the first variation of I onto $\mathcal{H}(M)$, with respect to the L^2 inner product on tensor fields on M, we then get

$$[\operatorname{grad}(I|\mathcal{H}(M))]_{ij}(g) = -R_{ij} + \frac{1}{n} r\, g_{ij} + \frac{1}{2}(R-r) g_{ij}$$

as the gradient field of $I|\mathcal{H}(M)$ (cf.[15]).

Unfortunately the evolution equation

$$\frac{\partial}{\partial t} g_{ij} = [\operatorname{grad}(I|\mathcal{H}(M))]_{ij}(g) ,$$

which would define the gradient flow of $I|\mathcal{H}(M)$, will not have solutions in general even for a short time, since it implies a backward heat equation for R. Nevertheless, let α and β be real numbers and consider the evolution equation

$$\frac{\partial}{\partial t} g_{ij} = \alpha\left(-R_{ij} + \frac{1}{n} r\, g_{ij}\right) + \beta(R-r) g_{ij} \tag{2}$$

which also would define a flow on $\mathcal{H}(M)$. Then the following holds.

THEOREM 5([7, 15]). *If* $\alpha > 0$ *and* $\beta < \alpha/2(n-1)$, *then the evolution equation* (2) *has a unique solution for a short time on any compact n-manifold* M *with any initial metric* g *at* $t = 0$.

It should be noted that the evolution equation (2) with $\alpha > 0$ and $\beta < \alpha/2(n-1)$ is almost, but not strictly parabolic; the invariance of (2) under the action of the diffeomorphism group of M leads to the degeneracy. The proof of Theorem 5 employs the existence theorem of Hamilton[6](cf.[5] also), which is based on the Nash-Moser inverse function theorem, for weakly parabolic systems.

3.3. To prove the long time existence of solutions for the initial value problem for the parabolic Einstein equation (1), we need several a priori estimates on the evolution of curvatures. To derive these estimates, we may deal with the unnormalized evolution equation

$$\frac{\partial}{\partial t} g_{ij} = -2R_{ij} \tag{3}$$

which is easier to handle. A solution of (3) differs from a solution of (1) only by a change of scale in space and a change of parametrization in time (cf.[7]).

This being remarked, assume that the unnormalized equation (3) has a solution g on a time interval $0 \leqq t < T$. Then, from the evolution equation for the Riemannian curvature tensor Rm, we can see, among others, that if $\hat{R} > 0$ (resp. $\mathring{R} > 0$) at $t = 0$ then it remains so on $0 \leqq t < T$.

Moreover, it is also observed that Conjectures I and II can be proved by the same arguments as in Hamilton[7], once we can establish the estimate: *We can find a constant* $\varepsilon > 0$ *independent of time* t *such that on* $0 \leqq t < T$ *we have* $<\hat{R}(\omega), \omega> \geqq \varepsilon R|\omega|^2$ *for any skew-symmetric 2-tensors* ω *(resp.* $<\mathring{R}(\zeta), \zeta> \geqq \varepsilon R|\zeta|^2$ *for any traceless symmetric 2-tensors* ζ*) provided that* $\hat{R} > 0$ *(resp.* $\mathring{R} > 0$*) at* $t = 0$.

Unfortunately, it seems very hard to prove these most desirable estimates by (at least naive) maximum principle arguments applied to the evolution equations for the curvature operators.

3.4. At present, we can only illustrate how these estimates, if proved, will fit into the arguments in [7] instead.

For example, let $G = (G_{ij})$ denote the traceless Ricci curvature tensor of (M, g), that is

$$G_{ij} = R_{ij} - \frac{1}{n} R g_{ij} ,$$

which measures the deviation of (M, g) from being an Einstein manifold. If $R > 0$ at $t = 0$, then for any $\gamma \geqq 0$ the quantity $|G|^2/R^{2-\gamma}$ satisfies the evolution equation

$$\frac{\partial}{\partial t}\left(\frac{|G|^2}{R^{2-\gamma}}\right) = \Delta\left(\frac{|G|^2}{R^{2-\gamma}}\right) + \frac{2(1-\gamma)}{R}\left\langle \nabla R, \nabla\left(\frac{|G|^2}{R^{2-\gamma}}\right)\right\rangle - \frac{2}{R^{4-\gamma}}|R\cdot\nabla G - \nabla R\cdot G|^2 - \frac{\gamma(1-\gamma)}{R^{4-\gamma}}|\nabla R|^2|G|^2 + A$$

where

$$A = -\frac{2}{R^{3-\gamma}}\left((2-\gamma)|G|^4 - \frac{\gamma}{n}R^2|G|^2 + 2R\langle \mathring{R}(G), G\rangle\right) .$$

Note that, choosing $\gamma > 0$ small enough, the desirable estimate for $\mathring{R}$ now implies $A \leqq 0$ and hence, by the maximum principle,

$$|G|^2 \leqq CR^{2-\gamma}$$

for some constant $C < \infty$. When this is established, the rest of the proof in [7] goes through almost unchanged, and Conjecture II can be proved.

The argument for Conjecture I is similar but more involved (cf.4.3).

§4. Manifolds with bounded curvature ratios

4.1. If we further assume a priori pinching estimate on the Riemannian curvature tensor, then we can carry out our program. More precisely, the following theorem can be proved.

THEOREM 6 (Huisken [9], Bourguignon-Lafontaine-Margerin [12], Nishikawa [15]). *Let* (M, g) *be a compact Riemannian n-manifold. Let* R *and* Z *denote the scalar curvature and the concircular curvature tensor of* (M, g), *respectively. Suppose that* $R > 0$ *and*

$$|Z| < \delta(n)\cdot R \quad \textit{with} \quad \delta(n) = 2[(n-2)(n-1)n(n+1)]^{-1/2} .$$

(1) *Then the initial value problem for the evolution equation* (1) *has a unique solution* $g(t, x)$ *with* $g(0, x) = g(x)$, $x \in M$ *for all time* $t \in [0, \infty)$.

(2) *As* $t \to \infty$ *the metrics* $g_t(x) = g(t, x)$ *converge to a metric of constant positive curvature in the* C^∞ *topology.*

The pinching constant given in [15] is very strict. The better constant $\delta(n)$ in the theorem is due to [9, 12].

4.2. The pinching assumption $|Z| < \delta(n)\cdot R$ implies the positivity of the curvature operator $\hat{R}$. In fact, it leads to the desirable estimate for $\hat{R}$ as follows.

Let $\hat{Z}$ denote the linear operator defined on the space Λ^2 of skewsymmetric 2-tensors by $\hat{Z}(\omega) = (Z_{ijk\ell}\omega^{k\ell})$, $\omega = (\omega_{ij}) \in \Lambda^2$. Then, from the definition of Z, we have

$$< \hat{R}(\omega),\ \omega > = \frac{2R}{n(n-1)} < \omega,\ \omega > - < \hat{Z}(\omega),\ \omega > .$$

On the other hand, since $\hat{Z}$ defines a symmetric tracefree operator on Λ^2, an elementary algebraic argument gives

$$\begin{aligned} < \hat{Z}(\omega),\ \omega >^2 &\leqq \frac{N-1}{N}\,[\mathrm{Trace}\ \hat{Z}^2] < \omega,\ \omega >^2 \\ &= \frac{(n-2)(n+1)}{n(n-1)}\,|Z|^2|\omega|^2 , \end{aligned}$$

where $N = \dim \Lambda^2 = n(n-1)/2$.

Note that we can find a $\varepsilon > 0$ such that

$$|Z| \leqq (1-\varepsilon N)\,\delta(n)\cdot R$$

holds on M, for M is compact and $|Z| < \delta(n)\cdot R$. Substituting this into the equations above, we then get

$$< \hat{R}(\omega),\ \omega > \geqq \varepsilon R|\omega|^2$$

for any $\omega \in \Lambda^2$, which is the desirable estimate for $\hat{R}$.

4.3. This being remarked, let us now consider the quantity $|Z|^2/R^{2-\gamma}$ for a solution of the unnormalized evolution equation (3). If $R \neq 0$, then for any $\gamma \geqq 0$ it satisfies

$$\begin{aligned} \frac{\partial}{\partial t}\Big(\frac{|Z|^2}{R^{2-\gamma}}\Big) &= \Delta\Big(\frac{|Z|^2}{R^{2-\gamma}}\Big) + \frac{2(1-\gamma)}{R} < \nabla R,\ \nabla\Big(\frac{|Z|^2}{R^{2-\gamma}}\Big) > \\ &\quad - \frac{2}{R^{4-\gamma}}\,|R\cdot\nabla G - \nabla R\cdot G|^2 - \frac{\gamma(1-\gamma)}{R^{4-\gamma}}\,|\nabla R|^2|Z|^2 + B \end{aligned}$$

where

$$\begin{aligned} B = -\frac{2}{R^{3-\gamma}}\Big\{&\Big((2-\gamma)|Z|^2 + \frac{4R^2}{n(n-1)}\Big)|R_{ij}|^2 \\ &+ R\big(R^{ijk\ell}R_{ij}{}^{pq}R_{pqk\ell} + 4R^{ijk\ell}R^{p}{}_{i}{}^{q}{}_{k}R_{pjq\ell}\big)\Big\} . \end{aligned}$$

Then, by using the relation between $\hat{R}$ and the quantity P in 2.4 and estimating the second term of B carefully, we can see $B \leqq 0$ whenever $|Z| < \delta(n)\cdot R$. This implies, by the maximum principle (with

$\gamma = 0$), that the condition $|Z| < \delta(n)\cdot R$ continues to hold along the solution. Moreover, choosing $\gamma > 0$ small enough, we can also see

$$|Z|^2 \leqq CR^{2-\gamma}$$

for some constant $C < \infty$.

This is a clue for the proof of Theorem 6. For more details, see [9, 12, 15].

4.4. Let δ be a positive number with $0 < \delta \leqq 1$. A Riemannian manifold (M, g) is said to be *locally δ-pinched* if there exists a positive function $A : M \to \mathbb{R}$ such that at every point $x \in M$ its sectional curvature K satisfies $\delta A(x) \leqq K \leqq A(x)$. Then, since a manifold of constant positive curvature is isometric to a spherical space form S^n/Γ, as an immediate corollary of Theorem 6 we have the following differentiable pinching theorem.

THEOREM 7 (Ruh[17]). *There exists $\delta = \delta(n)$ with $\frac{1}{4} < \delta < 1$ such that any compact locally δ-pinched Riemannian n-manifold M is diffeomorphic to a spherical space form S^n/Γ.*

It should be noted that, since the evolution equation (1) is invariant under the action of the diffeomorphism group of M, any isometry which exists in the initial metric in Theorem 6 is preserved as the metric evolves. Therefore, in Theorem 7, if G is the isometry group of M, then G is isomorphic to a closed subgroup of the isometry group of the spherical space form $\overline{M} = S^n/\Gamma$, and the diffeomorphism is equivariant with respect to the action of G on M and $\overline{M}$ respectively.

4.5. By making use of an idea in Siu[18] and that of [1], Bando[2] (cf. [8] also) proved that if (M, g) is a compact, simply connected, irreducible Riemannian n-manifold with *nonnegative* curvature operator $\hat{R}$, then either M admits a metric for which $\hat{R}$ is positive, or M is diffeomorphic to a Riemannian symmetric space of compact type. Indeed, he proved that unless the curvature operator $\hat{R}$ becomes strictly positive along the solution of the evolution equation (3), the metric has a restricted holonomy group whose Lie algebra is the image of the curvature operator, and this gives the result.

A complex analogue of our problem can be seen, for instance, in Siu[18]. See also Ruh's note in this conference.

REFERENCES

[1] S. Bando, On the classification of three-dimensional compact Kaehler manifolds of nonnegative bisectional curvature, J. Differential Geometry, 19(1984), 283-297.

[2] S. Bando, Curvature operators and eigenvalue problems, informal notes (in Japanese).

[3] M. Berger, Sur les variétés à opérateur de courbure positif, Comptes rendus, 253(1961), 2832-2834.

[4] J.P. Bourguignon and H. Karcher, Curvature operators: pinching estimates and geometric examples, Ann. Sci. Ec. Norm. Sup., 11 (1978), 71-92.

[5] D.M. Deturck, Deforming metrics in the direction of their Ricci tensors, J. Differential Geometry, 18(1983), 157-162.

[6] R.S. Hamilton, The inverse function theorem of Nash and Moser, Bull. Amer. Math. Soc., 7(1982), 65-222.

[7] R.S. Hamilton, Three-manifolds with positive Ricci curvature, J. Differential Geometry, 17(1982), 255-306.

[8] R.S. Hamilton, Four-manifolds with positive curvature operator, preprint.

[9] G. Huisken, Ricci deformation of the metric on a Riemannian manifold, preprint.

[10] T. Kashiwada, On the curvature operator of second kind, preprint.

[11] N. Koiso, On the second derivative of the total scalar curvature, Osaka J. Math., 16(1979), 413-421.

[12] C. Margerin, Some results about the positive curvature operators and point-wise $\delta(n)$-pinched manifolds, informal notes.

[13] D. Meyer, Sur les variétés riemanniennes à opérateur de courbure positif, Comptes rendus, 272(1971), 482-485.

[14] Y. Muto, On Einstein metrics, J. Differential Geometry, 9(1974), 521-530.

[15] S. Nishikawa, Deformation of Riemannian metrics and manifolds with bounded curvature ratios, to appear in Proc. Symp. Pure Math., 1985.

[16] K. Ogiue and S. Tachibana, Les variétés riemanniennes dont l'opérateur de courbure restreint est positif sont des sphères d'homologie réelle, Comptes rendus, 289(1979), 29-30.

[17] E.A. Ruh, Riemannian manifolds with bounded curvature ratios, J. Differential Geometry, 17(1982), 643-653.

[18] Y.T. Siu, Complex-analyticity of harmonic maps, vanishing and Lefschetz theorems, J. Differential Geometry, 17(1982), 55-138.

[19] S. Tachibana, A theorem on Riemannian manifolds of positive curvature operator, Proc. Japan Acad., 50(1974), 301-302.

Quasiconformal mappings and manifolds of negative curvature

Pierre Pansu
Centre de Mathématiques
Ecole Polytechnique
F-91128 Palaiseau Cédex
(U.A. 169 du CNRS)

In 1928, H.Grötzsch [19] observed that the classical Liouville and Picard theorems - nonexistence of entire functions which are bounded or, more generally, omit more than two points - extended to a class of non holomorphic maps. A holomorphic bijection between plane domains is conformal with respect to Euclidean metric : its differential at each point is a similitude, i.e., an isometry times a homothety. H. Grötzsch considered maps whose differential sits at a bounded distance from similitudes. If, at some point x, the differential takes a circle to an ellipse with axes a and b, he defined the distorsion $Q(x)$ as the least number such that

$$1/Q \leq a/b \leq Q .$$

He furthermore allowed a discrete set of singular points where the map is a ramified covering. He showed that the Picard theorem extends to the class of maps defined on the whole plane which have bounded distorsion.

Nowadays, these maps are called quasiregular maps. A smooth diffeomorphism with bounded distorsion is called quasiconformal, and the word "quasiregular" includes maps which are not 1-1 and whose differential may vanish.

It is clear that one can define quasiconformality, at least for smooth diffeomorphisms between Riemannian manifolds in any dimension. This notion plays a crucial role in G.D. Mostow's rigidity theorem for compact Riemannian manifolds of constant sectional curvature -1 and dimension ≥ 3. This theorem states that two such manifolds, if diffeomorphic, have to be isometric. The argument involves quasiconformal mappings in the following way. A diffeomorphism δ between two quotients D/Γ and D/Γ' of the unit disk D in $\mathbb{R}^n$ lifts to a quasiconformal mapping Δ of D. Δ extends by continuity to a quasiconformal homeomorphism f of the unit sphere S^{n-1}. It satisfies

$$f \cdot g = \delta_*(g) \cdot f \quad , \quad g \in \Gamma \qquad (*)$$

where δ_* denotes the isomorphism induced by δ on the fundamental groups Γ and Γ'. Equation (*), together with some regularity of f implies that f is in fact conformal. The corresponding hyperbolic isometry of D descends to an isometry between the quotients.

In this talk, we investigate whether the first steps of the argument carry over to general simply connected manifolds of negative curvature. In this class, there is a natural notion of "ideal boundary" [7], thus two questions arise :

do quasiconformal mappings extend to this boundary ?
is the extension quasiconformal in some sense ?

In section 1, we discuss a generalization of Schwarz' Lemma which shows that a quasiconformal mapping f between manifolds of bounded negative curvature is a quasiisometry, i.e., satisfies metric inequalities

$$- C + d(x,y)/L \leq d(fx,fy) \leq Ld(x,y) + C$$

where C and L are large constants. Such a map extends to ideal boundaries. Our proof relies on conformal distances constructed by means of capacities. Section 2 contains capacity estimates which lead to a nice application : locally symmetric spaces do not admit very pinched metrics. The precise definition of quasiregular maps is delayed until section 3, where the conformal structure on the boundary of homogeneous Riemannian manifolds of negative curvature is described. In sections 4 and 5, we organise material found in several papers by Pekka Tukia and Dennis Sullivan, in particular, we observe that one can speak of conformal and quasiconformal mappings on the boundary of a manifold of negative curvature as soon as there is a cocompact isometry group.

I learned most of the material described here in conversations with Mikhail Gromov. I hope that, in the present paper, it is apparent how much I owe him. Section 2 was completed in Japan during the Symposium. It is a pleasure to thank the Taniguchi Foundation for its generous support.

1. Schwarz Lemma and conformal distances

The Schwarz Lemma claims that, if a holomorphic function f on the unit disk fixes the origin, i.e., $f(0) = 0$, and if $|f(z)| \leq 1$ for all z with $|z| \leq 1$, then $|f'(0)| \leq 1$. The normalization $f(0) = 0$ can be avoided by expressing the result in terms of the *hyperbolic* metric d of the unit disk.

1 *Schwarz Lemma*.- Let the unit disk be equipped with its hyperbolic metric d of constant curvature -1 . Then any holomorphic map of the disk to itself is distance decreasing, i.e.,

$$d(f(x),f(y)) \leq d(x,y)$$

In this form, the lemma belongs to Riemannian geometry. This is even more clear with L. Ahlfors' extension of Schwarz' Lemma to surfaces of variable curvature.

2 *Theorem* [1] .-Let S be a surface endowed with a Riemannian metric of curvature ≤ -1 . Any holomorphic function on the disk with values in S is distance decreasing.

It is apparent in L. Ahlfors' paper [1] that Schwarz' Lemma only depends on the isoperimetric behaviour of the target surface S . Let us define the *isoperimetric profile* $I(v)$ of a Riemannian manifold M as follows. For a real number $v \leq \text{Volume}(M)$, let $I(v)$ be the infimum of the volumes of the hypersurfaces in M which bound a compact domain of volume v .

$$I(v) = \inf \{\text{vol}(\delta D) : D \text{ compact}, \text{vol}(D) = v\} .$$

Remember that the Classical Isoperimetric Inequality states that the isoperimetric profile of Euclidean space R^n has the form

$$I_{eucl}(v) = (c_n\ v)^{n-1/n}$$

for a sharp constant c_n .

3 *Theorem* (M. Gromov [18], compare I.G. Reshetnjak [49]).- Let N be a complete n-dimensional Riemannian manifold whose isoperimetric profile I satisfies the following two inequalities :

(i) for small v , N is at least as good as Euclidean space, i.e., if one writes

$$I(v)^{n/n-1} = c_n v(1 + \tau(v)) ,$$

then the integral

$$\int_0 \tau(v)^{-} \, dv/v$$

should be finite.

(ii) for large v , N is strictly better than Euclidean space, i.e., the integral

$$\int^{+\infty} I(v)^{-n/n-1} \, dv$$

should be finite.

Let M be a Riemannian manifold with sectional curvature $\geq -a^2$. Then any conformal (resp. quasiregular) immersion of M to N is Lipschitz (resp. Hölder continuous) with constants which depend only on a , the function I and the deviation from conformality.

Remark.- This theorem is sharp under the condition that a sharp isoperimetric inequality is used. For example, in order to conclude from it that isometries are the only conformal self maps of a rank one symmetric space M (which in fact is known, see [28]), one needs that, among domains in M of a given volume, balls have minimum boundary volume. This is known yet only when the sectional curvature is constant.

4 *Corollary*.- Let M , M' be complete simply connected Riemannian manifolds with bounded negative sectional curvature, i.e.,

$$-a^2 \leq K \leq -b^2 < 0$$

Then any quasiconformal diffeomorphism of M onto M' is a *quasiisometry*, i.e., satisfies inequalities of the form

$$-C + d(m,m')/L \leq d(fm,fm') \leq L\, d(m,m') + C$$

for some constants C and L which depend only on a, b and the deviation from conformality.

Proof.- It is known that the assumption on M implies a linear isoperimetric inequality (see [16] chap. 6), thus condition (ii) is satisfied. Condition (i) follows from a very general principle : as far as small volumes are concerned, all Riemannian manifolds behave almost like Euclidean space, see [4]. Finally, the Hölder condition for f and f^{-1} , as well as any relation of the type $d(fx,fy) \geq \Phi(d(x,y))$, for some homeomorphism Φ of $\mathbb{R}_+$, implies that f is a quasiisometry.∎

Let us now return to H. Grötsch's original motivation for the study of quasiregular mappings. Schwarz' Lemma implies Liouville's theorem as follows. A bounded entire function f is interpreted as a quasiregular map f from $\mathbb{C}$ to the disk. Let us compose f with homotheties of $\mathbb{C}$, and restrict them to the unit disk in $\mathbb{C}$. With respect to hyperbolic metrics, all these maps should be uniformly Hölder, thus equicontinuous. Considering larger and larger homotheties at a point z shows that $f'(z)$ vanishes, thus f is constant. The same argument gives Picard's theorem, once one observes that the plane minus two points admits a conformal complete metric with bounded negative curvature, the *quasihyperbolic metric* which will be defined in § 10.

However, the isoperimetric method does not provide any extension of Picard's theorem to dimensions ≥ 3. The following theorem by S. Rickman requires some more value distribution theory.

5 *Theorem* ([50]).- For $n \geq 3$, $q \geq 3$, there is a lower bound $K(n,q)$ for the deviation from conformality of quasiregular maps on $\mathbb{R}^n$ which omit at least q values.

It is striking that there exist such maps in dimension 3 [52]. The preceding theorem has a quantitative version ([51]) : A quasiregular map on the disk which omits enough points $a_1,\ldots,a_q$ is Lipschitz (as far as large distances are concerned) with respect to a complete conformal metric on the complement of these points. This metric is not the quasihyperbolic metric.

Question.- What are the invariance properties of this metric ?

The Schwarz Lemma, as a tool to prove nonexistence of holomorphic maps from $\mathbb{C}^n$ to certain complex manifolds, has had many developments (see [25]). The idea is to construct functorial pseudo-distances on complex manifolds, for which holomorphic maps are distance decreasing. Such a distance has to be identically zero for $\mathbb{C}^n$ thus no holomorphic map can exist from $\mathbb{C}^n$ to a manifold for which the pseudo-distance is a distance ("hyperbolic" complex manifolds). There are analogues in projective [26] and affine geometry [66].

In conformal geometry, functorial pseudo-distances can be constructed using capacities. Conformal capacity has been used in the plane for a long time. Its introduction in higher dimensions is due to Ch. Loewner [34].

6 Definition.- A condenser in a manifold is a triple (C,B_0,B_1) where C is open, B_0 and B_1, the "plates", are closed and contained in the closure of C. We shall admit $B_0 = \infty$ when C is unbounded. Assume that the manifold is Riemannian. The conformal capacity of a condenser is the infimum of the volumes of the conformal metrics on C for which the plates stay at distance 1 apart, i.e.,

$$\mathrm{dist}(B_1,B_0) \geq 1 .$$

7 Definition.- Let M be Riemannian manifold. We define two conformal pseudo-distances α and β as follows : for two points x and y,

$$\alpha(x,y)^{-n} = \inf \{\mathrm{cap}(M,B_0,B_1) : B_0 \text{ (resp. } B_1) \text{ is connected, unbounded and contains } x \text{ (resp. } y)\}$$

$$\beta(x,y) = \inf \{\mathrm{cap}(M,B,\infty) : B \text{ is compact, connected and contains } x \text{ and } y)\}$$

Both these quantities enter as tools in a number of papers (for references, see the papers of M. Vuorinen [68] and H. Tanaka [54]). They have been studied for their own sake by J. Ferrand [31] and I.S. Gál [8] respectively.

8 Properties.-
i) α and β are conformal invariants
ii) a quasiconformal homeomorphism is bi-Lipschitz with respect to α and β.
iii) a quasiregular map is Lipschitz with respect to β.

9 Example.- For two point homogeneous spaces, the conformal pseudo--distances and the symmetric Riemannian metric d are functionally dependant, i.e., $\beta = \Phi(d)$, but the function Φ is more or less unknown. In the constant curvature case, F.W. Gehring [9] has shown that, in the definition of α or β, the minimizing condenser is the one whose plates are geodesic segments (known as the Teichmüller - resp. Grötsch - condenser). This is unknown for other rank one (non compact) symmetric spaces. This yields inequalities of the form

$$A\, d \leq \Phi(d) \leq A\, d + B$$

where A and B are constants depending only on the dimension, see [68]. A similar inequality is obtained for α thanks to the identity

$$\alpha(x,y)^{-n} = 2^{n-2}\, \Phi(\log(1+2t+\sqrt{1+t})) \quad \text{where} \quad 2t = 1/\cosh(\tfrac{1}{2}d(x,y))-1 .$$

In dimension 2, more is known. Indeed, the Schwarz- Christoffel formula for the conformal mapping of the half-space onto the Teichmüller condenser leads to a representation of Φ by means of an elliptic integral, see [30].

It is obviously important to know for which manifolds the functions α and β really are distances, i.e., are positive. This has been studied by M. Vuorinen in the case of Euclidean domains. Then things are easier since one can use direct comparison with the ball. This is the reason why one introduces the following distances.

10 Definition [12].- Let Ω be an open, connected subset of Euclidean space. The quasihyperbolic metric k_Ω on Ω is obtained from Euclidean metric by a conformal factor equal to the inverse of the Euclidean distance to the boundary $\delta\Omega$. The distance j_Ω is defined by

$$j_\Omega(x,y) = \log(1 + |x-y|/d(\{x,y\},\delta\Omega))$$

11 Theorem [68].- For a domain Ω in $\mathbb{R}^n$, the pseudo-distances α, β, k_Ω and j_Ω are linked by the following inequalities.

$$j_\Omega \le k_\Omega$$

$$\beta \le \Phi_1(k_\Omega) \le A\, k_\Omega + B$$

$$\text{if } \delta\Omega \text{ is connected, } j_\Omega \le C\,\beta$$

$$\exp(D\, k_\Omega + E) \le \alpha \le \Phi_2(j_\Omega)$$

where A, B, C, D, E are constants, Φ_1 and Φ_2 are homeomorphisms of $\mathbb{R}_+$ depending only on the dimension.

These inequalities show that, for most Euclidean domains the conformal invariants α and β are distances. This is not that clear for general Riemannian manifolds. Again, isoperimetric inequalities give a useful criterion concerning β.

12 Theorem ([43], compare with [16]). Assume that the n-dimensional complete Riemannian manifold M satisfies a strictly stronger isoperimetric inequality than Euclidean n-space, i.e., condition (ii) in Theorem 3. Then the conformal pseudo-distance β is a distance.

If furthermore M has bounded geometry, i.e., bounded sectional curvature and a positive injectivity radius, and if for example one has an isoperimetric inequality of type

$$\mathrm{vol}(\delta D) \ge \text{const. } \mathrm{vol}(D)^{1-\tau} \quad \text{with} \quad \tau < 1/n ,$$

then, for all $x, y \in M$,

$$\beta(x,y) \ge A\, d(x,y)^{1-n\tau} - B$$

for some constants A and B.

13 As a consequence, all simply connected manifolds of bounded negative sectional curvature, all non almost abelian solvable Lie groups (N. Varopoulos [64] and [65]) have a conformal distance. They do not admit quasiregular maps from Euclidean space, or, more generally, from any manifold with vanishing β. Any quasiconformal mapping between two such manifolds is a quasiisometry, as defined in the introduction.

14 For manifolds of bounded negative curvature, Theorem 12 is reasonably sharp. Indeed, the reverse inequality

$$\beta(x,y) \le A\, d(x,y) + B$$

holds. To check this, one merely needs compute the L^n norm of the gradient for some function of the distance to the geodesic segment through x and y.

On the other hand, let M be the 3-dimensional Heisenberg group, i.e., the simply connected nilpotent nonabelian group in dimension 3. Then an isoperimetric inequality holds with exponent $\tau = 1/4$ [43], thus Theorem 12 yields $\beta \ge d^{1/4}$. One easily sees that, conversely,

$$\liminf_{d(x,y)\to+\infty} \beta(x,y)/d(x,y) = 0 .$$

Indeed, given a left-invariant metric g, split g orthogonally as

$$g = g_H + g_Z$$

where Z is the direction of the center and H its orthogonal complement. Define a group automorphism δ_t by

$$\delta_t = t\, \mathrm{id}_H + t^2\, \mathrm{id}_Z$$

so that δ_t is a homothety by a factor t of g onto the metric

$$g^t = g_H + t^2\, g_Z .$$

Let σ be a geodesic segment, tangent to H, with extremities x and

y . Then $d(\delta_t x,\delta_t y) = t\, d(x,y)$, whereas

$$\beta(\delta_t x,\delta_t y) = t\, \beta^t(x,y) \le t\, \mathrm{Cap}^t(M,\sigma,\infty) .$$

When t goes to infinity, the metrics g^t "converge" to a Carnot metric d_∞ (see section 3) with Hausdorff dimension equal to 4. The capacity $\mathrm{Cap}^t(M,\sigma,\infty)$ converges to a corresponding capacity $\mathrm{Cap}_\infty^\infty(M,\sigma,\infty)$ which vanishes, since σ has codimension 3 with respect to d_∞ .

Question.- Determine the asymptotics of β on nilpotent groups.

To show that the conformal invariant α is non trivial, one merely needs produce a condenser with finite capacity.

15 Proposition.- (see section 2) Let M be an n-dimensional simply connected Riemannian manifold with pinched negative curvature

$$-1 \le K < -1 + 1/n .$$

Its conformal pseudo-distance α is a distance.

On the other hand, it is very likely that the invariant α vanishes for a smaller pinching, in particular for quaternionic hyperbolic spaces (compare Corollary 21 of section 2). This would show that no general inequality holds between α and β .

16 In [63] chapter17, J. Väisälä introduces property P1 for a domain Ω in $\mathbb{R}^n$: Ω has property P1 at a boundary point b if, for all connected subsets E and F of Ω containing b in their closure, the capacity $\mathrm{cap}(\Omega,E,F)$ is infinite. He then shows that, if Ω has property P1 and if Ω' is an other domain which is finitely connected at each boundary point, then every quasiconformal mapping of Ω onto Ω' extends continuously to the boundary. A Riemannian manifold has vanishing invariant α (the class H_∞ of [31]) if and only if it has property P1 at ∞ . It would be interesting to have natural examples of manifolds with this property.

One can find other kinds of conformally invariant distances in the litterature. The Yamabe conjecture - in a conformal class of Riemannian metrics, find one, often unique, with constant scalar curvature - produces a conformally invariant Riemannian metric. It has been studied exactly for this purpose by Ch. Loewner and L. Nirenberg [35] on Euclidean domains. Another method consists in starting with some Riemannian metric and normalizing it by a clever conformal factor, a suitable power of the length $|W|$ of the Weyl tensor, see [42]. Both these tricks have pseudoconformal analogues, see [23] and [6]. These metrics behave badly under quasiconformal mappings. Indeed, there are quasiconformal mappings which are not locally Lipschitz, and thus, not Lipschitz under any Riemannian metric.

To study a conformally flat manifold M , one may be tempted to imitate the construction of the Kobayashi holomorphic or projective distances, i.e., define

$$\sigma(x,y) = \inf \{ d(z,t) : z\ ,\ t \text{ in the disk } D \subset \mathbb{R}^n ,\ \text{there exists a conformal embedding } f : D \hookrightarrow M \text{ such that } f(z) = x\ ,\ f(t) = y \} ,$$

and make a distance out of the function σ . For Euclidean domains, inequalities as in Theorem 11 hold, but it is unclear for which conformally flat manifolds this distance vanishes.

2. Capacity estimates

In this section, we prove two inequalities concerning conformal distances. This amounts to showing that certain conformal capacities do not vanish. The proof applies to capacities with arbitrary exponents, which is also of interest. (Remember that capacity originates from electrostatic capacity in physics, which, in dimension 3, has exponent 2, see [47]).

17 Definition.-Let (C,B_0,B_1) be a condenser in a Riemannian manifold. Let p be a positive real number. The p-capacity is

$$\inf \{ \int_C |du|^p : u \text{ smooth on } C \text{, extends continuously to values } 0 \text{ on } B_0 \text{ and } 1 \text{ on } B_1 \}$$

Conformal capacity is obtained for p equal to the dimension (see [10] for the equivalence with definition 6).

Imitating J. Ferrand's definition of conformal distance α, we introduce a notion of critical exponent for a simply connected Riemannian manifold of non positive sectional curvature.

18 Definition.- Let M be an open Riemannian manifold. Its critical exponent $p(M)$ is the least exponent $q > n-1$ such that there exist condensers with connected and unbounded plates and which have a finite q-capacity.

If furthermore M has non positive sectional curvature, it has an Eberlein-O'Neill boundary δM. One can consider condensers of type (M,x,y) where x and y are points at infinity. We define the modified exponent $\underline{p}(M)$ as the infimum of exponents q such that there exist points $x, y \in \delta M$ with $Cap_q(M,x,y) < +\infty$.

It is likely that $p(M) = \underline{p}(M)$. On the other hand, $p(M) > \dim M$ implies that the conformal distance α does not vanish. Thus Proposition 15 follows from the following Lemma.

19 Lemma.- Let M be simply connected and have bounded negative curvature $-a^2 \leq K \leq -b^2 < 0$. Then both $p(M)$ and $\underline{p}(M) \leq (n-1)a/b$.

Proof.- We exhibit a condenser which has finite q-capacity for all $q > (n-1)a/b$. We choose B_0 and B_1 to be two opposite rays on a geodesic τ. Let m be a point on τ between B_0 and B_1. Let u be any function on M which is constant on rays through m. We claim that du is L^q-integrable outside a neighbourhood of m. Indeed, Rauch's comparison theorem and $K \leq -b^2$ yield

$$du \leq e^{-br}$$

on the sphere S_r of center m and radius r, whereas $K \geq -a^2$ implies

$$vol(S_r) \leq e^{(n-1)ar} .$$

Thus

$$\int_M |du|^q \leq \int^{+\infty} e^{-qbr}\, vol(S_r)\, dr \leq \int^{+\infty} e^{((n-1)a-qb)r}\, dr$$

is finite if $q > (n-1)a/b$. ▪

In fact, if we denote the volume entropy by

$$h_{vol}(M) = \limsup_{r\to+\infty} \log vol(S_r)/r ,$$

we have proven the following inequality

$$K \leq -b^2 \Rightarrow \underline{p}(M) \leq h_{vol}(M)/b .$$

For example, volume entropy for a rank one symmetric space with sectional curvature normalized by $-4 \leq K \leq -1$ and dimension n is equal to $n+k-2$, where $k = 2$ for complex hyperbolic spaces, $k = 4$ for quaternionic hyperbolic spaces, $k = 8$ for Cayley hyperbolic plane. Thus these space have $\underline{p} \leq n+k-2$. This inequality is sharp.

20 Lemma.- Let M be a rank one symmetric space with sectional curvature normalized by $-4 \leq K \leq -1$. For each $n-1 < q < n+k-2$, there exists a positive constant $c_{n,q}$ such that, if h denotes a horofunction attached to a point at infinity x, and B is any closed subset of M, then

$$\mathrm{Cap}_q(M,B,x) \geq c_{n,q}\ \mathrm{length}(h(B))\ .$$

21 Corollary.- For such a symmetric space, $p = n + k - 2$.
Indeed, for $q < n + k - 2$, x , $y \in \delta M$, u smooth on M with $u(x) = 1$, $u(y) = 0$, h a horofunction attached to x or y one has

$$2\int_M |du|^q \geq c_{n,q}\, 2^{-q}[\mathrm{length}(h\{u>½\}) + \mathrm{length}(h\{u<½\})] = +\infty\ .\blacksquare$$

22 Corollary.- Let M be a compact quotient of a rank one symmetric space of dimension $n \geq 2k$. Then M does not admit any metric with pinching better than $(n-1/n+k-2)^2$.
Notice that the sharp result - 1/4 is the best possible pinching - is already known in dimension 4 (M. Ville [67]).

The invariants p and p can be defined in a combinatorial way for nets. It is likely that they are quasiisometry invariants (see the work of M. Kanai [24] for 2-capacities). If this is true, then the conclusion of Corollary 22 extends to all compact manifolds whose fundamental group is isomorphic to a cocompact subgroup of $U(m,1)$, $Sp(m,1)$ or F_4^{-20} .
Question (Gromov).- Compare $p(M)$ with the exponents for which L^q or $L^{q,\infty}$-cohomology of M in degree 1 vanishes.

Proof of Lemma 20 .- Let us foliate the symmetric space by parallel horospheres N centered at x - the levels of the horofonction h . Let u be a function on M which takes value 1 on B and extends by continuity to value 0 at x . By the coarea formula, it suffices to uniformly estimate

$$\int_N |du_{|N}|^q\ .$$

For the horospheres which hit B - a set of measure length $h(B)$ - the function $u_{|N}$ takes the value one on N , thus the integral is greater that some capacity $\mathrm{Cap}_q(N,\mathrm{point},\infty)$. This capacity does not depend on the particular point, since N is homogeneous. It does not depend on the particular horosphere, since they are all pairwise isometric. It is non zero for two reasons.
i) Since the exponent $q > n-1$, the Sobolev embedding of $W^{1,q}$ into $C^{1-(n-1)/q}$ allows one to replace the point by a ball of finite size as a plate of the condenser.
ii) The horosphere N is isometric to a nilpotent Lie group with left-invariant metric, whose isoperimetric profile satisfies $I(v) \geq$ const. $v^{p-1/p}$ where $p = n + k - 2$ (N. Varopoulos [65]), thus Theorem 12 applies.∎

For M a symmetric space and x , y points at infinity, we show that $\mathrm{Cap}_q(M,x,y) = +\infty$ for all $q < n + k - 2$. Question.- What happens when q is equal to the critical exponent ?

3. Regularity properties of quasiconformal mappings.

Early, it has turned out to be necessary to consider quasiregular mappings which are not of class C^1 . In Teichmüller's theory (see [55], [3], [5]) one obtains as solutions of a variational problem mappings which are smooth except at a finite number of points. Furthermore, in the deformation theory of Riemann surfaces, there definitely occur quasiconformal mappings which are nowhere smooth, as we shall see below. We give two equivalent definitions of quasiregular maps. A quasiconformal mapping is a quasiregular homeomorphism.

23 Analytic definition.- (C.B. Morrey [37]) A continuous map f between Riemannian manifolds of dimension $n \geq 2$ is K-quasiregular if it admits a differential df in the sense of distributions which is a locally L^n-integrable function and satisfies

$$|df|^n \leq K \,\mathrm{Jac}(f) .$$

The number K is only one of the various ways to measure the deviation from conformality, i.e., the distance between the differential df and the similitudes. In terms of the eigenvalues $\mu_1^2,\ldots,\mu_n^2$ of the endomorphism ${}^t df \cdot df$, one has

$$K = \mu_n^n/\mu_1\ldots\mu_n .$$

An equally satisfactory quantity is

$$Q = \mu_n/\mu_1$$

which satisfies

$$\log Q \leq \log K \leq (n-1)\log Q .$$

For a linear map A between Euclidean spaces, the number Q has a metric interpretation : Given a ball B , its image is pinched between two balls $B(s)$ and $B(S)$ - i.e.,

$$B(s) < AB < B(S)$$

such that $Q = S/s$.

More generally, if f is a continuous, discrete, open map between Riemannian manifolds, x is a point and r is small enough, one can define the ratio $Q_f(x,r) = S/s$ where S is the minimum radius of a ball centered at $f(x)$ which contains $fB(x,r)$, and s is the maximum radius of a ball centered at $f(x)$ which is contained in $fB(x,r)$.

24 Metric Definition (M.A.Lavrentiev [29]) A continuous map between Riemannian manifolds is quasiregular if it is orientation preserving, open, discrete, and if

$$Q_f(x) = \limsup_{r\to 0} Q_f(x,r)$$

is bounded.

There is a third characterization of quasiconformality by means of capacities, [46], [2], [36]. The fact that, in dimensions ≥ 2, these definitions coincide is a series of theorems by I.N. Pesin [45], J.A. Jenkins [22], F.W. Gehring - J. Väisälä [14], O. Martio - S. Rickman - J. Väisälä [36]. This is the conclusion of longstanding efforts to determine to which class of regularity quasiregular maps exactly belong. This regularity is expressed by the following properties.

25 Properties.- In dimensions ≥ 2, quasiregular mappings are absolutely continuous on lines, i.e., in a coordinate patch, a quasiregular map is absolutely continuous on almost every line. As a consequence, they send Lebesgue null sets to null sets.
Quasiregular mappings have a differential almost everywhere, which is L^n integrable.

These properties have turned out to be essential in G.D. Mostow's rigidity theorem for compact manifolds of constant sectional curvature.

26 Theorem [39].- If two compact Riemannian manifolds of dimension $\geq$ 3 , with constant sectional curvature -1 , are diffeomorphic, then they are isometric.

27 Here is a sketch of the proof. A diffeomorphism between two such manifolds lifts to a quasiconformal mapping f of the universal covers, i.e., the unit disk in $\mathbb{R}^n$. Let us denote by Γ and Γ' the fundamental groups of the compact manifolds. They act conformally on the disk. The diffeomorphism induces an isomorphism $i : \Gamma \approx \Gamma'$ and, for $g \in \Gamma$, one has

$$f \cdot g = i(g) \cdot f$$

The quasiconformal mapping f extends to the unit sphere (Property P1 of § 16) and the extension, still denoted by f ,is a quasiconformal homeomorphism of the (n-1)-sphere (Schwarz' reflection principle). We now show, following P. Tukia [60], that f is a conformal mapping of the sphere. This is due to the fact that the action of Γ on the sphere is highly transitive, and necessitates little regularity of f . Still, it fails when $n = 2$. Choose the upper half-space model, and normalize f so that $f(0) = 0$ and $f(\infty) = \infty$. Consider the 1-parameter group of homotheties h_t . Since $\Gamma < O(n,1)$ is cocompact, there exist elements $g_t \in \Gamma$ such that $h_t^{-1} \cdot g_t = k_t$ are bounded in $O(n,1)$ Then one can write $i(g_t) = h_s \cdot j_t$ with j_t and the ratio s/t bounded. The conjugacy condition now reads

$$h_t^{-1} \cdot f \cdot h_t = h_{s/t} \cdot j_t \cdot f \cdot k_t^{-1} .$$

Choose subsequences such that s/t , k_t and j_t converge. If f is differentiable at 0 ,then in the limit $k \cdot f \cdot j = df(0)$ is linear. From there on, it is easy to show that f is conformal.∎

28 In [40], [41], G.D. Mostow generalized the rigidity theorem to all locally symmetric spaces without 2-dimensional factors. The argument in the rank one case also relies on the theory of quasiconformal mapping, but in a slightly extended context. Indeed, the first steps are the same. A symmetric space of rank one is a simply connected Riemannian manifold M with negative sectional curvature. As such, it admits an "ideal boundary", defined by means of asymptotic geodesics [7]. The lift of a diffeomorphism - in fact, of any homotopy equivalence - is a quasiisometry(as defined in the introduction). It extends to a homeomorphism of the ideal boundary (a fact which can be traced back to M. Morse [38]). This extension is not quasiconformal with respect to any Riemannian metric on δM . Indeed, this fails to be true even for isometries : the analogue of the homotheties in the upper half space model for hyperbolic geometry is a 1-parameter group δ_t of isometries whose action on δM can be written, in suitable coordinates x_i , y_j ,

$$\delta_t(x_i) = t\, x_i \ , \quad \delta_t(y_j) = t^2\, y_j \ .$$

The plane with equations $dy_j = 0$ at the origin is part of a distribution of planes V which is invariant under the isometry group. The boundary extensions of isometries are conformal on the subbundle V and only there.

29 Let us define a family of distances on δM adapted to the situation. Fix a point $x \in M$. There is a unique Euclidean metric g_x on the subbundle V which is invariant under the isometries fixing x . It allows one to define the length of curves tangent to V , and we set, for two points p , q in δM ,

$$d_x(p,q) = \inf \{\text{length } c : c \text{ joins } p \text{ to } q \text{ in the boundary, } c \text{ is tangent to } V\}$$

This number is finite since the distribution V is non integrable, and defines a distance on δM . When x varies, the distance d_x changes conformally, i.e., a small d_x-ball is very close to a d_y-ball. Thus we have defined a conformal structure (in a generalized sense) on the boundary δM .

Now the boundary extension of a quasiisometry of M is a quasiconformal mapping with respect to any of the metrics d_x . Here we take the metric definition for quasiregular maps, which is meaningful for arbitrary metric spaces. The class of maps obtained coincides with G.D. Mostow's "quasiconformal mappings over a division algebra" [40]. These maps are absolutely continuous on a suitable class of "lines" [41] and almost everywhere differentiable [44] in a sense which we explain below. Thus P. Tukia's argument, as well as G.D. Mostow's, extends to prove the rigidity theorem in rank one.∎

Let M be a rank one symmetric space with isometry group G . To a choice of a point x in M and a boundary point $p \in \delta M$, there corresponds an Iwasawa decomposition $G = KAN$ where

K is the stabilizer of x
N is simply transitive on $\delta M - p$
A is a one-parameter group of translations along the geodesic through x and p .

In the constant curvature case, N is abelian and A consists of homotheties. In the other cases, N is two-step nilpotent, its Lie algebra splits as

$$n = V + [n,n]$$

and the element δ_t of A acts on n by multiplication by t on V and t^2 on $[n,n]$. Thus the ideal boundary of a rank one symmetric space identifies with a nilpotent Lie group. The results of absolute continuity and differentiability of quasi- conformal mappings will in fact apply to the whole class of Carnot groups, which we define now.

30 Definition.- A Carnot group is a simply connected nilpotent Lie group whose Lie algebra n splits as

$$n = V_1 \oplus \ldots \oplus V_r \quad \text{where} \quad [V_i, V_j] = V_{i+j} \quad .$$

A Carnot group N admits a one-parameter group of homotheties

$$\delta_t \in \mathrm{Aut}(N) \ , \ \delta_t \ \text{is multiplication by} \ t^i \ \text{on} \ V_i \ .$$

By a norm , we mean a left-invariant distance on N which is homogeneous of degree one under the group of homotheties. Particular norms are the Carnot metrics : given a Banach space structure on V_1 , one can define the length of curves in N which are tangent to the left-invariant subbundle of TN generated by V_1 . One defines quasiconformal mappings using the metric definition. The class obtained does not depend on the particular choice of norm.

A continuous map f between Carnot groups N and N' equipped with homotheties $\{\delta_t\}$ and $\{\delta'_t\}$ is said to be δ-differentiable at x if the limit

$$Df(x)\mu = \lim_{t \to 0} \delta'^{-1}_t (f(x)^{-1} f(x\delta_t\mu))$$

exists for all μ .

A line is an orbit of a left-invariant vector field which is tangent to V_1 .

For a smooth function u on N, let

$$du(x) = \sup \{wu(x) : w \in V_1 \ , \ |w| = 1 \} \ .$$

We define the p-capacity of a condenser (C,B_0,B_1) as the infimum of the integrals (with respect to Haar measure)

$$\int_C |du|^p$$

over all smooth functions u on C which tend to 0 on B_0 and to 1 on B_1 . Conformal capacity is obtained for p equal to the group's Hausdorff dimension

$$p = \Sigma \dim V_i$$

31 Theorem [44].- A quasiconformal homeomorphism between open subsets of Carnot groups admits almost everywhere a δ-differential which is a group isomorphism intertwining the two one-parameter groups of homotheties.

It is absolutely continuous on almost every line [41] and, as a consequence, it send null-sets to null-sets.

1-quasiconformal mappings preserve conformal capacities, and K-quasiconformal mappings multiply them at most by K (for a suitable measurement K of the deviation from conformality).

In other words, a big part of the analytic theory of quasiconformal mappings in Euclidean space can be carried out on Carnot groups. However, it seems to be harder to obtain capacity estimates. For example, the condenser between two concentric balls has positive capacity c(r) , depending only on the ratio r of the radii. This is sufficient to prove that 1-quasiconformal mappings are Lipschitz and to obtain some modulus of continuity for a general quasiconformal mapping.

However, one needs further information - still unknown - on the function c(r) to conclude that quasiconformal mappings are Hölder continuous. It is also unclear whether the condenser whose plates are two arbitrary curves has a non-zero capacity.

32 A new feature of the nilpotent theory is that, in general, there are no quasiconformal mappings at all. The reason is that there is too little choice for differentials. Indeed, these should live in the group $Aut_\sigma(N)$ of automorphism of N which commute with the homotheties. In the abelian case, this is the whole linear group, and every smooth diffeomorphism is locally quasiconformal. The Iwasawa component of U(n,1) is the Heisenberg group. The group $Aut_\sigma(N)$ consists of homotheties times symplectic 2n-2 by 2n-2 matrices ; a smooth diffeomorphism is locally quasiconformal if and only if it is a contact transformation, i.e., it preserves the plane distribution V [25]. This still produces an infinite dimensional group of quasiconformal mappings. In contrast, when N is the Iwasawa component of Sp(n,1), $n \geq 2$, the group $Aut_\sigma(N)$ consists of homotheties and a compact group Sp(n-1)Sp(1). In this case, any quasiconformal mapping is 1-quasiconformal.

33 Corollary [44].- The elements of Sp(n,1) are the only global quasiconformal self-maps of the boundary of quaternionic hyperbolic n-space.

In the same vein, if $Aut_\sigma(N)$ consists of homotheties only - a case which definitely occurs, see [44] - then any (even local) quasiconformal mapping of N is the restriction of a translation or homothety.

It would be interesting to have a local version of the preceding corollary. This amounts to prove that local 1-quasiconformal mappings are smooth. Proofs of this fact in the Euclidean case are due to I.M. Reshetnjak [48] and F.W. Gehring [11]. They rely on non-linear elliptic regularity theory. In the nilpotent case, the corresponding equations are hypoelliptic and the necessary regularity is not yet available.

34 There is much room left for further generalizations, since the metric definition for quasiregular maps can be taken using arbitrary metric spaces.

i) The solvable Lie groups which admit left-invariant metrics with strictly negative curvature have been classified by E. Heintze [21]. They are of the form AN where N is a nilpotent Lie group and A a one-parameter group of contracting automorphisms. The data of N, A together with a norm is a very natural generalization of a Banach space with its homotheties. One can speak of differentiability and of quasiconformal mappings, a class which will not depend on the particular norm. A new feature is that these groups are not length spaces (cf [16]), i.e., the distance between two points is not the length of any curve joining them. In fact, a group N admits a length norm if and only if it is a Carnot group.

ii) A question of Gromov : as far as I know, nothing is known about quasiconformal mappings between separable Hilbert spaces. What about Liouville theorem ? Notice that traditional methods use integration, and thus do not extend to infinite dimensions.

iii) Lift the assumption of discreteness and you can speak of quasiregular maps between manifolds of different dimensions. Is Liouville theorem still true ?

4. Quasiconformal groups

35 Definition [13].- A *quasiconformal group* on a metric space is a group of uniformly quasiconformal homeomorphisms.

36 We are concerned with the following question, originally due to F.W. Gehring and B.P. Palka : when is a quasiconformal group of the standard sphere (resp., ideal boundary of a manifold of negative curvature) quasiconformally conjugate to a group of conformal transformations ?

A general method to adress this question is due to D. Sullivan [53]. He observes that every quasiconformal group leaves invariant at least one measurable conformal structure. Indeed, at each point, the space of conformal structure on the tangent space identifies with $Gl(n)/CO(n)$, which admits an invariant metric with non-positive sectional curvature. The set of pull-backs of the standard structure by the elements of the group is bounded, thus one can attach to it a unique point, the center of the smallest ball which contains it, for example (see [59]). These data form a measurable conformal structure, almost everywhere invariant under the group. Notice that this argument carries over to Carnot groups, since all what is needed is the a. e. differentiability of quasiconformal mappings. In this case, by a measurable conformal structure, we mean the data at each point of a Euclidean metric on the left-invariant subbundle generated by V_1. The metric should depend measurably on the point.

37 In dimension 2, the sphere has only one conformal structure, by Riemann's mapping theorem. (The extension to measurable conformal structures is due to C.B. Morrey [37]). Thus the invariant conformal structure is quasiconformally conjugate to the standard one, and we are done [57]. As a consequence, all quasiconformal groups in dimension 2 are known.

38 The argument fails in higher dimensions, and in fact, P. Tukia has constructed domains in $\mathbb{R}^n$ which admit a transitive, connected quasiconformal group which is not isomorphic to any subgroup of $O(n,1)$ [58]. Thus one needs extra assumptions. A typical argument is as follows : assume that the group contains an element g which is expanding at the point x . Assume that the invariant conformal structure is smooth at x . Normalize is so that it coincides with the standard structure at x . Using iterates of g , one sees that some neighborhood U of x can be mapped conformally to smaller and smaller neighborhoods of x , which are very close to a standard disk. One concludes that U itself is conformally equivalent to a standard disk. In fact, as proved by P. Tukia, this argument works under no regularity assumption on the invariant conformal structure.

39 *Theorem* [59].- Let μ be a measurable conformal structure on the sphere. If its conformal group is cocompact in the space of triples of distinct points (for example, if it comes from a cocompact group of quasiisometries of the disk), then μ is the standard conformal structure.

Clearly, this applies to Carnot groups too.

D. Sullivan has a different result : let μ be invariant under a discrete subgroup Γ of $O(n,1)$. Then μ is standard under a weaker assumption on Γ : that it approaches almost every point horospherictly [53]. The assumption in P.Tukia's theorem is conical approach a.e.

Since the connected subgroups of $O(n,1)$, $U(n,1)$,.. are known, this leads to a a method to decide where two homogeneous Riemannian manifolds are quasiconformally equivalent. For the case of Euclidean domains, see the forthcoming work by F.W. Gehring and G. Martin.

40 Corollary.- Let N be a Carnot group with homothety group A , let M denote the group AN endowed with a left-invariant Riemannian metric of negative sectional curvature. Let M' be a simply connected Riemannian manifold with negative sectional curvature and cocompact isometry group. Assume that M and M' are quasiconformally equivalent. Then Isom(M) and Isom(M') are cocompact subgroups in a common topological group.

Notice that, if Isom(M') is discrete, we may conclude that both Isom(M) and Isom(M') are subgroups of a simple Lie group O(n,1) , U(n,1)...

5. Global characterizations of quasiconformal mappings

41 I want to emphasize the fact that the conformality or quasiconformality of a homeomorphism of a manifold M can be checked from its behaviour under conjugacy with conformal mappings of M . This applies only when the conformal group of M is large enough. Therefore, in the sequel, S denotes either the boundary of a rank one symmetric space (i.e., a sphere with an exotic conformal structure) or a Carnot group N . We denote by G its "conformal group", i.e., a simple group O(n,1) , U(n,1) , Sp(n,1) , F_4^{-20} in the symmetric case, the group MAN where M is maximal compact in $Aut_s(N)$ in the general case. Let us begin with a consequence of the preceding discussion.

42 Corollary (see [62] for an elementary proof in the case of Euclidean space) A quasiconformal group on S which contains a cocompact subgroup of the conformal group G consists only of 1-quasiconformal mappings (conformal in the symmetric case).

One of the applications of the methods of section 1, especially Theorem 3, is to equicontinuity properties of "normalized" quasiconformal mappings (see also [63], chap. 20). Given balls D_1, $D_2 \ll D_3$ and a point $x \in D_1$, we say that a homeomorphism f of the sphere S is normalized if

$$f(x) = x \quad \text{and} \quad D_2 < f(D_1) < D_3$$

then the normalized quasiconformal mappings of S with a given distorsion are equicontinuous. Any quasiconformal mapping can be normalized - depending only on its distorsion - by multiplying it with suitable elements of the conformal group. Thus one can state : if Q denotes the set of quasiconformal mappings on S with deviation from conformality less than K , then Q < GB where B is a compact subset of the homeomorphism group of the sphere. A kind of converse is true.

43 Proposition.- Let Γ be a cocompact subgroup of the conformal group G . A homeomorphism f of S is quasiconformal if and only if fΓ < ΓB where B is a compact set of homeomorphisms of S .

44 Corollary.- Let Γ be cocompact in G . A homeomorphism f is 1-quasiconformal (conformal in the symmetric case) if and only if Γ is cocompact in the closed subgroup of Homeo(S) generated by Γ and f .

45 Remark.- Since the boundary of a simply connected manifold of negative curvature can be reconstructed functorially from any discrete cocompact group of isometries, this corollary shows that the conformal group G can be recovered from any discrete cocompact subgroup. By the way, this is Mostow's rigidity theorem : any isomorphism between

lattices extends to an isomorphism between the Lie groups. This leads us to a notion of conformal mappings between the boundaries of the universal covers of arbitrary compact manifolds with negative curvature.

46 Definition.- Let M , M' be simply connected manifolds with negative sectional curvature. Let Γ , Γ' be cocompact groups of isometries. A homeomorphism $f : \delta M$ $\delta M'$ is said to be conformal if the multiplicative set generated by Γ , f , f^{-1} and Γ' is contained in $\Gamma B \cup \Gamma' B'$ where B, $B' <$ Homeo(δM), Homeo($\delta M'$) are compact. The map f is called quasiconformal if $f\Gamma < \Gamma' B$ where $B <$ Homeo($\delta M'$) is compact.

47 Theorem (P. Tukia [61]).- With the above definition, a quasiconformal mapping extends to a quasiisometry of M onto M', well-defined modulo maps with bounded displacement. Conversely, a quasiisometry extends to a quasiconformal mapping.

48 Question.- Does the conformal group of δM preserve some conformal class of distances ? In case the group Γ is smooth with respect to some differentiable structure on δM , is there any relation between the quasiconformal group as defined above and the quasiconformal group attached to the smooth structure ? The case of symmetric spaces already shows that there is no inclusion.

49 Invariants of patterns of points.- It is easy to construct conformal invariants of a finite number of points on the standard sphere : given k distinct points, take the various simplices they generate in hyperbolic space, choose a combination of volumes, mutual angles and distances. I claim that any such invariant is quasi-invariant under quasiconformal mappings of the sphere. This just follows from compactness modulo conformal normalization.

Conversely, quasiconformal mappings of the standard sphere can be defined using any conformal invariant of 4 points $(.,.,.,.)$ which tends to infinity when exactly two points become close to each other. Indeed, given a homeomorphism f , use a conformal mapping so that f fixes three given points a, b and c . Then the range of variation of $f(d)$ is controlled by the 4-points invariant $(a,b,c,f(d))$, and a modulus of continuity at d is given by the invariant $(a,b,f(d),f(e))$. Thus the set of normalized homeomorphisms which almost preserve $(.,.,.,.)$ is compact, and so uniformly quasiconformal. In particular, a group G' of homeomorphisms of S^n strictly containing $O(n,1)$ cannot preserve such an invariant. Does this imply some dynamical property of the action of G' on quadruples of distinct points ?

There are two famous examples of conformal invariants of several points. First, the cross-ratio on the two-sphere. There, the invariant has values in the two-sphere itself. Second, the volume of the hyperbolic simplex generated by $n+2$ points on the n-sphere, which enters M. Gromov's proof of Mostow's rigidity, see [56], chap. 6. It characterizes conformal mappings. Indeed, a hyperbolic simplex is regular (i.e., has maximal symmetry) if and only if it has maximal volume [20]. A homeomorphism which preserves regular simplices preserves the pattern generated by one of them by reflection across the faces, which is dense in the sphere, so the homeomorphism extends to an isometry of hyperbolic space. I do not know whether quasiconformal mappings can be characterized by this invariant of $n+2$ points.

50 One may wonder whether the conformal mappings on the boundary of a manifold of negative curvature as defined in § 46 preserve some kind of capacity. One can characterize the class of semi-open function on δM whose derivative is L^n-integrable, since they have adequate compactness properties, see [31], but it is unclear whether one can

reconstruct the whole Royden algebra of continuous functions with L^n-integrable derivative, together with its norm

$$\|u\|_{L^\infty} + \|du\|_{L^n} .$$

It is known (see [32], [33]) that this algebra completely determines the conformal structure.

Question.- Can one concoct a Royden algebra for δM out of the Royden algebra of M ? Some answer must exist already for the disk.

References

[1] L. AHLFORS, Zum theorie der Überlagerungsflächen, Acta Math. 5, 157-194 (1936).

[2] L. AHLFORS, On quasiconformal mappings, J. d'Analyse Math. 3, 1-58 (1954).

[3] L. AHLFORS, On quasiconformal mappings, J. d'Analyse Math. 4, 204-208 (1954).

[4] P. BERARD - D. MEYER, Inégalités isopérimétriques et applications, Ann. Sci. de l'E. N. S. de Paris 13, 513-542 (1982).

[5] L. BERS, Quasiconformal mappings and Teichmüller's theorem, 89-119 in "Analytic functions", Princeton Math. Series Vol 24, Princeton Univ. Press, Princeton (1960).

[6] D. BURNS Jr - S. SHNIDER, Geometry of hypersurfaces and mapping theorems in $\mathbb{C}^n$, Comment. Math. Helvetici 54, 199-217 (1979).

[7] P. EBERLEIN - B. O'NEILL, Visibility manifolds, Pacific J. Math. 46, 45-109 (1973).

[8] I. S. GAL, Conformally invariant metrics on Riemann surfaces, Proc. Nat.Acad. Sci.USA 45, 1629-1633 (1959).

[9] F.W. GEHRING, Symmetrization of rings in space, Trans. Amer. Math. Soc. 101, 499-519 (1961).

[10] F.W. GEHRING, Extremal length definitions for the conformal capacity of rings in space, Michigan Math. J. 9, 137-150 (1962).

[11] F.W. GEHRING, Rings and Quasiconformal mappings in space, Trans. Amer. Math. Soc. 103, 353-393 (1962).

[12] F.W. GEHRING - B.G. OSGOOD, Uniform domains and the quasi-hyperbolic metric, J. d'Analyse Math. 36, 50-74 (1979).

[13] F.W. GEHRING - B.P. PALKA, Quasiconformally homogeneous domains, J. d'Analyse Math. 30, 172-199 (1976)

[14] F.W. GEHRING - J. VÄISÄLÄ, On the geometric definition for quasiconformal mappings, Comment. Math. Helvetici 36, 19-32 (1961).

[15] M. GROMOV, Hyperbolic manifolds, groups and actions, 183-213 in "Riemann surfaces and related topics, Stony Brook 1978", Ann. of Math. Studies Vol.97, Princeton University Press, Princeton (1981).

[16] M. GROMOV, Structures métriques pour les variétés riemanniennes, chap. 6, notes de cours rédigées par J. Lafontaine et P. Pansu, CEDIC-Fernand-Nathan, Paris (1981).

[17] M. GROMOV, Filling Riemannian manifolds, J. Diff. Geo. 18, 1-148 (1983).

[18] M. GROMOV, Pseudo-holomorphic curves in symplectic manifolds, to appear in Inventiones Mathematicae.

[19] H. GRÖTZSCH, Über die Verzerrung bei schlichten nichtkonformen Abbildungen und über eine damit zusammenhängende Erweiterung des Picardschen Satzes, Ber. Verh. Sächs. Akad. Wiss. Leipzig 80, 503-507 (1928).

[20] U. HAAGERUP - H.J. MUNKHOLM, Simplices of maximal volume in hyperbolic n-space, Acta Math. 147, 1-11 (1981).

[21] E. HEINTZE On homogeneous manifolds of negative curvature, Math. Ann. 211, 23-24 (1974).

[22] J.A. JENKINS, A new criterion for quasiconformal mappings, Ann. Of

Math. 65, 208-214 (1957).
[23] D. JERISON - J.M. LEE, The Sobolev inequality on the Heisenberg group and the Yamabe problem on CR manifolds, Séminaire Goulaouic-Meyer-Schwarz, Ecole Polytechnique, Palaiseau (1984).
[24] M. KANAI, Rough isometries and the parabolicity of Riemannian manifolds, Preprint Keio Univ. Yokohama (1985).
[25] S. KOBAYASHI, Hyperbolic manifolds and holomorphic mappings, Dekker, New-York (1970).
[26] S. KOBAYASHI, Invariant distances for projective structures, Symp. Math. 26, 153-161 (1982).
[27] A. KORANYI - H.M. REIMANN, Quasiconformal mappings on the Heisenberg group, Preprint Bern (1984)
[28] N.H. KUIPER, Einstein spaces and connections I, Proc. Kon. Akad. v. Wetensch. Amsterdam 12, 506-521 (1950).
[29] M.A. LAVRENTIEV, Sur une classe de représentation continue, Math. Sb. 42, 407-423 (1935).
[30] O.LEHTO - K.I. VIRTANEN, Quasikonforme Abbildungen, Grundlehren der Math... Band 126, Springer Verlag Berlin...(1965).
[31] J. LELONG-FERRAND, Invariants conformes globaux sur les variétés riemanniennes, J. of Diff. Geo. 8, 487-510 (1973).
[32] J. LELONG-FERRAND, Etude d'une classe d'applications liées à des homomorphismes d'algèbres de fonctions, et généralisant les quasiconformes, Duke Math. J. 40, 163-185 (1973)
[33] L.G. LEWIS, Quasiconformal mappings and Royden algebras in space, Trans. Amer. Math. Soc. 158, 481-492 (1971).
[34] Ch. LOEWNER, On the conformal capacity in space, J. Math. Mech. 8, 411-414 (1959).
[35] Ch. LOEWNER - L. NIRENBERG, Partial Differential Equations invariant under Conformal or Projective Transformations, 245-272 in"Contributions to Analysis", ed. L.Ahlfors, I.Kra, B. Maskit, L. Nirenberg, Academic Press, New-York (1974).
[36] O. MARTIO - S. RICKMAN - J. VÄISÄLÄ, Definitions for quasiregular mappings, Ann. Acad. Sci. Fennicae Ser. A.I 488, 1-40 (1969).
[37] C.B. MORREY, On the solution of quasilinear elliptic partial differential equations, Trans. Amer. Math. Soc. 43, 126-166 (1938).
[38] M. MORSE, Recurrent geodesics on a surface of negative curvature, Trans. Amer. Math. Soc. 2 84-100 (1921).
[39] G.D. MOSTOW, Quasiconformal mappings in n-space and the rigidity of hyperbolic space forms, Publ. I.H.E.S. 34, 53-104 (1968).
[40] G.D. MOSTOW, Strong rigidity of discrete subgroups and quasiconformal mappings over a division algebra, 203-209 in proc. of "Discrete subgroups of Lie groups and application to moduli, Bombay 1973".
[41] G.D. MOSTOW, Strong rigidity of locally symmetric spaces, Annals of Math. Studies Vol 78, Princeton Univ. Press, Princeton (1973).
[42] M. OBATA, The conjectures on conformal transformations of Riemannian manifolds, Bull. Amer. Math. Soc. 77, 269-270 (1971).
[43] P. PANSU, An isoperimetric inequality on the Heisenberg group, in proc. of "Differential geometry on homogeneous spaces, Torino (1983).
[44] P. PANSU, Métriques de Carnot-Carathéodory et quasiisométries des espaces symétriques de rang un, Preprint Ecole Polytechnique, Palaiseau (1984).
[45] I.N. PESIN, Metric properties of Q-quasikonformal mappings, Math. Sb. 40 (82), 281-295 (1956).
[46] A. PFLUGER, Quasikonforme Abbildungen und logarithmische Kapazität, Ann. Institut Fourier 2, 29-80 (1951).
[47] G. POLYA - G. SZEGÖ, Isoperimetric inequalities in mathematical physics, Ann. of Math. Studies Nr 27, Princeton Univ. Press (1951).
[48] I.G. RESHETNJAK, On conformal mappings of a space, Dokl. Akad. Nauk. SSSR 130, 1196-1198 (1960) = Soviet. Math. Dokl. 1, 153-155 (1960).
[49] I.G. RESHETNJAK, Stability of conformal mappings in multi-dimensio-

nal spaces, Sibirsk. Mat. Zh. 8, 91-114 (1967).
[50] S. RICKMAN, On the number of omitted values of entire quasiregular mappings, J. d'Analyse Math. 37, 100-117 (1980).
[51] S. RICKMAN, Quasiregular mappings and metrics on the n-sphere with punctures, Comment. Math. Helvetici 59, 136-148 (1984).
[52] S. RICKMAN, The analogue of Picard's theorem for quasiregular mappings in dimension three, to appear in Acta Mathematica.
[53] D. SULLIVAN, On the ergodic theory at infinity of an arbitrary discrete group of hyperbolic motions, 465-496 in "Riemann surfaces and related topics, Stony Brook 1978", Ann. of Math. Studies Vol.97, Princeton Univ. Press, Princeton (1981).
[54] H. TANAKA, Boundary behaviours of quasiregular mappings, 405-408 in "Complex Analysis, Joensuu 1978", Lecture Notes Vol. 747, Springer Verlag Berlin... (1979).
[55] O. TEICHMÜLLER, Extremale quasikonforme Abbildungen und quadratische Differentialen, Abh. Preuss. Akad. Wiss. 22, 1-197 (1940).
[56] W. THURSTON, Geometry and topology of 3-manifolds, Preprint Princeton (1978).
[57] P. TUKIA, On two-dimensional quasiconformal groups, Ann. Acad. Sci. Fennicae Ser. A I 5, 73-78 (1980)
[58] P. TUKIA, A quasiconformal group not isomorphic to a Möbius group, Ann. Acad. Sci. Fennicae Ser. A.I 6, 149-160 (1981).
[59] P. TUKIA, On quasiconformal groups, Preprint Helsinki (1982).
[60] P. TUKIA, Differentiability and rigidity of Möbius groups, preprint Univ. Helsinki (1982).
[61] P. TUKIA, Quasiconformal extension of quasisymmetric mappings compatible with a Möbius group, Acta Math. 154, (1985).
[62] P. TUKIA - J. VÄISÄLÄ, A remark on 1-quasiconformal maps, Ann. Acad. Sci. Fennicae ser. A I 10, 839-842 (1985).
[63] J. VÄISÄLÄ, Lectures on n-dimensional quasiconformal mappings, Lecture Notes in Math. Vol. 229, Springer Verlag, Berlin...(1970)/
[64] N. VAROPOULOS, Chaînes de Markov et inégalités isopérimétriques, C. R. Acad. Sci. Paris 298 (10), 233-236 (1984).
[65] N. VAROPOULOS, Théorie du potentiel sur les groupes nilpotents, C. R. Acad. Sci. Paris 301 (5), 143-144 (1985).
[66] E. VESENTINI, Invariant metrics on convex cones, Ann. Scu. Norm. Sup. Pisa Ser. 4 3, 671-696 (1976).
[67] M. VILLE, Sur le volume des variétés riemanniennes pincées, to appear in Bull. Soc. Math. de France (1985).
[68] M. VUORINEN, Growth inequalities for quasiregular mappings, preprint Univ. Helsinki (1979).

HELICAL IMMERSIONS

Kunio Sakamoto
Department of Mathematics
Tokyo Institute of Technology
Ohokayama, Meguro-ku, Tokyo
JAPAN

§1. Introduction

Let $f:M \longrightarrow \overline{M}$ be an isometric immersion of a Riemannian manifold M into a Riemannian manifold $\overline{M}$. If for each geodesic γ of M the curve $f\circ\gamma$ in $\overline{M}$ has constant Frenet curvatures which are independent of the chosen geodesic, then f is said to be helical. Let κ_i denote the i-th Frenet curvature of $f\circ\gamma$. If $\kappa_i \neq 0$ $(i \leq d-1)$ and $\kappa_d = 0$, then the helical immersion is said to be of order d.

Example 1. Planes and spheres in Euclidean space are helical submanifolds. Great and small spheres in a sphere are also helical. The former is of order 1 and latter is of order 2. These are trivial examples.

Example 2 (Veronese submanifolds). An isometric immersion $S^2(1/3) \longrightarrow S^4(1)$ defined by

$$u^1 = yz/\sqrt{3}, \quad u^2 = zx/\sqrt{3}, \quad u^3 = xy/\sqrt{3},$$

$$u^4 = (x^2 - y^2)/2\sqrt{3}, \quad u^5 = (x^2 + y^2 - 2z^2)/6$$

is a helical minimal immersion of order 2, where $\{x,y,z\}$ (resp. $\{u^1,u^2,\dots,u^5\}$) is the canonical coordinates of the Euclidean space E^3 (resp. E^5) in which $S^2(1/3)$ (resp. $S^4(1)$) is naturally imbedded. Here we note that the above functions are harmonic homogeneous polynomials so that when they are restricted to $S^2(1/3)$, they are eigenfunctions corresponding to the second eigenvalue of the Laplace operator on $S^2(1/3)$. More generally, for a compact rank one symmetric space, we can obtain a helical immersion into a unit sphere by using the same method (cf. §2 and [27]).

In [2], A. Besse constructed helical minimal immersions of strongly harmonic manifolds into a unit sphere, which are called standard minimal immersions. This construction will be explained in §2. Conversely, if a complete Riemannian manifold M admits a helical minimal immersion into a sphere, then M is a strongly harmonic manifold (cf. §3). So we can say that the theory of helical minimal immersions into a sphere is a submanifold version of the strongly harmonic manifold theory. Thus the theory of helical immersions may be useful when we consider the well-known conjecture that strongly harmonic manifolds are compact symmetric spaces of rank one.

Let $f: M \longrightarrow S(1)$ be a helical minimal immersion into a unit sphere. There are two problems. The first is to prove that M is isometric to one of compact rank one symmetric spaces. The second is to prove that f is congruent to a standard minimal immersion. In [25], K. Tsukada solved the second problem. The first problem was studied in [9, 11, 14, 16 ~ 21] and the low order cases ($d \leq 5$) were solved. Since M is a strongly harmonic manifold, it will be nice to show that the first standard minimal immersion is of low order. However, at present, the only known result is that the order of the first standard minimal immersion is even except for the totally geodesic case (see [21]).

Helical immersions into S(1) are characterized by $\bar{\delta}(f(x),f(y)) = F(\delta(x,y))$ for every x and y in M, where $\bar{\delta}$ (resp. δ) denotes the distance function of the ambient space S(1) (resp. M) and F is a certain even function on $\mathbb{R}$. So we can say that the extrinsic distance is a function of the intrinsic distance. By using this fact, we can prove that M is a Blaschke manifold (cf. §3 and [2, 17]).

In §4, we shall study a certain special case of a helical minimal imbedding $f: M \longrightarrow S(1)$. Since M is a Blaschke manifold, geodesics are simply closed geodesic loops with the same period. We consider the condition that for every geodesic $\gamma:[0,2L] \longrightarrow M$ (2L is the period), the tangent line of γ at $\gamma(0)$ is parallel to the tangent line at $\gamma(L)$ in the Euclidean space E in which S(1) is canonically imbedded. This condition is equivalent to that $F''(L) = \pm 1$. Under this condition, we can prove that the order of f is 2 or odd, and M is isometric to one of real, complex, quaternion and Cayley projective space or a sphere respectively.

Note. Helical immersions into a Euclidean space were studied in [4, 5, 22, 26]. D. Ferus and S. Schirrmacher [8] studied submanifolds in a Euclidean space whose geodesics are curves with constant Frenet curvatures, where we note that Frenet curvatures may depend on the individual geodesic.

§2. Standard minimal immersions

Let M^n be a Riemannian manifold and $X \in T_xM - \{0\}$. Let γ be the geodesic $\gamma: s \longmapsto \exp_x \frac{s}{\|X\|} X$ and $\{J_i\} i=2,\dots,n$ be Jacobi fields along γ such that $J_i(0)=0$ for every i and $\{J'_i(0)\} i=2,\dots,n$ forms an orthonormal base of the subspace $\{X\}^\perp$ in T_xM. Then we define $\theta: TM \longrightarrow \mathbb{R}$ by $\theta(0)=1$ and $\theta(X) = \|X\|^{-n+1} \det(J_2(\|X\|),\dots,J_n(\|X\|))$, where the determinant should be understood with respect to the parallel frame field which coincides with $\{J_i'(0)\}i=2,\dots,n$ at x. If for every $x \in M$ there exists a positive real number ε_x and a function $\Theta_x:[0,\varepsilon_x) \longrightarrow \mathbb{R}$ such that $\theta(X) = \Theta_x(\|X\|)$ for every $X \in T_xM$ satisfying $\|X\|<\varepsilon_x$, then M is called a locally harmonic manifold. Moreover if M is complete and $\varepsilon_x=\infty$ for every x, then M is said to be globally harmonic. Using the canonical geodesic involution on the tangent bundle, we easily see that Θ_x does not depend on x (cf. p. 156 [2]). Hence the subscript x of Θ_x will be omitted. Let M be a compact Riemannian manifold. It has a unique kernel of the heat equation $K:M\times M\times\mathbb{R}_+^* \longrightarrow \mathbb{R}$. If there exists a function $\Psi:\mathbb{R}_+\times\mathbb{R}_+^* \longrightarrow \mathbb{R}$ such that $K(x,y,t)=\Psi(\delta(x,y),t)$ for every $x, y\in M$ and $t \in \mathbb{R}_+^*$, then M is said to be strongly harmonic. Since $K(x,y,t)=\sum_\alpha e^{-\mu_\alpha t}\sum_i \phi_i^\alpha(x)\phi_i^\alpha(y)$, where $\{\phi_i^\alpha\}i=1,\dots,N_\alpha$ is an orthonormal base of the eigenspace corresponding to the α-th eigenvalue μ_α of the Laplace operator on M, we see that for each α there exists a function $\hat{F}_\alpha:\mathbb{R}_+ \longrightarrow \mathbb{R}$ such that $\sum_i \phi_i^\alpha(x)\phi_i^\alpha(y)=\hat{F}_\alpha(\delta(x,y))$ for every $x, y \in M$.

It was shown in [2] that a strongly harmonic manifold is globally harmonic and in [13] D. Michel proved the converse in the simply connected case (see also [3]). The only known examples of strongly harmonic manifolds are compact symmetric spaces of rank one. However, the conjecture that a strongly harmonic manifold is a symmetric space (and hence of rank one) is still open as far as I know.

For each α, we define ϕ_α by $\phi_\alpha(x)=c_\alpha\cdot(\phi_1^\alpha(x),\dots,\phi_{N_\alpha}^\alpha(x)) \in \mathbb{R}^{N_\alpha}$, where $c_\alpha=(\mathrm{vol}M/N_\alpha)^{\frac{1}{2}}$. Besse [2] showed

Proposition 2.1. *If we change the metric* g *of* M *for* $\hat{g}=(\mu_\alpha/n)g$, *then* ϕ_α *is a helical minimal immersion of* M *into a unit hypersphere* $S(1)$ *in* $\mathbb{R}^{N_\alpha}$.

Proof. We sketch the proof (for the detail, see p. 175 ~ p. 178 [2]). Let < , > denote the inner product of $\mathbb{R}^{N_\alpha}$. From the definition, we have $<\phi_\alpha(x),\phi_\alpha(y)>=c_\alpha^2\hat{F}_\alpha(\delta(x,y))$, where δ is the distance function of the metric g. It follows that $\|\phi_\alpha(x)\|^2=c_\alpha^2\hat{F}_\alpha(0)$ and $\hat{F}_\alpha(0)\mathrm{vol}(M,g)=N_\alpha$. Therefore we have $\|\phi_\alpha(x)\|^2=1$, i.e., $\phi_\alpha(x) \in S(1)$ for each $x\in M$. Next we show that ϕ_α is an isometric immersion. Let $\gamma(s)$ be a geodesic with the arc-length parameter s with respect to g. Let $X=\dot{\gamma}(0)$ and $x=\gamma(0)$. Differentiating twice $\|\phi_\alpha(\gamma(s))\|^2=1$ and $<\phi_\alpha(x),\phi_\alpha(\gamma(s))>=c_\alpha^2\hat{F}_\alpha(s)$ with respect to s, we easily see that $\|\phi_{\alpha*}(X)\|^2+<\phi_\alpha(x),(\mathrm{Hess}\phi_\alpha)(X,X)>=0$ and $<\phi_\alpha(x),(\mathrm{Hess}\phi_\alpha)(X,X)>=c_\alpha^2\hat{F}''_\alpha(0)$. It follows that $\|\phi_{\alpha*}(Y)\|^2 =-c_\alpha^2\hat{F}''_\alpha(0)\|Y\|_g^2$ for any $Y \in T_xM$. Let $\{X_i\}i=1,\dots,n$ be an orthonormal base in T_xM. Then we have $<\phi_\alpha(x),\Sigma(\mathrm{Hess}\phi_\alpha)(X_i,X_i)>=-<\phi_\alpha,\Delta\phi_\alpha>=-\mu_\alpha$ and hence $\hat{F}''_\alpha(0)=-\mu_\alpha/nc_\alpha^2$. We have proved $\|\phi_{\alpha*}(Y)\|^2=(\mu_\alpha/n)\|Y\|_g^2$ for every $Y \in T_xM$. We recall Takahashi's theorem [22]. Takahashi showed that an isometric immersion f into $S(1) \subset E$ satisfies $\Delta f = \mu f$ for some μ iff f is minimal. Thus we see that $\phi_\alpha : (M, \hat{g}) \longrightarrow S(1)$ is minimal. Finally, we show that ϕ_α is helical. There is a function F_α such that $\langle \phi_\alpha(\gamma(s)), \phi_\alpha(\gamma(t)) \rangle = F_\alpha(|t-s|)$ for every close enough t and s, where $\gamma(s)$ is a unit speed geodesic of (M,g). From this fact, we can show that the Frenet curvatures of the curve $\phi_\alpha \circ \gamma$ in $S(1)$ are determined by derivatives of F_α at 0. q.e.d.

The helical minimal immersion $\phi_\alpha : (M, \hat{g}) \longrightarrow S(1)$ is called the <u>α-th standard minimal immersion</u>.

§3. <u>Helical immersions into</u> S(1)

Let $f: M \longrightarrow S(1)$ be a helical immersion of a complete Riemannian manifold into $S(1)$. Let $\iota: S(1) \hookrightarrow E$ be the canonical inclusion into a Euclidean space E and $\phi = \iota \circ f$.

<u>Proposition</u> 3.1 ([17]). <u>The isometric immersion</u> $\phi: M \longrightarrow E$ <u>is a helical immersion of order</u> d^* <u>where</u> $d^* = d$ <u>if</u> d <u>is even and</u> $= d+1$ <u>if</u> d <u>is odd</u>. <u>Denote by</u> H <u>and</u> D <u>the second fundamental form and van der Waerden-Bortolotti covariant differentiation of</u> f <u>respectively</u>. <u>Let</u> γ <u>be a unit speed geodesic</u>. <u>Then Frenet frame</u> $\{\tau_i\} i=1,\ldots,d^*$ <u>of</u> $\tau = \phi \circ \gamma$ <u>is given as follows</u>: $\tau_1 = \phi_* X$, τ_i <u>is a linear combination of</u> τ, $H(X,X)$, $(D^2H)(X,\ldots,X)$, $\ldots$, $(D^{i-2}H)(X,\ldots,X)$ <u>if</u> i <u>is even and linear combination of</u> $(DH)(X,X,X)$, $(D^3H)(X,\ldots,X)$, $\ldots$, $(D^{i-2}H)(X,\ldots,X)$ <u>if</u> i <u>is odd</u> (≥ 3), <u>where</u> $X = \dot{\gamma}$ <u>and the coefficients of the linear combinations are determined by the Frenet curvatures</u> $\lambda_1, \ldots, \lambda_{d^*-1}$ <u>of</u> τ.

<u>Sketch of proof</u>. Let ∂ and $\overline{\nabla}$ be the covariant differentiation on E and $S(1)$ respectively. By Gauss equation, we have $\partial_X X = \overline{\nabla}_X X + \langle \partial_X X, \tau \rangle \tau = H(X,X) - \tau$. Since $f \circ \gamma$ is a curve of constant Frenet curvatures in $S(1)$, we have $\|H(X,X)\|^2 = \kappa_1^2$: constant for every $X \in UM$ (unit sphere bundle of M). Thus $\lambda_1^2 = 1 + \kappa_1^2$ and $\tau_2 = \lambda_1^{-1}(-\tau + H(X,X))$. Next, using Weingarten equation, we get $\partial_X \tau_2 + \lambda_1 \tau_1 = \lambda_1^{-1}(\partial_X H(X,X) - X) + \lambda_1 X = \lambda_1^{-1}(DH)(X,X,X)$. From the condition that the second Frenet curvature κ_2 of $f \circ \gamma$ is constant, we have $\|(DH)(X,X,X)\| = \kappa_1 \kappa_2$ for every $X \in UM$. Thus $\lambda_2 = \lambda_1^{-1} \kappa_1 \kappa_2$ and $\tau_3 = (\lambda_1 \lambda_2)^{-1} (DH)(X,X,X)$. In this way, we can prove the assertion inductively. q.e.d.

Let f_i, $i=1,2,\ldots,d^*$ be the solution of a differential equation

$$
(3.1) \qquad
\begin{aligned}
f_1' &= 1 - \lambda_1 f_2, \\
f_i' &= \lambda_{i-1} f_{i-1} - \lambda_i f_{i+1} \qquad (2 \leq i \leq d^*-1), \\
f_{d^*}' &= \lambda_{d^*-1} f_{d^*-1}
\end{aligned}
$$

with initial conditions $f_i(0)=0$ for all i. It is easily verified that f_i is an odd (resp. even) function if i is odd (resp. even). Moreover, define F by $F=1-(c_2f_2+c_4f_4+\cdots+c_{d^*}f_{d^*})$, where $c_j=\lambda_2\lambda_4\cdots\lambda_{j-2}/\lambda_1\lambda_3\cdots\lambda_{j-1}$ $(c_2=1/\lambda_1)$. We have

Proposition 3.2 ([17]). *For arbitrary* $x\in M$ *and* $X\in U_xM$, *we obtain*

$$\phi(\exp_x sX) = F(s)\phi(x) + f_1(s)\phi_* X + \xi(s;X) + \zeta(s;X)$$

where $\xi(s;X)$ *and* $\zeta(s;X)$ *are normal vectors at* x *defined by*

$$\xi(s;X) = \sum_{j:\mathrm{even}} f_j(s)\tilde{\tau}_j \qquad (\tilde{\tau}_j = \tau_j - \langle\tau_j,\phi(x)\rangle\phi(x)),$$

$$\zeta(s;X) = \sum_{j:\mathrm{odd}\geq 3} f_j(s)\tau_j$$

and $\{\tau_j\}j=1,\dots,d^*$ *is the Frenet frame of the curve* $\phi(\exp_x sX)$ *at* x.

Proof. Let $\sigma(s)=\phi(\exp_x sX)$. We have only to solve the Frenet equation

$$\sigma' = \sigma_1, \quad \sigma_1' = \lambda_1\sigma_2, \quad \sigma_i' = -\lambda_{i-1}\sigma_{i-1} + \lambda_i\sigma_{i+1} \quad (2 \leqq i \leqq d^*-1),$$
$$\sigma_{d^*}' = -\lambda_{d^*-1}\sigma_{d^*-1}$$

with initial conditions $\sigma(0)=\phi(x)$, $\sigma_1(0)=\phi_* X$, $\sigma_i(0)=\tau_i$ $(2\leqq i\leqq d^*)$. Then we have $\sigma(s)=\phi(x)+f_1(s)\phi_* X+\sum_{j\geq 2} f_j(s)\tau_j$. Noting that $-\langle\tau_j,\phi(x)\rangle=c_j$ (cf. [17]), we obtain from Prop. 3.1 the desired equation. q.e.d.

Clearly we have

Corollary 3.3. *If* $f:M \longrightarrow S(1)$ *is helical*, *then we have* $\langle\phi(x),\phi(y)\rangle = F(\delta(x,y))$ *for every* x, $y\in M$.

Let $A_{\xi(s;X)}$ denote the Weingarten map corresponding to the normal vector $\xi(s;X)$. Let $V\in T_xM$ such that $V\perp X$. Define $\xi_X(s;V)$ and $(D\xi)(s;V;X)$ by $\xi_X(s;V)=\|V\|\, d/d\omega\, \xi(s;\cos\omega\cdot X+\sin\omega\cdot V/\|V\|)|_{\omega=0}$ and $(D\xi)(s;V;X)=\nabla_V^{\perp}\xi(s;X^*)$ respectively where X^* is the parallel vector field along a curve $\beta(t)$ tangent to V at x such that $X^*(0)=X$ and $\nabla^{\perp}$ denotes the normal connection of f. We define $\zeta_X(s;V)$ and $(D\zeta)(s;V;X)$ in the similar way.

Lemma 3.4. *Jacobi fields* J_V *and* J_V^* *along a unit speed geodesic* γ $(=\exp_x sX)$ *such that* $J_V(0)=0$, $\nabla_X J_V=V$ *and* $J_V^*(0)=V$, $\nabla_X J_V^*=0$ *respectively are given by*

$$J_V(s) = f_1(s)V + \xi_X(s;V) + \zeta_X(s;V), \tag{3.2}$$

$$J_V^*(s) = F(s)V - A_{\xi(s;X)}V - A_{\zeta(s;X)}V + f_1(s)H(V,X) + (D\xi)(s;V;X) + (D\zeta)(s;V;X), \tag{3.3}$$

where we omitted ϕ_*.

Proof. Let V be a unit vector. Consider a variation $(s,t)\longmapsto$

$\exp_x s(\cos tX+\sin tV)$. Using Prop. 3.2 and from $J_V(s)=d/dt\ \phi(\exp_x s(\cos tX+\sin tV))|_{t=0}$, we have (3.2). Next, we consider a variation $(s,t)\longmapsto \exp_{\beta(t)} sX^*(t)$. By Prop. 3.2,

$$\phi(\exp_{\beta(t)} sX^*(t)) = F(s)\beta(t) + f_1(s)X^*(t) + \xi(s;X^*(t)) + \zeta(s;X^*(t)).$$

Using Gauss and Weingarten equation, we have (3.3). q.e.d.

Lemma 3.5. Assume that $f:M\longrightarrow S(1)$ is a helical imbedding and M has a pole x. Then there is a divergent sequence $\{s_k\}_{k=1}^{\infty}$ such that $s_k>0$ and $\lim_{k\to\infty} f_i(s_k)=0$ for $i=1,\ldots,d^*$.

Proof. Let γ be a unit speed geodesic in M and $x=\gamma(0)$. Put $x_k=\gamma(k)$ $(k\in \mathbb{Z}_+)$. Since $\phi(x_k)\in S(1)$ for every k, a subsequence $\{\phi(y_k)\}$ of $\{\phi(x_k)\}$ converges. Put $t_k=\delta(y_k,x)$. Then $\lim F(t_k-t_{k-1})=\lim<\phi(y_k),\phi(y_{k-1})>=1$. Define a sequence $\{u_k\}$ by $u_k=t_k-t_{k-1}$ (≥ 1) for every $k\in \mathbb{Z}_+$. If $\{u_k\}$ is bounded, then a subsequence $\{u_k'\}$ of $\{u_k\}$ converges and $F(\lim u_k')=\lim F(u_k')=1$. By the assumption that f is an imbedding, we have a contradiction: $0=\lim u_k'\geq 1$. Thus $\{u_k\}$ has a subsequence $\{s_k\}$ which diverges and satisfies $\lim F(s_k)=1$. Since it is easily verified from the proof of Prop. 3.2 that $\Sigma f_i^2=2(1-F)$, we have the assertion. q.e.d.

Next we recall the definition of Blaschke manifolds ([2]). Let M be a compact Riemannian manifold. Let $x\in M$ and Cut(x) denote the cut-locus of x. If for every $y\in \mathrm{Cut}(x)$ the link

$$\{\dot{\gamma}(y)\ :\ \gamma \text{ is a minimal unit speed geodesic from } x \text{ to } y\}$$

is a great sphere of the unit tangent sphere at y of M, then M is called a Blaschke manifold at x. Moreover, if M is a Blaschke manifold at every point, then M is called a Blaschke manifold. It is known that M is a Blaschke manifold at x if and only if the distance from x to each point of Cut(x) is constant (cf. [2]). If M is a Blaschke manifold, then every geodesic is a simply closed geodesic loop with the same period and the dimension of links are independent of $x\in M$ and $y\in \mathrm{Cut}(x)$. Let e=(the dimension of links)+1. Then we see that e=1, 2, 4, 8 or n (cf. [2]). Corresponding to e=2, 4 or 8, n (=dim M) is equal to 2m, 4m or 16. It is well-known (M. Berger's theorem) that if M is a Blaschke manifold and e=n (resp. 1), then M is isometric to a sphere (resp. real projective space) of constant curvature.

Next we prove that if a complete Riemannian manifold M admits a helical immersion into S(1), then M is a Blaschke manifold. We see from [17] that if M is compact, then M is a Blaschke manifold and if M is noncompact, then every point of M is a pole. Thus it suffices to show that M is compact.

Theorem 3.6. Let M be a complete Riemannian manifold and admit a helical immersion into $S(1)$, then M is a Blaschke manifold.

Proof. At first, we note that we may always assume that a helical immersion $f:M \longrightarrow S(1)$ is an imbedding. This is shown as follows. Let $f(x)=f(y)$ for x, y $(\neq)$ in M. Then, by Cor. 3.3, $f(x)=f(z)$ for every z such that $\delta(x,z)=\delta(x,y)$. Since M is a Blaschke manifold or every point of M is a pole, the set $\{z \in M;\delta(x,z)=\delta(x,y)\}$ is a submanifold of M with positive dimension except when $M=S^n$ and x, y are antipodal points. It follows that if f is not injective, then $M=S^n$ and f is the composite $f=\hat{f}\circ\pi$, where $\pi;S^n \longrightarrow \mathbb{R}P^n$ is the projection and $\hat{f}:M \longrightarrow S(1)$ is a helical imbedding.

Assume that M is noncompact. Let $x \in M$ and $X \in U_xM$. Let γ be a unit speed geodesic such that $\gamma(0)=x$ and $\dot{\gamma}(0)=X$. Since M has no conjugate points, the set $\{J_V(s);V \in \{X\}^\perp\}$ spans the subspace $\{\dot{\gamma}(s)\}^\perp$ in $T_{\gamma(s)}M$. Thus there exists a symmetric transformation $S_X(s)$ acting on $\{X\}^\perp$ such that $J^*_V(s)=J_{S_X(s)V}(s)$ for every $V \in \{X\}^\perp$ and s $(\neq 0)$. Clearly $S_X(s)$ is smooth with respect to s. From Lemma 3.4 we have $S_X(s)=(F(s)I-A_{\xi(s;X)}-A_{\zeta(s;X)})/f_1(s)$. Using Jacobi equation, we get $<J_V(s),J_W(s)> =-<S'_X(s)^{-1}V,W>$ for every V, $W \in \{X\}^\perp$ and s $(\neq 0)$ (cf. [18, 21]). It follows that for $V \in \{X\}^\perp$, $<S_X(s)V,V>$ is a monotone decreasing function on $\mathbb{R}_+$. Moreover we have $\xi(0;X)=\zeta(0;X)=0$ and $\lim_{s\to+0}f_1(s)=+0$ because $f'_1(0)=1$ (cf. (3.1)). Thus we get $\lim_{s\to+0}<S_X(s)V,V>=+\infty$. On the other hand, we have, from Lemma 3.5, $\lim_{k\to\infty}F(s_k)=1$, $\lim_{k\to\infty}f_1(s_k)=0$ and $\lim_{k\to\infty}\xi(s_k;X)=\lim_{k\to\infty}\zeta(s_k;X)=0$ for some divergent sequence $\{s_k\}$, $(s_k>0)$. Since $<S_X(s)V,V>$ is monotone decreasing, we have $\lim_{s\to+\infty}<S_X(s)V,V>=-\infty$. Therefore there exists a unique $u_0 \in \mathbb{R}_+$ such that $<S_X(u_0)V,V>=0$ for each $X \in U_xM$ and V $(\neq 0) \in \{X\}^\perp$.

Let $V \in \{X\}^\perp \cap U_xM$ be fixed. Exchange X for $-X$. We have also $u_1 \in \mathbb{R}_+$ such that $<S_{-X}(u_1)V,V>=0$. Let β be a unit speed geodesic such that $\beta(0)=x$ and $\dot{\beta}(0)=V$. Consider a Jacobi field $J=J^*_V-J_{S_X(u_0)V}$ along γ, which satisfies $J(u_0)=0$, $J(0)=V$ and $\nabla_XJ=-S_X(u_0)V$. Since $<S_X(u_0)V,V>=0$, we have $\nabla_XJ \in \{V\}^\perp$. Therefore $\gamma(u_0)$ is a focal point of β along γ. Similarly we see that $\gamma(-u_1)$ is a focal point of β along γ, where we note that $S_{-X}(u_1)=-S_X(-u_1)$. But M has no conjugate points, and we have a contradiction. q.e.d.

Corollary 3.7. If a complete Riemannian manifold M admits a helical minimal immersion f into $S(1)$, then M is strongly harmonic.

Proof. By Takahashi's theorem [23] (see also [27]), we see that the height function $h_x;y \longmapsto <\phi(x),\phi(y)>$ satisfies $\Delta h_x=nh_x$ for arbitrary point $x \in M$. We rewrite this equation in the geodesic coordinates around x as follows:

$$(3.4)\qquad -F'' - F'\cdot\left(\frac{n-1}{s} + \frac{\theta_x'}{\theta_x}\right) = nF$$

where θ_x is defined by $\theta_x(s,X)=\theta(sX)$ for $s\in\mathbb{R}_+$ and $X\in U_xM$. Thus θ_x depends only on s. Since M is a Blaschke manifold, we can easily show that M is strongly harmonic (cf. §2 and [3, 13]). q.e.d.

§4. Helical minimal imbeddings with certain conditions

Let $f:M \longrightarrow S(1)$ be a helical minimal imbedding of a complete Riemannian manifold. Let $x\in M$, $X\in U_xM$ and $y=\exp_x LX\in \mathrm{Cut}(x)$ where L denotes the diameter of M. Let $\mathcal{H}_x(X)\subset T_xM$ be the subspace spanned by $\{\dot{\gamma}(x):\gamma$ is a geodesic from x to y$\}$ and $\mathcal{H}_x^*(X)$ the orthogonal complement of X in $\mathcal{H}_x(X)$. Cut(y) is a submanifold orthogonal to $\mathcal{H}_x(X)$ (cf. [2]). By Prop. 3.2, we see that $\mathrm{Cut}(x)=\{b\phi(x)+\xi(V):V\in U_xM\}$ and $y=b\phi(x)+\xi(Z)$ for any $Z\in U_xM\cap\mathcal{H}_x(X)$, where $b=F(L)$ and $\xi(V)=\xi(L;V)$. Taking account of Lemma 3.4 and the fact that the index of the conjugate point y equals e-1 (cf. [2]), we easily have $\mathcal{H}_x^*(X)=\{Z\in T_xM:X\perp Z$ and $\xi_X(Z)=0\}$ and $T_y\mathrm{Cut}(x)=\{\xi_X(Y):Y\in T_x\mathrm{Cut}(y)\}$ (cf. [19]).

Lemma 4.1. *For any* $V\in\{X\}^\perp$, *we have*

$$(4.1)\qquad \begin{aligned} &2a^2H(X,V) + 2a(D\zeta')(V;X)\\ &= -\xi_X(aV + A_{\eta(X)}V) + \eta_X(bV - A_{\xi(X)}V) - 2\zeta_X'(A_{\zeta'(X)}V), \end{aligned}$$

where $a=f_1'(L)$, $\zeta'(X)=\zeta'(L;X)$ *and* $\eta(X)=\xi''(L;X)$.

Proof. As in §3, we have

$$S_X(s)V = \frac{1}{f_1(s)}(F(s)V - A_{\xi(s;X)}V - A_{\zeta(s;X)}V)$$

for any s ($\neq mL$, $m\in\mathbb{Z}$). Substituting this equation into $J_V^*(s)=J_{S_X(s)V}(s)$ and using Lemma 3.4, we have

$$(4.2)\qquad \begin{aligned} &f_1(s)^2H(V,X) + f_1(s)(D\zeta)(s;V;X)\\ &= \xi_X(s;F(s)V - A_{\xi(s;X)}V) - \zeta_X(s;A_{\zeta(s;X)}V) \end{aligned}$$

for every s. Differentiate (4.2) twice at s=L. Noting that $F'=-f_1$ which is obtained from (3.1) and the definition of F, we have (4.1).

q.e.d.

Lemma 4.2. *Let M be a compact symmetric space of rank one and* $\phi_\alpha:M \longrightarrow S(1)$ *the α-th standard minimal imbedding. Then we have*

$$a = (-1)^\alpha\frac{n}{e}\binom{\alpha+\frac{e}{2}-1}{\alpha}\Big/\binom{\alpha+\frac{n}{2}-1}{\alpha}$$

where e=1, 2, 4, 8 *or* n *according to* $M=RP^n$, CP^m, QP^m, $\mathrm{Cay}P^2$ *or* S^n ($m=n/e$, $m\geqq 2$).

Proof. Since M is a compact symmetric space of rank one, Θ is given by

$$\Theta(s) = \left(\frac{2}{\nu s}\right)^{n-1}\left(\sin\frac{\nu s}{2}\right)^{n-1}\left(\cos\frac{\nu s}{2}\right)^{e-1}$$

where $\nu=\pi/L$. Substituting this equation into (3.4) and putting $t=\cos\nu s$, (3.4) reduces to

$$(4.3)\qquad (1 - t^2)\ddot{w} - \tfrac{1}{2}\{(n + e)t + (n - e)\}\dot{w} + \frac{n}{\nu^2}w = 0$$

with initial conditions $w(1)=1$ and $\dot{w}(1)=1/\nu^2$. Since the α-th eigen value of the Laplacian of M equals n, we have $\nu^2=2n/\alpha(2\alpha+n+e-2)$. It is well-known (cf. [10, 24]) that the solution of (4.3) is given by the Jacobi polynomial

$$(4.4)\qquad P_\alpha^{p,q}(t) = \frac{(-1)^\alpha}{2^\alpha}\frac{1}{\binom{p+\alpha}{\alpha}}\sum_{i=0}^{\alpha}(-1)^i\binom{p+\alpha}{i}\binom{q+\alpha}{\alpha-i}(1 - t)^{\alpha-i}(1 + t)^i$$

where $p=n/2-1$ and $q=e/2-1$. From (4.3) we have $a=-F''(L)=-\nu^2\dot{w}(-1)$ $=nw(-1)/e$. Moreover, from (4.4) we have $w(-1)=(-1)^\alpha\binom{q+\alpha}{\alpha}\Big/\binom{p+\alpha}{\alpha}$. q.e.d.

Prop. 3.2 implies that $a^2+\sum_{i\geq 3}f_i'(L)^2=1$ and hence $a^2\leq 1$. In the following theorem, we consider the extreme case $a^2=1$. If γ is a unit speed geodesic such that $\dot{\gamma}(0)=X$, then $\dot{\gamma}(L)=aX+\zeta'(X)$ (cf. Prop. 3.2). Therefore $a^2=1$ if and only if the tangent line of γ at $\gamma(0)$ is parallel in E to that at $\gamma(L)$. For instance, if the order of f is equal to 2, then we see easily that $f\circ\gamma$ is a small circle on S(1) and hence the above condition is satisfied.

Theorem 4.3. *Let* $f:M \longrightarrow S(1)$ *be a helical minimal imbedding of order* d *of a complete Riemannian manifold* M. *If* $a^2=1$, *then* d=2 *and* M $=\mathbb{R}P^n$, $\mathbb{C}P^m$, $\mathbb{Q}P^m$, $\mathbb{C}ayP^2$ *or* d *is odd and* $M=S^n$.

Proof. We assume that $e\neq n$. If e=n, then by Berger's theorem [2] (see §3) we see that M is isometric to a sphere.

First we note that the case a=1 does not occur. Because if a=1, then, from (3.1), $f_2(L)=f_4(L)=\cdots=f_{d^*}(L)=0$ and hence b=1. This contradicts the assumption that f is an imbedding. Let x be an arbitrary point of M. Let $X\in U_xM$ and $y=\gamma(L)$ where $\gamma(s)=\exp_x sX$.

Step 1. We shall prove that $R(Y,X)X=(\lambda_1^2/4)Y$ for every $Y\in T_x\mathrm{Cut}(y)$ where R is the curvature tensor of M. By Prop. 3.2,

$$(4.4)\qquad \begin{aligned}&F(s+t)x + f_1(s+t)X + \xi(s+t;X) + \zeta(s+t;X)\\ &= F(t)\gamma(s) + f_1(t)\dot{\gamma}(s) + \xi(t;\dot{\gamma}(s)) + \zeta(t;\dot{\gamma}(s)).\end{aligned}$$

Differentiating the both hand sides with respect to s at s=0, we have $A_{\xi(t;X)}X=(F(t)-f_1'(t))X$ and $\xi'(t;X)=f_1(t)H(X,X)+(D\zeta)(t;X)$ where $(D\zeta)(t;X)$ $=\nabla_X^\perp\zeta(t;\dot{\gamma})$. Thus $A_{\xi(X)}X=(b-a)X$ and $\eta(X)=aH(X,X)+(D\zeta')(X)$. Since $\zeta'(X)=0$, we get $\eta(X)=-H(X,X)$ and $\eta_X(Z)=-2H(X,Z)$ for every $X\in U_xM$, $Z\in\mathscr{K}_x^*(X)$. $\mathscr{K}_x(X)$ is an eigen space $\{Z\in T_xM: A_{\xi(X)}Z=(b-a)Z\}$, because $\xi(X)=\xi(Z)$ for every $Z\in\mathscr{K}_x(X)\cap U_xM$. Put $V=Z\in\mathscr{K}_x^*(X)$ in (4.1), then $\xi_X(A_{\eta(X)}Z)=0$. Thus we conclude that $A_{H(X,X)}$ leaves $\mathscr{K}_x^*(X)$ and $T_x\mathrm{Cut}(y)$ invariant. Moreover, putting $V=Y\in T_x\mathrm{Cut}(y)$ in (4.1), we obtain $2H(X,Y)=\xi_X(Y+A_{H(X,X)}Y)$. It follows from $<\xi_X(V),H(X,W)>+<\xi(X),H(V,W)>=(b-a)<V,W>$ for any V, $W\in\{X\}^\perp$ that $2\|H(X,Y)\|^2=1+<H(X,X),H(Y,Y)>$ where $Y\in T_x\mathrm{Cut}(y)\cap U_xM$. On the other hand, f is isotropic, i.e., $\|H(X,X)\|^2=\kappa_1^2$ for every $X\in UM$, which is equivalent to the eqation $\mathfrak{S}_{U,V,W}<H(X,U),H(V,W)>=\kappa_1^2\mathfrak{S}_{U,V,W}<X,U><V,W>$ for any X, U, V, $W\in T_xM$. It follows that $\|H(X,Y)\|^2=\lambda_1^2/4$ and $<H(X,X),H(Y,Y)>=\lambda_1^2/2-1$ for $Y\in T_x\mathrm{Cut}(y)\cap U_xM$. Using Gauss equation, we have $R(Y,X)X=(\lambda_1^2/4)Y$ for $Y\in T_x\mathrm{Cut}(y)$.

<u>Step</u> 2. We shall prove $R(Z,X)X=\lambda_1^2Z$ for $Z\in\mathscr{K}_x^*(X)$. Since $A_{H(X,X)}Y=(\lambda_1^2/2-1)Y$, we have $4\|H(X,Y)\|^2=(\lambda_1^4/4)\|\xi_X(Y)\|^2$ for every $Y\in T_x\mathrm{Cut}(y)$ U_xM. It follows that $\|\xi_X(Y)\|=2/\lambda_1$. Consider a submersion $\Pi:U_xM\longrightarrow \mathrm{Cut}(x)$ defined by $V\mapsto bx+\xi(V)$. We identify the tangent space of U_xM at X with $\{X\}^\perp$ and introduce a metric G in U_xM which is defined by $G(U,W)=(4/\lambda_1^2)<U,W>$ for U, $W\in\{X\}^\perp$. The fiber of Π through X is $\mathscr{K}_x(X)\cap U_xM$ and every fiber is an (e-1)-dimensional totally geodesic submanifold in U_xM. Since $\mathscr{K}_x^*(X)$ is orthogonal to $T_x\mathrm{Cut}(y)$ with respect to G, $T_x\mathrm{Cut}(y)$ is the horizontal space at X. For $Y\in T_x\mathrm{Cut}(y)\cap U_xM$, we have

$$\Pi_*Y = d/d\omega\,\Pi(\cos\omega X + \sin\omega Y)|_{\omega=0} = \xi_X(Y).$$

It follows that $\|\Pi_*Y\|=\|\xi_X(Y)\|=\|Y\|_G$. Therefore we have proved that $\Pi:(U_xM,G)\longrightarrow(\mathrm{Cut}(x),\text{metric induced from } M)$ is a Riemannian submersion. Here we apply Escobales' theorem [6] to our submersion Π ($e\geq 2$) which states that if $S^{n-1}(1)\longrightarrow B$ is a Riemannian submersion with connected totally geodesic fibers, then B is isometric to one of $\mathbb{C}P^m$, $\mathbb{Q}P^m$ with the canonical metric whose maximal curvature equals 4 and S^2, S^4 and S^8 of constant curvature 4. Using (4.4), we have $\zeta'(X)=(D\xi)(X)=0$. It was shown in [19] that if $A_{(D\xi)(X)}=0$ for every $X\in UM$, then every cut locus is totally geodesic in M and $\mathscr{K}_\gamma^*(\dot\gamma)$ is a parallel (e-1)-dimensional plane field along any geodesic γ. Therefore we conclude that, for each x, Cut(x) is a totally geodesic submanifold in M and isometric to a compact symmetric space of rank one with maximal curvature λ_1^2. Let $Y_0\in T_x\mathrm{Cut}(y)$ $\cap U_xM$ and $y_0=\exp_xLY_0$. Then $\mathscr{K}_x(Y_0)\subset T_x\mathrm{Cut}(y)$ by virtue of the above conclusion. Thus $\mathscr{K}_x(X)\subset T_x\mathrm{Cut}(y_0)$ which implies that the sphere $\Sigma(x,y)=\{\exp_x sZ: Z\in\mathscr{K}_x(X),\ 0\leq s\leq L\}\subset\mathrm{Cut}(y_0)$ is of constant curvature λ_1^2. Since $\Sigma(x,y)$ is totally geodesic in M, we have $R(Z,X)X=\lambda_1^2Z$ for every $Z\in\mathscr{K}_x^*(X)$.

<u>Step</u> 3. Using the fact that $\mathscr{K}_\gamma^*(\dot\gamma)$ is parallel along γ, it is easy to show $\nabla R(X;X,\cdot)X=0$ for any $X\in TM$. From this result we can show that M is locally symmetric (cf. p. 65 [2]). Since M is a Blaschke manifold

(see §3), we see that M is a compact symmetric space of rank one. By Takahashi's theorem [23] (cf. §2), n (=dimM) is an eigen value of the Laplacian. Let n be the α-th eigen value. Consider the α-th standard minimal immersion $\phi_\alpha : M \longrightarrow S(1)$ where we note that $\hat{g}=g$ in Prop. 2.1. Since F and F_α (of ϕ_α) satisfies the same differential equation (3.4) with the same initail conditions, we have $F=F_\alpha$. It follows from Lemma 4.2 that

$$-1 = a = (-1)^\alpha \frac{n}{e} \frac{e/2(e/2+1)\cdots(e/2+\alpha-1)}{n/2(n/2+1)\cdots(n/2+\alpha-1)}.$$

Therefore we see that α is odd if n=e and α=1 if n≠e. On the other hand, Mashimo [12] and Tsukada [25] proved that the order d of the α-th standard minimal imbedding ϕ_α is equal to α if $M=S^n$ and 2α if $M=RP^n$, CP^m, QP^m, $CayP^2$. Using this result, we have the assertion. q.e.d.

References

[1] A. Allamigeon, Propriétés globales des espaces de Riemann harmoniques, Ann. Inst. Fourier 15, 91-132 (1965).

[2] A. Besse, Manifolds all of whose geodesics are closed, Ergebnisse der Mathematik, Bd. 93. Berlin, Heidelberg, New York, Springer 1978.

[3] J. Cheeger and S. T. Yau, A Lower bound for the heat kernel, Comm. Pure and Applied Math. 34, 465-480 (1981).

[4] B.-Y. Chen and P. Verheyen, Sous-variétés dont les sections normales sont des géodésiques, C. R. Acad. Sci. Paris Sér. A293, 611-613 (1981).

[5] B.-Y. Chen and P. Verheyen, Submanifolds with geodesic normal sections, Math. Ann. 269, 417-429 (1984).

[6] R. Escobales, Riemannian submersions with totally geodesic fibers, J. Differential Geometry 10, 253-276 (1978).

[7] D. Ferus, Symmetric submanifolds of Euclidean space, Math. Ann. 247, 81-93 (1980).

[8] D. Ferus and S. Schirrmacher, Submanifolds in Euclidean space with simple geodesics, Math. Ann. 260, 57-62 (1982).

[9] S. L. Hong, Isometric immersions of manifolds with plane geodesics into Euclidean space, J. Differential Geometry 8, 259-278 (1973).

[10] Y. Komatsu, Special functions, in Japanese, Asakura, 1967.

[11] J. A. Little, Manifolds with planar geodesics, J. Differential Geometry 11, 265-285 (1976).

[12] K. Mashimo, Order of the standard isometric immersions of CROSS as helical geodesic immersions, Tsukuba J. Math. 7, 257-263 (1983).

[13] D. Michel, Comparison des notions de variétés Riemanniennes globalement harmoniques et fortement harmoniques, C. R. Acad. Sci. Paris, Ser. A282, 1007-1010 (1976).

[14] H. Nakagawa, On a certain minimal immersion of a Riemannian manifold into a sphere, Kodai Math. J. 3, 321-340 (1980).

[15] H. S. Ruse, A. G. Walker and T. J. Willmore, Harmonic spaces, Consiglio Nacionale delle Ricerche Monographie Mathematiche 8, Roma, Edizioni Cremonese 1961.

[16] K. Sakamoto, Planar geodesic immersions, Tohoku Math. J. 29, 25-56 (1977).

[17] K. Sakamoto, Helical immersions into a unit sphere, Math. Ann. 261, 63-80 (1982).

[18] K. Sakamoto, On a minimal helical immersion into a unit sphere, Advanced Studies in Pure Math. 3, 193-211 (1984).

[19] K. Sakamoto, Helical minimal immersions of compact Riemannian manifolds into a unit sphere, Trans. Amer. Math. Soc. 288, 765-790 (1985).

[20] K. Sakamoto, Helical minimal imbeddings of order 4 into spheres, J. Math. Soc. Japan 37, 315-336 (1985).

[21] K. Sakamoto, The order of helical minimal imbeddings of strongly harmonic manifolds, to appear in Math. Z.

[22] K. Sakamoto, Helical immersions into a Euclidean space, to appear in Michigan Math. J.

[23] T. Takahashi, Minimal immersions of Riemannian manifolds, J. Math. Soc. Japan 18, 203-215 (1968).

[24] F. G. Tricomi, Vorlesungen über orthogonal reihen, Springer, 1955.

[25] K. Tsukada, Helical geodesic immersions of compact rank one symmetric spaces into spheres, Tokyo J. Math. 6, 267-285 (1983).

[26] P. Verheyen, Submanifolds with geodesic normal sections are helical, to appear in Rendiconto del Seminario Mathematics dell' reniversita e del Politecmico di Torino.

[27] N. R. Wallach, Minimal immersions of Riemannian manifolds, Symmetric spaces, edited by W. M. Boothby and G. L. Weiss, M. Dekker, New York, 1972.

On topological Blaschke conjecture III

Hajime SATO
Mathematical institute
Tôhoku University
Sendai 980/ JAPAN

§1. Statement of results

In this note, a sequel to the preceding papers I[8] and II[9], we study the homotopy type of Blaschke manifolds whose cohomology ring is equal to that of the quaternion projective space $\mathbb{H}P^n$.

Our main result is the following.

Theorem. Let (M,g) be a Blaschke manifold whose cohomology ring is equal to that of the quaternion projective space $\mathbb{H}P^n$ for $n \geq 1$. Then M is homotopy equivalent to $\mathbb{H}P^n$.

By a discussion concerning the K-theory of $\mathbb{H}P^n$ as in I[8], Theorem will imply the following;

any Blaschke manifold is homeomorphic to a compact rank one symmetric space,

which is also a conjecture of Gluck-Warner-Yang[3].

Our Theorem is a consequence of the following.

Theorem A. The structure group of a smooth fibration of a unit sphere by great 3-spheres reduces to $Sp(1) = S^3$.

As a corollary of Theorem A, we have the following.

Theorem B. The base manifold of a smooth fibration of S^{4k+3} by great 3-spheres is homotopy equivalent to $\mathbb{H}P^k$.

In II[9], Thorem A is proved for smooth fibrations of S^7 by great 3-spheres. So we study smooth fibrations of S^{4k+3} by great 3-spheres for $k \geq 2$.

§2. Unknotting lemma

Let $\mathbb{H}$ denote the non-commutative field of quaternions.

In $S^{4k+3} = \{(q_0,q_1,\dots,q_k);\ \Sigma|q_i|^2 = 1,\ q_i \in \mathbb{H}\}$,

we have the two spheres

$$S_0^{4k-1} = \{(q_0,q_1,\dots,q_k) \in S^{4k+3};\ q_k = 0\}$$

$$S_0^{7} = \{(q_0,q_1,\dots,q_k) \in S^{4k+3};\ q_i = 0 \quad \text{for } k-1 \geq i \geq 1\} .$$

Then

$$S_0^{4k-1} \cap S_0^{7} = S_0^{3} = \{(q_0,q_1,\dots,q_k) \in S^{4k+3};\ q_i = 0 \text{ for } i \geq 1\} .$$

Now let Σ^{4k-1} and Σ^7 be two PL-submanifolds in S^{4k+3} such that Σ^{4k-1} is PL-homeomorphic to S^{4k-1} and Σ^7 is PL-homeomorphic to S^7 . Assume that

$$\Sigma^{4k-1} \cap \Sigma^7 = S_0^3 .$$

<u>Lemma 1</u>. There exists a PL-isotopy f_t $(0 \leq t \leq 1)$ of S^{4k+3} keeping S_0^3 fixed with $f_0 = \text{id.}$ and $f_1(\Sigma^{4k-1}) = S_0^{4k-1}$.

Proof. Since the codimension of Σ^{4k-1} in S^{4k+3} is greater than two, by Zeeman's unknotting theorem[4], we have a PL-isotopy h_t of S^{4k+3} moving Σ^{4k-1} onto S_0^{4k-1}. We want to keep S_0^3 fixed under the isotopy. We extend the isotopy

$$h = \{h_t\} : S^{4k+3} \times I \rightarrow S^{4k+3} \times I$$

to a PL-homeomorphism $\tilde{h}$ of $D^{4k+4} = S^{4k+3} \times I \cup_{S^{4k+3} \times \{0\}} D_0^{4k+4}$ onto itself by letting $\tilde{h}$ be the identity on D_0^{4k+4}. Then the image

$$\tilde{h}(S_0^3 \times I \cup D_0^4)$$

is a 4-disc and the pair $(D^{4k+4}, \tilde{h}(S_0^3 \times I \cup D_0^4))$ is a disc knot. By the unknotting theorem again, we can move $\tilde{h}(S_0^3 \times I \cup D_0^4)$ by an isotopy of D^{4k+4} keeping the boundary fixed to the standard 4-disc.

Since D_0^{4k+3} is a regular neighborhood of the center O of D_0^{4k+4}, we can further assume that the resultant PL-homeomorphism of D^{4k+4} fixes D_0^{4k+4} in D^{4k+4}. Thus we obtain a pseudo-isotopy of S^{4k+3} moving Σ^{4k-1} to S_0^{4k+3} keeping S_0^3 fixed. This is equivalent to the existence of an isotopy of S^{4k+3} with the same properties ([4], Theorem 9.1). This completes the proof.

Let V be an open regular neighborhood of Σ^{4k-1} in S^{4k+3}. Then, as a corollary of Lemma 1, we obtain the following.

<u>Lemma 2</u>. $S^{4k+3} - V$ is PL-homeomorphic to $S^3 \times D^{4k}$.

After the isotopy of Lemma 1, we have the embedding of $\Sigma^7 - (\Sigma^7 \cap V)$ in $S^{4k+3} - V$ such that $\partial(\Sigma^7 - (\Sigma^7 \cap V)) = \partial(S_0^7 - (S_0^7 \cap V_0))$, where V_0 is a regular neighborhood of S_0^3 in S_0^7. The space $S_0^7 - (S_0^7 \cap V_0)$ is PL-homeomorphic to $S^3 \times D_0^4$.

§3. Trivialization

In the situation of §2, we assume the following.

The space $S^{4k+3} - V$ is the total space of the trivial fiber bundle over a closed 4k-disc E with fiber S^3 and $\Sigma^7 - (\Sigma^7 \cap V)$ is the restriction of the bundle over a closed 4-disc F embedded in E.

By the isotopy of Lemma 1, we have two trivializations of the bundle $\partial(\Sigma^7 - (\Sigma^7 \cap V)) = \partial(S_0^7 - (S_0^7 \cap V_0)) \to \partial F$; one is obtained by restricting a trivialization of the S^3-bundle over ∂F and the other is obtained as the total space $S^3 \times \partial D_0^4$ over ∂D_0^4 by choosing a PL-homeomorphism from $\partial(\Sigma^7 - (\Sigma^7 \cap V))$ to $S^3 \times \partial D_0^4$.

Let PL(4) denote the structure group of PL 4-block bundle (Rourke-Sanderson[6]). Then we know that ([6], 1.4, 1.5, 2.10)

$$\pi_3(PL(4)) = \pi_3(GL(4)) .$$

<u>Lemma 3</u>. The homotopy classes of the two trivializations of $\partial(\Sigma^7 - (\Sigma^7 \cap V))$ coincide, i.e., the difference in $\pi_3(PL(4))$ is zero.

Proof. Since both trivializations are the restriction of the trivialization of an S^3-bundle over S^{4k-1}, it is sufficient to show that the trivializations coincide over S^{4k-1}. The difference results from the restriction to the boundary $S^3 \times S^{4k-1}$ of a PL-homeomorphism of $S^3 \times D^{4k}$ onto itself. But, by the PL-analogue (Kato[5]) of a result in [7], we know that, if an element d in $\pi_{4k-1}(PL(4))$ lies in the image of the restriction homomorphism

$$\tilde{\pi}_0(PL(S^3 \times D^{4k})) \rightarrow \tilde{\pi}_0(PL(S^3 \times S^{4k-1})) ,$$

then $d = 0 \in \pi_{4k-1}(PL(4))$. This completes the proof.

§4. Embedding submanifolds by surgery

Let $p: S^{4k+3} \rightarrow B$ be a smooth fibration of S^{4k+3} by geat 3-spheres. The base space B is a simply connected smooth manifold. By the spectral sequence argument, we know that the integral cohomology ring of B is equal to that of $\mathbb{H}P^k$.

The following is a consequence of the surgery technique due to Browder-Novikov (See Browder-Hirsch[1] and Wall[10,Corollary 11.3.4]).

Lemma 4. We can embed a (4k-4)-dimensional closed PL-manifold A in B so that the inclusion is a homotopy equivalence between A and $B - \{pt.\}$.

Then the inverse $p^{-1}(A)$ is a closed PL-submaifold of S^{4k+3} and is homotopy equivalent to S^{4k-1}. By the h-cobordism theorem, $p^{-1}(A)$ is PL-homeomorphic to S^{4k-1}. Let N be an open regular neighborhood of A in B. The h-cobordism theorem implies that $B - N$ is PL-homeomorphic to the 4k-disc. By the unknotting theorem, we can choose a PL-4-sphere C in B such that $A \cap C = \{pt.\}$ and the intersection $F = C \cap (B - N)$ is PL-homeomorphic to the 4-disc.

The inverse $p^{-1}(C)$ is a PL-submanifold of S^{4k+3} and is PL-homeomorphic to S^7. The intersection

$$p^{-1}(A) \cap p^{-1}(C)$$

is PL-homeomorphic to S^3.

§5. Differential of the map g

Let $G(4k+4,4)$ denote the Grassmann manifold consisting of 4-planes in $\mathbb{R}^{4k+4}$. Let $p: S^{4k+3} \to B$ be a smooth fibration of S^{4k+3} by great 3-spheres as before. The base space B is a simply connected smooth manifold. By taking the 4-plane determined by the fiber, we have the smooth map

$$g: B \to G(4k+4,4) ,$$

which is an embedding(II[9, Propositon 3]). Let b_0 be a base point of B. We may suppose that $g(b_0) = x_0$, where x_0 is the canonical 4-plane $\mathbb{R}^4$ in $\mathbb{R}^{4k+4}$.

Let $M(4k,4)$ denote the set of real $(4k \times 4)$-matrices. We may naturally identify $T_{x_0}G(4k+4,4)$ with $M(4k,4)$ and the exponential map

$$\mathrm{Exp}: M(4k,4) \to G(4k+4,4)$$

is given by

$$\mathrm{Exp}(V) = \{\begin{pmatrix} I_4 \\ V \end{pmatrix}\}$$

where $\{\ \}$ denote the 4-plane determined by the column $(4k+4)$-vectors.

Let W be a small neighborhood of b_0 in B. For $b \in W$, we may write $g(b)$ as

$$g(b) = \{\begin{pmatrix} I_4 \\ (\mathrm{Exp})^{-1}(g(b)) \end{pmatrix}\} .$$

Since g is defined from a smooth fibration of S^{4k+3} by great 3-spheres, the map

$$\varphi : S^3 \times W \to S^{4k+3}$$

defined by

$$\varphi(q,b) = g(b)q \;/\; |g(b)q|$$

is an embedding for any $q \in S^3$, where $g(b)q$ is the product of the $(4k \times 4)$-matrix with the $(4k \times 1)$-matrix. The differential

$$\varphi_{*(q,0)}: T_qS^3 \times T_0W \to T_{\varphi(q,0)}S^{4k+3}$$

is non-singular for any q . This is equivalent to that the differential of the map

$$\tilde{\varphi} : \mathbb{R}^4 \times W \to \mathbb{R}^{4k+4}$$

defined by

$$\tilde{\varphi}(q,b) = g(b)(q)$$

is non-singular at $(q,0)$. Define a linear map

$$\tilde{g}_*: \mathbb{R}^4 \to M(4k,4k)$$

by

$$\tilde{g}_*(q)(x) = g_*(x)q \quad \text{for} \quad x \in \mathbb{R}^{4k} = T_{b_0}B .$$

Then the differential $\tilde{\varphi}_{*(q,0)}$ is equal to the multiplication by the matrix

$$\begin{pmatrix} I_4 & 0 \\ (\mathrm{Exp})^{-1}(g(b)) & \tilde{g}_*(q) \end{pmatrix} .$$

Consequently $\tilde{g}_*(q)$ is non-singular and we obtain the following, which is also proved in (II[9, Proposition 11] .

<u>Lemma 5</u>. For $x \neq 0 \in \mathbb{R}^{4k} = T_{b_0}(B)$,

$$\mathrm{rank}(g_*(x)) = 4 .$$

Let

$$g_t: W \to G(4k+4,4) ,$$

for $0 \leq t \leq 1$, be a deformation of $g|_W$ such that $g_0 = g|_W$,

$$g_{t_*}(b_0) = g_*(b_0) : T_{b_0}(B) \to M(4k,4)$$

and

$$g_1 = (\mathrm{Exp})g_*(b_0)(\mathrm{Exp})^{-1} .$$

Then we can retake a smaller neighborhood W of b_0 in B such that the map

$$\varphi_t: S^3 \times W \to S^{4k+3}$$

defined by $\varphi_t(q,b) = g_t(b)(q) \,/\, |g_t(b)(q)|$ is a smooth embedding for any $0 \leq t \leq 1$. Note that

$$\varphi_1(q,b)=((\mathrm{Exp})g_*(b_0)(\mathrm{Exp})^{-1})(b)(q)/|((\mathrm{Exp})g_*(b_0)(\mathrm{Exp})^{-1})(b)(q)|.$$

By the isotopy extension theorem, we have an ambient isotopy ψ_t of S^{4k+3} covering φ_t.

Let $\mathbb{R}^{4k} = \mathbb{R}^{4k-4} \oplus \mathbb{R}^4$ be the canonical decomposition. We choose the PL-submanifolds A and C of §4 so that

$$A \cap W \subset (\mathrm{Exp})g_*(b_0)(\mathbb{R}^{4k-4}),$$

$$C \cap W \subset (\mathrm{Exp})g_*(b_0)(\mathbb{R}^4).$$

Identify W with the subset $(\mathrm{Exp})^{-1}(W)$ in $\mathbb{R}^{4k}$ by $(\mathrm{Exp})^{-1}$. Then the map $\varphi_1 \colon S^3 \times W \to S^{4k+3}$ satisfies $\varphi_1(q,x) = g_*(x)(q)$. Write φ_1 simply by g_*. Then the map $g_* \colon S^3 \times W \to S^{4k+3}$ is the restriction of the bilinear map

$$\mathbb{R}^4 \times \mathbb{R}^{4k} \to \mathbb{R}^{4k+4} \ ; \quad (q,x) \to g_*(x)(q),$$

which we also denote by g_*. We may take W to be an ε-disc in $\mathbb{R}^{4k}$.

Denote by Σ^{4k-1} and Σ^7 the images of the isotopy $\psi_1(p^{-1}(A))$ and $\psi_1(p^{-1}(C))$ in S^{4k+3} respectively. We have proved the following.

Proposition 6. The PL-submanifolds Σ^{4k-1} and Σ^7 in S^{4k+3} satisfy the following equalities;

$$g_*(S^3 \times (\mathbb{R}^{4k} \cap W)) = \Sigma^{4k-1} \cap g_*(S^3 \times W)$$
$$g_*(S^3 \times (\mathbb{R}^4 \cap W)) = \Sigma^7 \cap g_*(S^3 \times W).$$

§6. Reduction on 4-skeleton

For a small ε-disc W in $\mathbb{R}^{4k}$, we have the map

$$g_* \colon S^3 \times W \to S^{4k+3}$$

defined by

$$g_*(q,x) = \{\begin{pmatrix} q \\ g_*(x)q \end{pmatrix}\} .$$

By Lemma 5, $g_*(x)$ is a $(4k \times 4)$-matrix whose rank is 4 if $x \neq 0$.

Consider the map

$$\tilde{g} : S^3 \to GL(4k)$$

defined by

$$\tilde{g}(q)(x) = g_*(q,x) .$$

Since the homomorphism

$$\pi_3(GL(4k-4)) \to \pi_3(GL(4k))$$

induced by the natural inclusion is surjective, we have the isotopy

$$\tilde{g}_t : S^3 \to GL(4k) \qquad \text{for} \quad 0 \le t \le 1$$

such that $\tilde{g}_0 = \tilde{g}_*$ and $\tilde{g}_1(q)$ is equal to the form

$$(*) \qquad \begin{pmatrix} a(q) & 0 \\ 0 & b(q) \end{pmatrix} ,$$

where $a(q) \in O(4k-4)$, $b(q) \in O(4)$.

By the isotopy extension theorem, we have an ambient isotopy $\tilde{\Psi}_t$ ($0 \le t \le 1$) of S^{4k+3} extending the embedding

$$\tilde{g}_t : S^3 \times W \to S^{4k+3}$$

defined by $\tilde{g}_t(q,x) = \tilde{g}_t(q)(x)$.

After the isotopy $\tilde{\Psi}_t$, we have two submanifolds $\tilde{\Psi}_1(\Sigma^{4k-1})$ and $\tilde{\Psi}_1(\Sigma^7)$ in S^{4k+3} such that

$$\tilde{\Psi}_1(\Sigma^{4k-1}) \cap \tilde{\Psi}_1(\Sigma^7) = S_0{}^3 ,$$

and there exists an open neighborhood U of $S_0{}^3$ such that

$$U \cap \tilde{\Psi}_1(\Sigma^{4k-1}) = U \cap S_0{}^{4k-1} ,$$

$$U \cap \tilde{\Psi}_1(\Sigma^7) = U \cap S_0{}^7 .$$

By Lemma 1, we have an isotopy f_t $(0 \le t \le 1)$ of S^{4k+3} keeping $S_0{}^3$ fixed such that

$$f_1(U) = U ,$$

$$f_1(U \cap \tilde{\Psi}_1(\Sigma^{4k-1}) = U \cap \tilde{\Psi}_1(\Sigma^{4k-1}) ,$$
$$f_1(U \cap \tilde{\Psi}_1(\Sigma^7) = U \cap \tilde{\Psi}_1(\Sigma^7) .$$

Then f_1 is the mapping of the trivial D^{4k}-bundle U over S_0^3 onto itself defined by the multiplication of the matrix

$$\begin{pmatrix} a'(q) & 0 \\ 0 & b'(q) \end{pmatrix} ,$$

where $a'(q) \in O(4k-4)$, $b'(q) \in O(4)$, $q \in S_0^3$.

Let $\mathbb{R}^4$ be the last 4-dimensional linear space in $\mathbb{R}^{4k+4}$ and put

$$V = W \cap \mathbb{R}^4 .$$

Then V is a 4-disc of radius ε in $\mathbb{R}^4$. Let Π_k be the projection of $\mathbb{R}^{4k+4}$ to the (last) $\mathbb{R}^4$. For $q \in S^3$ and $x \in V$, $p^{-1}(V)$ is equal to $g_*(q,x)$. By Lemma 3, the map

$$b'(q)\Pi_k : f_1(\tilde{\Psi}(g_*(S^3, V - \{0\}))) \to S^3 ,$$

is the trivialization of the S^3-bundle over $V - \{0\}$ which extends to a trivialization of the S^3-bundle $f_1(\tilde{\Psi}(p^{-1}(C - V)))$ over $C - V$.

Consequently, we obtain the following.

<u>Lemma 7</u>. The transition function

$$t : \partial V \to O(4)$$

of the bundle $p : \Sigma^7 \to C = V \cup (C - V)$ is given by

$$t(x)(q) = b'(q)b(q)(x) ,$$

for $x \in \partial V$ and $q \in S^3$.

Since $b'(q)b(q) \in O(4)$ and $t(x) \in O(4)$, the bilinear map

$$\mu : \mathbb{R}^4 \times \mathbb{R}^4 \to \mathbb{R}^4$$

defined by $\mu(q,x) = b(q)b'(q)(x)$ for $(q,x) \in S^3 \times S^3_\varepsilon$ has no zero devisor.

By II[9], or by [2], [3], we have the following.

<u>Proposition 8</u>. Let

$$\lambda : \mathbb{R}^4 \times \mathbb{R}^4 \to \mathbb{R}^4$$

be a bilenear map without zero devisor. Then the induced map

$$\tilde{\lambda} : S^3 \to GL(4)$$

defined by $\tilde{\lambda}(x)(y) = \lambda(x,y)$, for $x \in S^3 \subset \mathbb{R}^4$, $y \in \mathbb{R}^4$, is homotopic to the map

$$\tilde{\lambda}_0 : S^3 \to Sp(1) \subset GL(4)$$

defined by the multiplication (on the left or right) of norm one quaternions.

Since $H^4(B;\mathbb{Z}) \cong \mathbb{Z}$ is generated by $[C]$, we have proved the following.

Proposition 9. The structure group of a smooth fibration $p: S^{4k+3} \to B$ of S^{4k+3} by great 3-spheres reduces to $Sp(1)$ on the 4-skeleton of B .

§7. Proof of theorems

PROOF of Theorem A. We have the map $b: B \to G(4k+4,4)$. We also have the natural inclusion

$$i: G(4k+4,4) \to BO(4) = \lim_{N\to\infty} G(N,4) .$$

Then the composition ib classifies the bundle $p: S^{4k+3} \to B$. Since B is 3-connected, we have the lifting

$$\hat{b} : B \to BSpin(4)$$

of ib unique up to homotopy. Since the group $Spin(4)$ is isomorphic to the product $Sp(1) \times Sp(1)$, the cofibration

$$Sp(1) \to Spin(4) \to Spin(4)/Sp(1)$$

is equal to the exact sequence of the groups

$$Sp(1) \to Spin(4) \to Sp(1) .$$

This induces the cofibration of classifying spaces

$$BSp(1) \to BSpin(4) \xrightarrow{j} BSp(1) .$$

Then the map $\hat{b}: B \to BSpin(4)$ lifts to a map $B \to BSp(1)$ if and only if the composition $j\hat{b}$ is homotopic to the trivial map. Since $BSp(1)$ is homotopy equivalent to the infinite-dimensional quaternion projective space $\mathbb{H}P^\infty$,

$$H^*(BSp(1);\, \mathbb{Z}) \cong \mathbb{Z}[\alpha]\ , \qquad \alpha \in H^4(B;\, \mathbb{Z})\ .$$

The map $\hat{jb}$ is homotopic to zero if and only if

$$(\hat{jb})^*(\alpha^k) = 0$$

for all k. Note that $(\hat{jb})^*(\alpha^k) = ((\hat{jb})^*(\alpha))^k$. By Proposition 9, we have $(\hat{jb})^*(\alpha) = 0$. Consequently we have $(\hat{jb})^*(\alpha^k) = 0$ for all k and the map $\hat{jb}$ is homotopic to zero. This completes the proof of Theorem A.

PROOF of Theorem B. From Theorem A, we obtain the classifying map $h:B \to BSp(1)$ of the bundle $p:S^{4k+3} \to B$. Note that

$$BSp(1) \simeq \mathbb{H}P^\infty = \lim_{N\to\infty} \mathbb{H}P^N\ .$$

Since $\dim B = k$, by the cellular approximation, we may suppose that

$$h(B) \subset \mathbb{H}P^k\ .$$

The map $h:B \to \mathbb{H}P^k$ induces an isomorphism of integral homology groups. Since B and $\mathbb{H}P^k$ are simply connected, by the Whitehead theorem, h is a homotopy equivalence, which completes the proof of Theorem B.

PROOF of Theorem. Let $m \in M$ be a point of a Blaschke manifold M. By the definition, the tangential cut locus $C_m \subset T_mM$ is isometric to the sphere S^{4n-1} of the constant radius and the map

$$\mathrm{Exp}_m\colon C_m \to C(m)$$

is a smooth fibration of S^{4n-1} by great 3-spheres(see[8]), where $c(m) \subset M$ is the cut locus of m. Let E be the 4-disc bundle associated to the bundle Exp_m. The Blaschke manifold M is diffeomorphic to the union

$$D^{4n} \cup E$$

glued by an orientation preserving diffeomorphism along the boundaries. From Theorem B, we know that $C(m)$ is homotopy equivalent to $\mathbb{H}P^{n-1}$ and the bundle Exp_m is isomorphic to the Hopf bundle under the homotopy identification. Since an orientation preserving diffeomorphism of $\partial D^{4n} = S^{4n-1}$ is homotopic to the identity map, M is homotopy equivalent to $\mathbb{H}P^n$. The proof is complete.

REFERENCES

[1] W. Browder and M. W. Hirsch, Surgery on piecewise linear manifolds and applications, Bull Amer. Math. Soc. 72(1966), 959-964.

[2] T. Buchanan, Zur Topologie der projektiven Ebenen über reelen Divisionsalgebren, Geometriae Dedicata 8(1979), 383-393.

[3] H. Gluck, F. Warner and C. T. C. Yang, Division algebras, fibrations of spheres and the topological determination of spaces by the gross behavior of its geodesics, Duke. Math. J. 50 (1983), 1041-1076.

[4] J. F. P. Hudson, Piecewise Linear Topology, Benjamin, New-York, Amsterdam, 1969.

[5] M. Kato, A concordance classification of PL homeomorphisms of $S^p \times S^q$, Topology 8(1969), 371-383.

[6] C. P. Rourke and B. J. Sanderson, Block bundles I, II, III, Ann. of Math. 87(1968), 1-28, 255-277, 431-483.

[7] H. Sato, Diffeomorphism groups of $S^p \times S^q$ and exotic spheres, Quart. J. Math. Oxford 20(1969), 255-276.

[8] H. Sato, On topological Blaschke conjecture I, Geometry of Geodesics and Related Topics, Advanced Studies in Pure Math. 3, Kinokuniya, Tokyo, 1984.

[9] H. Sato and T. Mizutani, On topological Blaschke conjecture II, Tôhoku Math. J. 36(1984), 159-174.

[10] C. T. C. Wall, Surgery on compact manifold, Academic Press, London, 1970.

Critical Points of Busemann Functions on Complete Open Surfaces

Katsuhiro Shiohama

Department of Mathematics
Faculty of Science
Kyushu University
Fukuoka, 812-JAPAN

§0. Introduction. The existence of total curvature on a connected, oriented, complete, noncompact and finitely connected Riemannian 2-manifold M imposes strong restrictions to the Riemannian structure which defines it. The purpose of the present article is to investigate relation between total curvature and the behavior of Busemann functions on such an M. From definition of a Busemann function it will be natural to anticipate that a Busemann function will possess similar properties to a distance function to a point on M.

The distance function to an arbitrary fixed point p on a finitely connected M admitting total curvature has the following properties (see [8], p313) : (1) There exists a compact set $K \subset M$ such that the distance function to p has no critical point outside K, (2) If $S(t) := \{x \in M : d(x,p) = t\}$ lies in $M - K$, then the number of components of $S(t)$ is equal to that of the endpoints of M and each component of $S(t)$ is homeomorphic to a circle.

If the Gaussian curvature G on M is nonnegative everywhere, then every Busemann function has such nice properties that all of its level sets are simultaneously compact or non-compact and have at most two components and that all leavels except the munimum set (if exists) are homeomorphic to either a circle or else a line (sse [5] and [6]). It should be noted that all these properties are the consequences of convexity of Busemann functions, and this convexity is guaranteed by the assumption that $G \geq 0$ (see [3]).

However in general case where M does not admit total curvature, the behavior of Busemann functions will not be able to controle, but still certain restrictions can be obtained under the assumption that M admits total curvature.

We are interested in the distribution of critical points of Busemann functions as well as the topology of their level sets. To state our main results some notations will be needed. M is by definition finitely connected iff it is homeomorphic to a compact 2-manifold without boundary from which k $(1 \leq k < \infty)$ distinct points are removed. Such an M is called to have k ends. A core C of a finitely connected complete open surface M with k ends is by definition a compact domain whose boundary ∂C consists of k disjoint simply closed curves each of which is a broken geodesic and each component of $M - \mathrm{Int}(C)$ is homeomorphic to $S^1 \times [0,\infty)$. Such a component of $M - \mathrm{Int}(C)$ will be called a tube of M relative to C. For a core C of M and for a ray γ on M there exists a unique tube $U(\gamma)$ relative to C which contains a subray of γ. For a core C of M and for a ray γ let $M(\gamma)$ be a new complete noncompact Riemannian 2-manifold without boundary which satisfies : (1) $M(\gamma)$ has onr end, (2) There exists an isometric embedding $\iota : U(\gamma)\ C \to M(\gamma)$, (3) each component of $M(\gamma) - \iota(U(\gamma)\ C)$ is homeomorphic to a disk. It follows from definition of $M(\gamma)$ that

$$\chi(M(\gamma)) = \chi(M) + k - 1,$$

where $\chi(M)$ is the Euler characteristic of M.

With these notations our main result will be stated as follows.

Main Theorem. Let M be finitely connected with k ends and admit total curvature. Let $C \subset M$ be a core of M and γ a ray on M. Then the following statements are valid.

(1) There exists a core $C_1 \supset C$ of M such that F_γ has no critical point outside $U(\gamma) \cup C_1$.

(2) If $c(M(\gamma)) \neq (2\chi(M(\gamma)) - 1)\pi$, then F_γ has no critical point outside C_1.

(3) If the set of all critical points of F_γ is unbounded, then $c(M(\gamma)) = (2\chi(M(\gamma)) - 1)\pi$.

(4) If $c(M(\gamma)) < (2\chi(M(\gamma)) - 1)\pi$, then each level set of F_γ lying outside C_1 consists of k components each of which is homeomorphic to a circle.

(5) If $c(M(\gamma)) < (2\chi(M(\gamma)) - 1)\pi$, then each level set of F_γ lying outside $U(\gamma) \cup C_1$ consists of $k-1$ components, each of which is homeomorphic to a circle and that lying in $U(\gamma) - C_1$ is homeomorphic to a line.

A basic observation for the proof of our Main Theorem will be done on surfaces having one end, and we shall prove in §3 the following

Theorem 3.2. Assume that M has one end and admits total curvature $c(M)$. If there exists a Busemann function on M whose critical points contain an unbounded sequence, then $c(M) = (2\chi(M) - 1)\pi$. If $c(M) \neq (2\chi(M) - 1)\pi$, then there exists a compact set such that any Busemann function on M has no critical point outside the compact set.

It should be noted that the converse of the first statement of Theorem 3.2 is not true. For instance, let M^2 be a complete open surface in R^3 of positive Gaussian curvature having the total curvature π. Then, every Busemann function on M^2 is convex and non-exhaustion and has no minimum, and hence has no critical point. For other examples of Riemannian planes admitting total curvature π, see §6, [7].

The idea of the proof of Theorem 3.2 came out of the investigation on the exhaustion property of Busemann functions on M with one end. It was announced in [7], Main Theorem : (2) that *if the total curvature* $c(M)$ *of* M *with one end satisfies* $c(M) > (2\chi(M) - 1)\pi$, *then every Busemann function on* M *is exhaustion.* However the proof of Theorem 5.3,(2) in [7] in incomplete and therefore the Main Theorem. (2) has not been proved yet. I would like to establish a perfect proof of it in §2.

§1. Preliminaries. We shall give definitions and state known results obtained in [7] and used here. Let $\gamma : [0, \infty) \to M$ be a ray, where geodesics are all parametrized by arclenghs. The Busemann function $F_\gamma : M \to R$ for γ is defined by

$$F_\gamma(x) := \lim_{t\to\infty}[t - d(x,\gamma(t))], \quad x \in M,$$

where d is the distance function induced from the Riemannian structure of M. The right hand side converges uniformly on every compact set. Form $|F_\gamma(x) - F_\gamma(y)| \leq d(x,y)$ for all $x,y \in M$ it follows that F_γ is Lipshitz continuous, and hence differentiable except a set of measure zero. A function $f : M \to R$ is said to be exhaustion iff $f^{-1}((-\infty,c])$ is compact for all $c \in R$. A critical point of F_γ is defined as follows : First of all, a ray $\sigma : [0,\infty) \to M$ is said to be asymptotic to γ iff there exists a sequence $\{\sigma_j : [0,b_j] \to M\}$ of minimizing geodesics such that the sequence $\{\dot\sigma_j(0)\}$ of initial tangent vectors tends to $\dot\sigma(0)$ and such that the sequence $\{\sigma_j(b_j)\}$

of their terminal points is a monotone divergent sequence on $\gamma([0,\infty))$. A point $p \in M$ is said to be a <u>critical point of</u> F_γ iff for any vector $v \in M_p$ there is a ray $\sigma : [0,\infty) \to M$ asymptotic to γ and $\sigma(0) = p$ such that $\langle\dot{\sigma}(0),v\rangle \geq 0$, where $\langle\ ,\ \rangle$ means the inner product on M_p induced from the metric. A point $p \in M$ is by definition <u>non-critical for</u> F_γ iff there exists an open half-space in M_p which contains all unit vectors tangent to rays asymptotic to γ and emanating from p.

Now let M have one end and $\gamma : [0,\infty) \to M$ a ray. We may choose a core C of M such that $\gamma(0) \in C$. Let $U = M - \mathrm{Int}(C)$ be a tube relative to C. Let $\pi : \tilde{U} \to U$ be the universal Riemannian covering of U and π the covering projection, and let $\hat{U} \subset \tilde{U}$ be the fundamental domain of U whose boundary consists of two rays $\hat{\gamma}_1, \hat{\gamma}_2 : [0,\infty) \to \tilde{U}$ and a broken geodesic $\hat{P}$ such that $\pi \circ \hat{\gamma}_1 = \pi \circ \hat{\gamma}_2 = \gamma$ and $\pi(\hat{P}) = P := \partial U$. Let $\hat{d}$ be the distance function on $\hat{U}$ induced from the metric. Any two points in $\hat{U}$ can be joined by a $\hat{d}$-segment whose length realizes the $\hat{d}$-distance between them. Let $\hat{\Gamma}(t)$ for every $t \geq 0$ be the set of all $\hat{d}$-segments in $\hat{U}$ joining $\hat{\gamma}_1(t)$ to $\hat{\gamma}_2(t)$. Namely, each $\hat{P}(t) \in \hat{\Gamma}(t)$ has length $L(\hat{P}(t)) = \hat{d}(\hat{\gamma}_1(t), \hat{\gamma}_2(t))$. Then, Theorem 4.2 in [7] states that if $\hat{P}(t) \cap \hat{P} \neq \emptyset$ for all $t \geq 0$ and for all $\hat{P}(t) \in \hat{\Gamma}(t)$, then $c(M) \leq (2\chi(M) - 1)\pi$. Assume that there exists a $T_0 > 0$ such that $\hat{P}(t) \cap \hat{P} = \emptyset$ holds for all $t > T_0$ and for all $\hat{P}(t) \in \hat{\Gamma}(t)$. Theorem 4.3 in [7] states that if there is such a $T_0 > 0$ and if the function $g_\gamma(t) := t - L(\hat{p}(t))/2$ is bounded above, then $c(M) \leq (2\chi(M) - 1)\pi$. For each $t > T_0$ the four curves $\hat{P}(t)$, $\hat{\gamma}_1([0,t])$, $\hat{P}$, $\hat{\gamma}_2([0,t])$ bounds a disk domain in $\hat{U}$. Let $\hat{P}(t)^+$(resp. $\hat{P}(t)^-$) be chosen in $\hat{\Gamma}(t)$ so as to satisfy that the angles at corners $\hat{\gamma}_1(t)$ and $\hat{\gamma}_2(t)$ of this disk domain with boundary $\hat{P}(t)^+$(resp. $\hat{P}(t)^-$) are the maximal (resp. minimal). Thus, if $\hat{\Omega}(t) \subset \hat{U}$ for $t > T_0$ is the disk domain bounded by $\hat{P}(t)^+$ and $\hat{P}(t)^-$, then every $\hat{P}(t) \in \hat{\Gamma}(t)$ lies in $\hat{\Omega}(t)$.

The author tried to prove in Theorem 5.3.[7] that <u>if M with one end admits total curvature and if g_γ is unbounded for some ray γ, then F_γ is exhaustion</u>. However the proof is incomplete and this is not proved yet. A direct proof of the Main Theorem, (2) is established in §2 with above notions.

<u>§2. Exhaustion Property of Busemann Functions</u>. We shall prove the following theorem which was announced in [7].

Theorem 2.1 (Main Theorem. (2):[7]). Let M have one end and admit total curvature $c(M)$ such that $c(M) > (2\chi(M) - 1)\pi$. Then every Busemann function on M is exhaustion.

For the proof of Theorem 2.1 we shall argue by deriving a contradiction. Suppose that there exists a non-exhaustion Busemann function F_γ on M satisfying the assumptions in Theorem 2.1. Fix an arbitrary core $C \subset M$. As is mentioned in §1, $c(M) > (2\chi(M) - 1)\pi$ implies that there exists a $T_0 > 0$ such that $\hat{P}(t) \cap \hat{P} = \emptyset$ holds for all $t > T_0$ and for all $\hat{P}(t) \in \hat{\Gamma}(t)$, and that the function $g_\gamma(t) := t - L(\hat{P}(t))/2$ is unbounded. It is obvious that g_γ is strictly monotone increasing for $t > 0$. Then Lemma 5.1 implies that there exists a $T_1 \geqq T_0$ such that the midpoint of every $P(t) \in \Gamma(t)$ for all $t > T_1$ is the cut point to $\gamma(t)$ along it, where we set $P(t) = \pi(\hat{P}(t))$ and $\pi(\hat{\Gamma}(t)) = \Gamma(t)$.

Since F_γ is non-exhaustion, there exists a constant $c \in R$ and a divergent sequence $\{q_j\}$ of points on U such that $F_\gamma(q_j) \leqq c$ for all $j = 1,2,\cdots$. It follows that q_j is not on any $P(t)$ for all t with $g(t) > c$. There exists a monotone divergent sequence $\{t_j\}$ such that q_j for each j is contained in $\mathrm{Int}(\Omega_j)$, where we set $\Omega_j := \pi(\hat{\Omega}(t_j))$. Let $\hat{P}_j^{\pm} := \hat{P}_j^{\pm}(t_j)$ and $P_j^{\pm} := \pi(P_j^{\pm})$. Recall that the boundary of Ω_j consists of P_j^+ and P_j^-.

The following Lemma 2.2 is useful for the proof of Theorem 2.1.

Lemma 2.2. Under the assumptions in Theorem 2.1, suppose that $F_\gamma : M \to R$ is a non-exhaustion Busemann function. Then there exists for each $j = 1,2,\cdots$, a number $s_j \geq t_j$ such that the distance function $d(\gamma(t),\cdot)$ to $\gamma(t)$ for all $t > s_j$ has a critical point y_t in $\mathrm{Int}(\Omega_j)$. In particular, F_γ has a critical point y_j in $\mathrm{Int}(\Omega_j)$.

Proof. The Cohn-Vossen theorem (see [3]) and the assumption imply $(2\chi(M) - 1)\pi < c(M) \leqq 2\pi\chi(M)$, and hence $\int_M G_+ dM < \infty$, where $G_+ = \mathrm{Max}\,\{G, 0\}$.

For any given $\varepsilon \in (0,\ c(M) - (2\chi(M)-1)\pi)$ there exists a core C of M such that (setting $c(C) = \int_C G dM$)

$$\int_M G_+ dM - \int_C G_+ dM < \varepsilon$$

and

$$|c(M) - c(C)| < \varepsilon .$$

By choosing a subsequence of $\{q_j\}$, if necessary, we may consider that every Ω_j is c ntained in the tube U relative to C. For a fixed j let $\hat{q}_j \in \hat{\Omega}_j$ be the point with $\pi(\hat{q}_j) = q_j$ and $\hat{m}_j$ the midpoint of $\hat{P}_j^-$. Let $\hat{\tau}_{i,t}$ for $i = 1,2$ and for $t > t_j$ be a $\hat{d}$-segment joining $\hat{m}_j$ to $\hat{\gamma}_i(t)$, and also $\hat{\sigma}_{i,t}$ a $\hat{d}$-segment joining $\hat{q}_j$ to $\hat{\gamma}_i(t)$. We then have inequalities : $\hat{d}(\hat{q}_j, \hat{\gamma}_i(t)) \geq d(q_j, \gamma(t)) \geq |F_\gamma(q_j) - F_\gamma(\gamma(t))| \geq t - c$ and $\hat{d}(\hat{m}_j, \hat{\gamma}_i(t)) \leq (t - t_j) + L(\hat{P}_j^-)/2 = t - g_\gamma(t_j)$. Choose $s_j \geq t_j$ such that $\hat{d}(\hat{q}_j, \hat{\gamma}_i(t)) > \hat{d}(\hat{m}_j, \hat{\gamma}_i(t))$ holds for all $i = 1,2$ and for all $t > s_j$. This choice is possible because g_γ is unbounded.

Now, $\hat{\Omega}_j$ is divided into three sub-domains by the subarcs of $\hat{\tau}_{1,t}$, $\hat{\tau}_{2,t}$ and $\hat{P}_j^+$. If $\hat{q}_j$ is contained in one of the two sub-domains which are bounded by geodesic triangles with corners $(\hat{m}_j, \hat{\gamma}_1(t_j), \hat{\tau}_{1,y} \cap \hat{P}_j^+)$ and $(\hat{m}_j, \hat{\gamma}_2(t_j), \hat{\tau}_{2,t} \cap \hat{P}_j^+)$, then the circumference of the above triangle is less than $2\hat{d}(\hat{q}_j, \hat{\gamma}_i(t))$, and hence there exists a geodesic loop α_t at $\hat{\gamma}_i(t)$ for some $i = 1,2$ such that α_t bounds a disk $\hat{D}$ which contains $\hat{q}_j$ in its interior. The Gauss-Bonnet theorem implies $c(\hat{D}) > \pi$, and $\hat{D} \subset \hat{U}$ by construction. This contradicts to the choice of C which conclude $\int_{\hat{U}} G_+ dM < \varepsilon < \pi$.

The above argument shows that $\hat{q}_j$ is contained in the subdomain of $\hat{\Omega}_j$ which is bounded by the subarcs of $\hat{P}_j^+$, $\hat{\tau}_{1,t}$ and $\hat{\tau}_{2,t}$, (see Figure 1). It follows from $L(\hat{\sigma}_{1,t}) + L(\hat{\sigma}_{2,t}) > L(\hat{\tau}_{1,t}) + L(\hat{\tau}_{2,t})$ that there exists a geodesic $\hat{Q}(t)$ joining $\hat{\gamma}_1(t)$ to $\hat{\gamma}_2(t)$ in $\hat{U}$ which has the minimum length among all curves with the same endpoints in the subdomain of $\hat{U}$ which is a disk and bounded by $\hat{\gamma}_1([0,t])$, $\hat{P}$, $\hat{\gamma}_2([0,t])$, $\hat{\sigma}_{1,t}$ and $\hat{\sigma}_{2,t}$. Note that $\hat{Q}(t)$ is not a $\hat{d}$-segment and that $\hat{q}_j \notin \hat{Q}(t)$ follows from $L(\hat{Q}(t)) < L(\hat{\sigma}_{1,t}) + L(\hat{\sigma}_{2,t})$, and also that $\hat{Q}(t)$ lies in the subdomain bounded by geodesic quadrangle $(\hat{\sigma}_{1,t}, \hat{\sigma}_{2,t}, \hat{\tau}_{2,t}, \hat{\tau}_{1,t})$. If $Q(t) = \pi(\hat{Q}(t))$, then $Q(t)$ bounds a core $C(t)$ containing C and the tube $U(t)$ relative to $C(t)$ contains q_j in its interior. Moreover the midpoint y_t of the geodesic loop $Q(t)$ is the cut point to $\gamma(t)$ along $Q(t)$. In fact, if otherwise supposed, then there exists a minimizing geodesic λ joining $\gamma(t)$ to y_t whose length is less than $L(\hat{Q}(t))/2$. A subarc of $Q(t)$ together with λ forms a geodesic biangle which bounds a disk. If q_j lies in this disk, then there exists a geodesic loop β_t at $\gamma(t)$ which bounds a disk containing q_j in its interior. The curvature integral over the disk exceeds π, contradicting to the choice of C. If q_j does not lie in this disk, then it is possible to find a curve joining $\hat{\gamma}_1(t)$ to $\hat{\gamma}_2(t)$ and lying in the domain bounded by $\hat{\gamma}_1([0,t])$, $\hat{P}$, $\hat{\gamma}_2([0,t])$, $\hat{\sigma}_{1,t}$ and $\hat{\sigma}_{2,t}$ whose length is less than $L(\hat{Q}(t))/2 + L(\lambda) < L(\hat{Q}(t))$, a contradiction to the minimizing property of $Q(t)$.

The above stated property for $Q(t)$ and the $\hat{d}$-minimizing property of $\hat{P}_j^{+}$ imply together with $\hat{P}_j^{+} \cap \hat{Q}(t) \neq \emptyset$ that $\hat{y}_t$ lies in $\mathrm{Int}(\hat{\Omega}_j)$, and hence $y_t \in \mathrm{Int}(\Omega_j)$. This proves the first statement of Lemma 2.2.

Finally, by choosing a suitable monotone divergent sequence $\{b_m\}$ we may consider that the midpoints of $Q(b_m)$ converges to a point $y_j \in \mathrm{Int}(\Omega_j)$ and that $\{Q(b_m)\}$ converges to a geodesic $\sigma : R \to M$ with $\sigma(0) = y_j$. From construction it follows that both $\sigma_+(t) := \sigma(t)$ and $\sigma_-(t) := \sigma(-t)$, $t > 0$ are rays asymptotic to γ. This fact means that y_j is a critical point of F_γ.

This completes the proof of Lemma 2.2.

The proof of Theorem 2.1. With the same notations as in the proof of Lemma 2.2 we apply the Gauss-Bonnet theorem for $\mathbf{C}(b_m)$ to obtain $\lim_{m\to\infty} c(\mathbf{C}(b_m)) = (2\chi(M) - 1)\pi$. This is because we may choose $\{Q(b_m)\}$ in such a way that the angles at corners $\gamma(b_m)$ tends to 0 if they are measured with respect to $\mathbf{C}(b_m)$. If we set $V :=$ $M - \mathbf{C}(b_m)$, then $C \subset \mathbf{C}(b_m)$ implies that $\varepsilon < \int_V G\,dM \leq \int_V G_+\,dM < \varepsilon$, a contradiction. This completes the proof of Theorem 2.1.

§3. Critical Points of Busemann Functions. Let $\gamma : [0,\infty) \to M$ be a ray. If $p \in M$ is a critical point of F_γ, then either there exists a geodesic $\sigma : R \to M$ with $\sigma(0) = p$ such that both $\sigma_+(t) := \sigma(t)$ and $\sigma_-(t) := \sigma(-t)$, $t \geq 0$ are rays asymptotic to γ or else there are three distinct rays $\sigma_1, \sigma_2, \sigma_3 : [0,\infty) \to M$ with $\sigma_1(0) = \sigma_2(0) = \sigma_3(0) = p$ such that every open half-space of M_p centains at least one of the three vectors $\dot{\sigma}_1(0)$, $\dot{\sigma}_2(0)$ and $\dot{\sigma}_3(0)$. In this case there is a small ball B around p in which F_γ takes minimum at a unique point p. More precisely, there exists a small $h > 0$ such that B contains a component of $F^{-1}((-\infty, h + F_\gamma(p)])$.

The following Lemma 3.1 is a direct consequence of the above observation and the proof will be omitted.

Lemma 3.1. For a ray $\gamma : [0,\infty) \to M$ let $C \subset M$ be a core of M and $U(\gamma)$ a tube relative to C and containing a subray of γ. Assume that $p \in M - C$ is a critical point of F_γ. Then, one of the following statements holds.

(1) There exists a geodesic $\sigma : R \to M$ with $\sigma(0) = p$ such that both $\sigma_+, \sigma_- : [0,\infty) \to M$ are rays asymptotic to γ.

(2) If $p \in U(\gamma)$ and if (1) does not occur, then there exists a $T(p) > 0$ such that for all $t > T(p)$ there exists a geodesic loop $Q(t)$ at $\gamma(t)$ which either is freely homotopic to $\partial U(\gamma)$ in $U(\gamma)$ or else bounds a disk in $\hat{U}$.

(3) If $p \in V$ for a tube $V \neq U(\gamma)$ relrative to C and if (1) does not occur, then there exists a $T(p) > 0$ such that for all $t > T(p)$ there exists a geodesic loop $Q(t)$ at $\gamma(t)$ which intersects ∂V, and a closed curve in V obtained by joinning subarcs of $Q(t)$ and ∂V at their intersections bounds a disk containing p in its iterior.

It is obvious that if $q \in M$ belongs to the minimum set of F_γ, then q is a critical point of F_γ.

The proof of Theorem 3.2. It follows from $-\infty \leq c(M) \leq 2\pi\chi(M)$ that $\int_M G_+ dM < \infty$. This makes it possible to choose a core $C \subset M$ such that

$$\int_C G_+ dM > \int_M G_+ dM - \pi/2 \qquad \cdots\cdot \qquad (*)$$

Let U be a unique tube relative to C.

For the proof of the first statement let $\{q_j\}$ be an unbounded sequence of critical points of F_γ such that $q_j \notin C$ for all $j = 1,2,\cdots$. Let $\{\varepsilon_j\}$ be a monotone decreasing sequence of positive numbers with $\lim \varepsilon_j = 0$.

If (1) in Lemma 3.1 occurs for q_j and if $\sigma_j : R \to M$ is the geodesic with $\sigma_j(0) = q_j$ as stated in (1), then there exists a geodesic triangle $\Delta(j)$ with vertices $\sigma_j(1)$, $\sigma_j(-1)$ and $\gamma(k_j)$ for a sufficiently large k_j such that the sum of all angles of $\Delta(j)$ is greater than $2\pi - \varepsilon_j$ if they are measured with respect to the compact domain bounded by $\Delta(j)$. Suppose that $\Delta(j)$ bounds a disk. Then the disk is contained in U and the curvature integral over the disk exceeds π, a contradiction to the inequality (*).

The same argument can be developped when (2) in Lemma 3.1 occurs for q_j. It turns out in this case that $Q_j(k_j)$ for a sufficiently large k_j is freely homotopic to ∂U (and hence bounds a core) and the angle of $Q_j(k_j)$ at $\gamma(k_j)$ is less than ε_j when it is measured with respect to the core bounded by it.

For each $j = 1,2, \cdots,$ let $C_j(k_j)$ be the core bounded by $\Delta(j)$ when (1) in Lemma 3.1 occurs for q_j and also $C_j(k_j)$ the core bounded by $Q_j(k_j)$ when (2) in Lemma 3.1 occurs for q_j. By choosing a subsequence, if necessary, we may consider that $\{C_j(k_j)\}$ is monotone increasing. The proof of the first statement is complete if $\bigcup C_j(k_j) = M$.

Suppose finally that $\bigcup C_j(k_j) \subsetneq M$. In this case we observe that $\inf_{j\to\infty} F_\gamma(q_j) = -\infty$ and that there is a core C_∞ of M such that $\sigma_j(R)$ intersects C_∞ for all j. Let U_∞ be the tube relative to C_∞ and consider the fundamental domain $\hat{U}_\infty$ of U_∞ in the universal Riemannian covering $\tilde{U}_\infty$ just as in the same way as $\hat{U}$ and U, and let $\hat{d}_\infty$ be the distance function on $\hat{U}_\infty$. When (1) in Lemma 3.1 occurs for q_j for a sufficiently large j, there are numbers $a_j < 0 < b_j$ and a $\hat{d}_\infty$-segment $\hat{\lambda}_j$ in $\hat{U}_\infty$ joining $\hat{\sigma}_j(a_j)$ to $\hat{\sigma}_j(b_j)$ such that $\hat{\lambda}_j \cup \hat{\sigma}_j([a_j,b_j])$ bounds a disk $\hat{D}_j$ in $\hat{U}_\infty$ and $\hat{D}_j$ has the properties that (1) the angle at the corner $\hat{\sigma}_j(b_j)$ of $\hat{D}_j$ is less than ε_j, (2) the angle at the corner $\hat{\sigma}_j(a_j)$ of $\hat{D}_j$ is not less than $\pi/2$, (3) all the other corners of $\hat{\lambda}_j$ are on ∂C_∞ and have angles not less than π (see Figure 2). When (2) in Lemma 3.1 occurs for q_j for a sufficiently larg j, such a disk domain in U_∞ can be bound by using a subarc of $Q_j(k_j)$. In any case this contradicts to the inequality (*).

This completes the proof of the first statement of Theorem 3.2.

For the proof ot the second statement of Theorem 3.2 we use the core C and $\{\varepsilon_j\}$ as used before. Suppose that there is no compact set which contains all critical points of all Busemann functions on M. Then, there exists a divergent sequence of points $\{q_j\}$ on M and rays $\{\gamma_j\}$ such that q_j for each j is a critical point of the Busemann function for γ_j. It follows from the previous argument that there exists for each j a core C_j of M such that $(2\chi(M) - 1)\pi < c(C_j) < (2\chi(M) - 1)\pi + \varepsilon_j$. It is possible to choose a subsequence $\{C_k\}$ of $\{C_j\}$ which is monotone increasing and exhausting M. Therefore, $c(M) = \lim c(C_k) = (2\chi(M) - 1)\pi$, a contradiction to the assumption.

This completes the proof of the second statement of Theorem 3.2.

Corollary to Theorem 3.2. Assume that M has one end and admits total curvature. If there exists a Busemann function whose minimum set is noncompact, then $c(M) = (2\chi(M) - 1)$.

§4. The proof of Main Theorem. For the proof of our Main Theorem we shall choose a core C of M for which (*) is satisfied. Let C_1 be a core of M containing C with the property that if U_1 and U are tubes of M relative to C_1 and C respectively such that $U_1 \subset U$, then $d(x_1,x) > L(\partial U)$ holds for all $x_1 \in \partial U_1$ and for all $x \in \partial U$.

The proof of (1). Suppose that there exists a critical point $p \in M - U(\gamma) \cup C_1$ of F_γ. There are tubes U_1 and U with $U_1 \subset U$ relative to C_1 and C with $C_1 \supset C$ respectively such that $p \in \mathrm{Int}(U_1)$. It follows from (1) and (3) in Lemma 3.1 that there exists a geodesic $\mu : [0,b] \to U$ such that $\mu(0)$, $\mu(b) \in \partial U$ and $\mu(c)$ $\mathrm{Int}(U_1)$ for some $c \in (0,b)$ and such that $\mu([0,b])$ and a subarc of ∂U form a closed curve which bounds a disk D_0 in U. It follows from $b = L(\mu) > 2L(\partial U) > 4d(\mu(0),\mu(b))$ that there exists a segment ν in D_0 joining $\mu(0)$ to a point $\mu(b')$ for some $b' \in (0,b]$ such that ν has the minimum length among all curves in D_0 joining $\mu(0)$ to points on $\mu([b-L(\partial U),b])$. Then $\mu([0,b']) \cup \nu$ bounds a disk D contained in D_0 such that the angles of all corners of ν have the following properties : The angle at $\mu(b')$ is not smaller than $\pi/2$ and the angle at $\mu(0)$ lies in $(0,\pi)$, and the other angles at corners in the interior of ν(which are some corners of ∂U) are all greater than or equal to π when they are measured with respect to D. Now we have $c(D) > \pi/2$, contradicting to (*).

The proofs of (2) and (3) have already been established because they are obtained just by combining the above arguments and Theorem 3.2. We only need to choose C_1 in (2) such that the tube $U_1(\gamma)$ $U(\gamma)$ relative to C_1 is as required in the proof of the second statement of Theorem 3.2.

Before going into the proofs of (4) and (5), we shall observe relations between critical points and the topology of level sets of a Busemann function.

It is elementary that through each point on M there passes at least one ray asymptotic to γ, along which F_γ has derivative 1. In view of this property of F_γ a critical point of F_γ appears in the follwing cases. If a level set of F_γ contains a simply closed curve which bounds a disk, then there exists a critical point of F_γ in the disk at which a local minimum is attained. If x is an isolated point of an a-level set $F_\gamma^{-1}(\{a\})$, then there is a neighborhood W of x on which F_γ takes local minimum at x, and hence x is a critical point. If an a-level set $F_\gamma^{-1}(\{a\})$ contains a non-trivial curve $\alpha : [0,1] \to F_\gamma^{-1}(\{a\})$ such that $F_\gamma^{-1}(\{a\})$ does not divide any small neighborhood of $\alpha(1)$, then there exists an $h \in (0,1)$ such that every point on $\alpha([h,1])$ is a critical point of F_γ. In the last case there exists a 1-parameter family $\sigma_s : R \to M$ of geodesics such that $s \in [h,1]$ and $\sigma_s^+(t) := \sigma_s(t)$ and $\sigma_s^-(t) := \sigma_s(-t)$ for $t > 0$ are both rays asumptotic to γ and $\sigma_s^+(0) = \sigma_s^-(0) = \alpha(s)$ for $s \in [h,1]$.

It follows from the above observations that if $c(M) \neq (2\chi(M)-1)\pi$ and if an a-level set of F_γ is contained entirely in $M - C_1$, then each component of $F_\gamma^{-1}(\{a\})$ is homeomorphic to either a circle or a line which is divergent in both directions.

The proof of (4). In view of the above observations together with Theorem 2.1 every level set of F_γ in $U_1(\gamma)$ is compact, connected and homeomorphic to a circle. Let U_1 be a tube relative to C_1 such that $U_1 \cap U_1(\gamma) = \emptyset$. The compactness of levels in U_1 is seen as follows. Suppose that $F_\gamma^{-1}(\{a\}) \cap U_1$ is homeomorphic to a line. Let $\{q_j\}$ be a divergent sequence of points on $F_\gamma^{-1}(\{a\}) \cap U_1$ and $\sigma_j : [0,\infty) \to M$ an asymptotic ray to γ with $\sigma_j(0) = q_j$. If $\sigma_j(c_j) \in \partial U$, then $\{c_j\}$ contains a monotone divergent sequence and $F_\gamma(\sigma_j(c_j)) = a + c_j$. This contradicts the continuity of F_γ on ∂U. The same principle shows that $F_\gamma^{-1}(\{a\}) \cap U_1$ is connected.

This completes the proof of (4).

The proof of (5). We only need to prove that each component of $F_\gamma^{-1}(\{a\})$ in $U(\gamma)$ is homeomorphic to a line.

It follows from Main Theorem ; (2) in [7] that $F_\gamma : M(\gamma) \to R$ is non-exhaustion, and hence there is a noncompact level of it. If $F_\gamma^{-1}(\{a\})$ is noncompact, then so is $F_\gamma^{-1}(\{a'\})$ for all $a' > a$. This implies together with the previous arguments that each component of $F_\gamma^{-1}(\{a\})$ in $U_1(\gamma)$ is homeomorphic to a line. Here a is chosen such that $a > \mathrm{Max}\{F_\gamma(x) : x \in C_1\}$. Thus the proof of (5) is complete if we verify that $F_\gamma^{-1}(\{a\})$ for all such a is connected.

Suppose that α and β are disjoint divergent curves in $F_\gamma^{-1}(\{a\})$. α bounds an open half-space $H(\alpha)$ in $U_1(\gamma)$. Every point $x \in H(\alpha)$ satisfies $F_\gamma(x) > a$. This fact and $\alpha \cap \beta = \emptyset$

imply $H(\alpha) \cap H(\beta) = \emptyset$. Choose points $x \in \alpha$ and $y \in \beta$ and a curve $c : [0,1] \to U_1(\gamma) - H(\alpha) \cup H(\beta)$. It follows that $F_\gamma(c(u)) < a$ for all $0 < u < 1$. If $a_1 = \mathrm{Min}\{F_\gamma(c(u)) ; 0 \leq u \leq 1\}$, then the set $\{c(u) ; F_\gamma(c(u)) = a_1\}$ is contained entirely in a component of $F_\gamma^{-1}(\{a_1\})$. Therefore there exists a homotopy $H : [0,1] \times [0,1] \to U_1(\gamma) - H(\alpha) \cup H(\beta)$ of curves joining x to y such that $H(u,0) = c(u)$ and $H(u,1) \in F_\gamma^{-1}(\{a\})$ for all $u \in [0,1]$ and such that $H(0,v) = x$ and $H(1,v) = y$ for all $v \in [0,1]$. This contradicts to the choice of $x \in \alpha$ and $y \in \beta$. Thus the proof of (5) is complete.

References

[1] Busemann, H. : The Geometry of Geodesics, Academic Press, New York, 1955.

[2] Cheeger, J-Gromoll, D. : The splitting theorem for manifolds of nonnegative Ricci curvature, J. Differential Geometry, Vol. 6(1971), 119-128.

[3] Cohn-Vossen, S. : Kürzeste Wege und Totalkrümmung auf Flächen, Compositio Math., Vol. 2(1935), 63-133.

[4] Cohn-Vossen, S. : Totalkrümmung und Geodätische Linien auf einfach zusammenhängenden offenen volständigen Flächenstücken, Recueil de Math., Moscow, Vol. 43(1936), 139-163.

[5] Greene, R-Shiohama, K. : Convex functions on complete noncompact manifolds ; Topological Structure, Invent. Math., Vol. 63(1981), 129-157.

[6] Shiohama, K. : Busemann functions and total curvature, Invent. Math., Vol. 53(1979), 281-297.

[7] Shiohama, K. : The role of total curvature on complete noncompact Riemannian 2-manifolds, Illinois J. Math., Vol.28(1984), 597-620.

[8] Shiohama, K. : Total curvature and minimal areas of complete open surfaces, Proc. Amer. Math. Soc. Vol. 94(1985), 310-316.

L-functions in geometry and some applications

Toshikazu Sunada*

Department of Mathematics, Nagoya University, Nagoya 464, Japan

This article attempts to survey some facts on L-functions which come up in geometry, combinatorics and dynamical systems, and to give applications thereof. Several results are essentially part of a larger investigation carried out in collaboration with T. Adachi and A. Katsuda. We should point out that some of the results concerning dynamical L-functions were independantly obtained by Parry and Pollicott [24]. The reader may usefully consult the item on zeta functions in [15] on general questions on L-functions. Recently, Kurokawa [17] [18] proposed a fairly general setting for L-functions belonging to arithmetic categories.

A classical L-function in number theory is a natural generalization of the celebrated Riemannian zeta function, which fits with theory of Galois extensions of number fields. Let K/k be a finite Galois extension of a number field k with Galois group Γ (for simplicity we assume that K/k is unramified), and let $\rho : \Gamma \longrightarrow U(n)$ be a representation of Γ. The L-function associated with K/k and ρ is defined by

$$L(s,\rho) = \prod_{p} \det(I_n - \rho((\frac{K/k}{p}))N(p)^{-s})^{-1},$$

where p runs over all prime ideals in k, and $(\frac{K/k}{p})$ denotes the conjugacy class of the Frobenius automorphism associated to p. In the case $K = k$, $L(s,\rho)$ is just what we call the Dedekind zeta function $\zeta_k(s)$. The fundamental properties of $L(s,\rho)$ are embodied in

<u>Proposition</u> A. 1) $L(s,\rho)$ converges absolutely and is holomorphic in the region $\mathrm{Re}\ s > 1$.

---

*Suppoted by The Ishida Foundation.

2) $L(s,\rho)$ has a meromorphic continuation to an open neighborhood of the closed region $\mathrm{Re}\, s \geq 1$ (in fact, $L(s,\rho)$ can be extended to the whole plane).

3) $L(s,\rho)$ is non-vanishing in $\mathrm{Re}\, s \geq 1$.

4) If ρ is irreducible and non-trivial, then $L(s,\rho)$ is holomorphic in $\mathrm{Re}\, s \geq 1$.

5) The Dedekind zeta function $\zeta_k(s) = L(s,\mathbb{1})$ has simple pole at $s = 1$, and holomorphic in $\mathrm{Re}\, s \geq 1$ except for $s = 1$.

See S. Lang [19] for the proof and backgrounds of the materials.

After the above definition of L-function, we now give an abstract setting for generalized L-functions. Let $\mathcal{P} = \{p\}$ be a countable set, and $N : \mathcal{P} \longrightarrow \mathbb{R}$ be a map. Let Γ be a group (we do not assume that Γ is finite), and suppose that we are given a map $\mathcal{P} \longrightarrow [\Gamma]$ $(p \longmapsto \langle p\rangle)$. Given a unitary representation $\rho : \Gamma \longrightarrow U(n)$, define an L-function associated to the data $(\mathcal{P},\Gamma,N)$ by

$$(1) \qquad L(s,\rho) = \prod_p \det(I_n - \rho(\langle p\rangle)N(p)^{-s})^{-1}.$$

In many cases, the group Γ comes up as the fundamental group of a topological space (or its quotient group), and $\mathcal{P}$ is a relevant set of closed paths in the space, so that the map $\mathcal{P} \longrightarrow \Gamma$ is a canonical one.

We shall call the L-functions $\{L(s,\rho);\ \rho \in \hat{\Gamma}\}$ associated to $(\mathcal{P},\Gamma,N)$ *nice* if the followings are satisfied:

(L-1). There exist a positive h such that $L(s,\rho)$ converges absolutely and is holomorphic in the region $\mathrm{Re}\, s > h$.

(L-2). $L(s,\rho)$ has a meromorphic continuation to an open neighborhood of the closed region $\mathrm{Re}\, s \geq h$.

(L-3). $L(s,\rho)$ is non-vanishing in $\mathrm{Re}\, s \geq h$.

(L-4). If ρ is irreducible and non-trivial, then $L(s,\rho)$ is holomorphic in $\mathrm{Re}\, s \geq h$.

(L-5). $L(s,\mathbb{1})$ has a simple pole at $s = h$, and holomorphic in $\mathrm{Re}\, s \geq h$ except for $s = h$.

The number h will be called the critical exponent of $L(s,\rho)$.

Remark. As we will see later, (L-4) and (L-5) does not hold for some important examples, but can be replaced by modified conditions.

Recall that the L-functions in number theory are introduced as a means of proving various density theorems. An abstract aspect of density theorem is explained in the following way.

Proposition B (The density theorem of Chebotarev type). Suppose that Γ is a finite group, and the L-functions associated to $(\mathcal{P},\Gamma,N)$ are nice and have the critical exponent $h > 0$. Then for any conjugacy class $[\sigma] \in [\Gamma]$

$$\#\{\, p \in P\ ;\ \langle p\rangle = [\sigma],\ N(p) < x \,\} \sim \frac{\#[\sigma]}{\#\Gamma}\, x^h/\log x^h,$$

as x goes to $+\infty$.

Outline of the proof. Without loss of generality, we may assume that $h = 1$. Let $\Lambda(s,\rho)$ be the logarithmic derivative of $L(s,\rho)$, so that

$$-\Lambda(s,\rho) = \sum_{p,k} \mathrm{tr}(\rho(\langle p\rangle^k))(\log N(p))\, N(p)^{-ks}.$$

Multiplying $\mathrm{tr}(\rho(\sigma^{-1}))$ on the both sides, and summing up over all irreducible representations of Γ, we get, by using the orthogonal relation for characters,

$$-\sum_{\rho} \mathrm{tr}\ \rho(\sigma^{-1})\ \Lambda(s,\rho) = \left(\frac{\#[\sigma]}{\#\Gamma}\right)^{-1} \sum_{\langle p\rangle^k \in [\sigma]} \log N(p)\ N(p)^{-ks}.$$

Thus the Dirichlet series defined by the right hand side is a meromorphic function with a single simple pole at $s = 1$ in an open domain containing $\mathrm{Re}\ s \geq 1$. Since the residue is one, applying the Tauberian theorem to this Dirichlet series, we obtain

$$\sum_{\substack{N(p)^k < x \\ \langle p\rangle^k \in [\sigma]}} \log N(p) \sim \frac{\#[\sigma]}{\#\Gamma}\, x,$$

combined with which, a routine method in analytic number theory leads to the assertion.

The above proposition suggests a way to apply the L-function idea to the study of homotopy or homology of closed paths in some special classes.

We now exhibit several examples of L-functions belonging to geometric categories.

I. Let N be a compact Riemann surface with constant negative curvature -1, so that one can find a Fuchsian group $\Gamma \subset SL(2,\mathbb{R})$ such that N can be written as a qutient space $\Gamma \backslash H$, where H being the upper half-plane with the Poincare metric. Let $\mathcal{P}$ be the set of the conjugacy classes of primitive elements in Γ. Here, an element γ is called primitive when γ is not a positive power of other element (or what is the same in this case, γ generates its centralizer in Γ). We then have a canonical map : $\mathcal{P} \longrightarrow [\Gamma]$. Define a map $N : \mathcal{P} \longrightarrow \mathbb{R}$ by setting $N(p) = \xi_2{}^2$ where ξ_2 denotes the maximum eigenvalue of a representative of p. It should be noted that if we denote by $\ell(p)$ the length of a (unique) prime closed geodesic whose homotopy class corresponds to the conjugacy class p, then $N(p) = \exp(\ell(p))$. Associated with the data $(\mathcal{P},\Gamma,N)$, the L-function $L(s,\rho)$ is defined as above. This L-function satisfies all the properties (L-1) —— (L-5) with $h = 1$. In fact, $L(s,\rho)$ is closely related to the celebrated Selberg zeta function, which is defined by

$$Z(s,\rho) = \prod_{k=0}^{\infty} \prod_{p} \det(I_n - \rho(<p>)\ N(p)^{-(s+k)}).$$

In his paper [29], Selberg proved that $Z(s,\rho)$ can be extended holomorphically to the whole plane and the zeros of $Z(s,\rho)$ are $0, -1, -2, \dots,$ and

$$\frac{1}{2}\ (1 \pm (1 - 4\lambda_n(\rho))^{1/2}\ , \quad n = 0,\ 1,\ 2,\ \dots .$$

where $\{\lambda_n(\rho)\}$ denotes the eigenvalues of the Laplacian acting on the sections of the flat vector bundle on N associated to the representation ρ. This, in particular, implies that the Riemann hypothesis is almost valid for $Z(s,\rho)$. Namely, all zeros whose real parts are in $(0,1)$ lie on the line $\mathrm{Re}\ s = 1/2$, except for a finite number that lie on real line (zeros corresponding to the eigenvalues $\lambda_n(\rho) \in (0,1/4)$). From this view, it is interesting to know when small eigenvalues come out. As for the existence of small eigenvalues, see R. Brooks [5], who proves, among other things, that if a compact Riemannian manifold N has the fundamental group $\pi_1(N)$ surjecting onto an infinite, amenable and residually finite group, then there exist coverings N_i of N with $\lambda_1(N_i) \longrightarrow 0$. In particular, if the first Betti number of N is positive (this is the case of Riemann surfaces with constant negative curvature), we may obtain arbitrary small eigenvalue by taking a finite covering of N (see also [33]).

Since $L(s,\rho) = Z(s+1,\rho)/Z(s,\rho)$, we easily observe that $L(s,\rho)$ is nice. The proof of Selberg's results relies on a trace formula which gives a beautiful relationship between the eigenvalues and the length spectrum of closed geodesics. For instance, the trace formula applied to the heat kernel function is

$$\sum_{n=0}^{\infty} \exp(-t\lambda_n(\rho)) = \mathrm{Vol}(N)\ (4\pi t)^{-3/2} e^{-t/4} \int_0^{\infty} \frac{be^{-b^2/4t}}{\sinh b/2}\, db$$
$$+ \frac{1}{2} \sum_{n=1}^{\infty} \sum_{p} (4\pi t)^{-/2} \frac{\ell(p)}{\sinh n\ell(p)/2}\, e^{-t/4}\, e^{-n^2\ell(p)^2/4t}.$$

Taking the transformation $(2s-1)\int_0^{\infty} e^{-s(s-1)t}\cdot dt$ of the both sides, we find that

$$\frac{Z'(s,\rho)}{Z(s,\rho)} = (2s-1)\sum_{k=0}^{\infty}\left\{ \frac{1}{s(s-1)+\lambda_k(\rho)} - \frac{n\cdot\mathrm{Vol}(N)}{4\pi}\frac{1}{s+k}\right\},$$

which leads to Selberg's theorem (see D. Hejhal [13] for the rigouros proof).

The following gives a relation between small eigenvalues and the higher order asymptotics of distributions of $N(p)$.

<u>Proposition</u> C (cf. [13], P. Sarnak [28]). Let $t_1, \ldots, t_q$ be all the zeros of $Z(s,\mathbb{1})$ in the interval $[\frac{1}{2}, 1)$, and $\mathrm{Li}(u) = \int_2^u$

$\frac{1}{\log y}\,dy$. Then as $x \longrightarrow \infty$,

$$\#\{\, p \in \mathfrak{P}\ ;\ N(p) < x\,\} = Li(x) + Li(x^{t_1}) + \cdots Li(x^{t_q}) + O(x^{3/4}(\log x)^2).$$

R. Gangolli [10] defined a generalization of the Selberg zeta function for a compact locally symmetric space with negative sectional curvature, which has the form

$$Z(s,\rho) = \prod_{p} \prod_{k_1,\ldots,k_{n-1}=0}^{\infty} \det(I_n - \rho(<p>)\exp(-(s+k_1r_1h^{-1}+\cdots+k_{n-1}r_{n-1}h^{-1})\ell(p))),$$

where r_i^2 are the positive eigenvalues of the curvature operator : $u \longmapsto R(u,v)v$, v being a unit tangent vector, and $h = \sum_{i=1}^{n-1} r_i$. This zeta function also satisfies the nice properties as the Selberg zeta function, and is also related to L-function of the form (1) by the equality

$$L(s,\rho) = \frac{\prod_i Z(s+s_i,\rho) \prod_{i_1<i_2<i_3} Z(s+s_{i_1}+s_{i_2}+s_{i_3},\rho)\cdots}{Z(s,\rho) \prod_{i_1<i_2} Z(s+s_{i_1}+s_{i_2},\rho)\cdots}, \quad (s_i = r_ih^{-1})$$

for which (L-1) —— (L-5) hold.

II. We now give a one-dimensional analogue of Selberg zete functions associated with finite *oriented* graphs, which, as we will see later, is viewed as a prototype of L-function of a topological graph associated with a dynamical system.

Let $(\mathbb{V},\mathbb{E})$ be an oriented graph with $\mathbb{V}$, the set of verteces, and $\mathbb{E} \subset \mathbb{V} \times \mathbb{V}$, the set of edges. For each edge $e = (u,v)$, we put $o(e) = u$, and $t(e) = v$. We assign a length $\ell(e) > 0$ to each edge e. A path in $(\mathbb{V},\mathbb{E})$ is an element of the form $c = (e_1, \ldots, e_n)$, $e_i \in \mathbb{E}$, such that $t(e_i) = o(e_{i+1})$, $i = 1, \ldots, n-1$. We then put $|c| = n$ and $\ell(c) = \ell(e_1) + \ldots + \ell(e_n)$. A path c is called closed if $t(e_n) = o(e_1)$, and called prime if, in addition, there is no divisor k of n such that $1 \le k < n$ and $e_{i+k} = e_i$ for any $i \in \mathbb{Z}/n\mathbb{Z}$. We say that two closed paths $c = (e_1, \dots, e_n)$ and $c' = (e'_1, \dots, e'_n)$ are equivalent if there is an integer k with $e'_i = e_{i+k}$ for any $i \in \mathbb{Z}/n\mathbb{Z}$. A prime cycle is an equivalence class of a prime closed path, which will be denoted by p. The length of p is defined in an obvious way, and denoted $\ell(p)$. We now define $N(p) = \exp(\ell(p))$.

Suppose that $(\mathbb{V},\mathbb{E})$ is a finite non-circuit graph and is irreducible. The fundamental group $\pi_1(\mathbb{V},\mathbb{E})$ is defined to be that of the 1-dimensional CW-complex $|(\mathbb{V},\mathbb{E})|$ associated with $(\mathbb{V},\mathbb{E})$. A closed path c and a prime cycle p yield, in a natural way, conjugacy classes in $\pi_1(\mathbb{V},\mathbb{E})$, which we denote by $\langle c\rangle$ and $\langle p\rangle$ respectively. We finally let $\rho : \pi_1(\mathbb{V},\mathbb{E}) \longrightarrow U(n)$ be a representation. With these data, we define the L-function of $(\mathbb{V},\mathbb{E})$. Then one can prove that for a positive h, $\{L(s,\rho)\}$ are nice, though the property 4) should be modified; see below. In fact, $L(s,\rho)$ has a meromorphic extension to the whole plane. To see this, we shall show that there exists a matrix $L_{s,\rho}$ depending holomorphically on s such that $L(s,\rho) = \det(I - L_{s,\rho})^{-1}$. Let $(\tilde{\mathbb{V}},\tilde{\mathbb{E}})$ be a tree associated to the universal covering of the CW-complex $|(\mathbb{V},\mathbb{E})|$, on which $\pi_1(\mathbb{V},\mathbb{E})$ acts as automorphisms. Define an operator L_s of the linear space $\mathrm{Map}(\tilde{\mathbb{V}},\tilde{\mathbb{C}})$ by setting

$$(L_s\varphi)(x) = \sum_{e\in\tilde{\mathbb{E}},\, o(e)=x} (\exp -s\ell(e))\, \varphi(t(e)).$$

If we denote by F_ρ the finite dimensional subspace of $\mathrm{Map}(\tilde{\mathbb{V}},\tilde{\mathbb{C}})$ consisting of functions φ such that $\varphi(\gamma x) = \rho(\gamma)\varphi(x)$ for any $\gamma \in \pi_1(\mathbb{V},\mathbb{E})$, then $L_s(F_\rho) \subset F_\rho$. We then put $L_{s,\rho} = L_s|F_\rho$. When $\rho = \mathbb{1}$, the trivial representation, $L_{s,\mathbb{1}}$ is identified with an endomorphism of $\mathrm{Map}(\mathbb{V},\mathbb{C})$. The Perron-Frobenius theorem says that if $s \in \mathbb{R}$, there exists a maximum positive simple eigenvalue $\lambda(s)$ for the operator $L_{s,\mathbb{1}}$ with an associated positive eigenfunction. Noting that $\lambda(s)$ is strictly decreasing and $\lim_{s\to\infty}\lambda(s) = 0$, $\lim_{s\to-\infty}\lambda(s) = \infty$, we may find a unique positive h such that $\lambda(h) = 1$.

Expressing the operator $L_{s,\rho}$ by a matrix with respect to an appropriate basis of F_ρ, we find that

$$\operatorname{tr}(L_{s,\rho})^k = \sum_{\gamma\in\pi_1} \sum_{x\in\mathcal{D}} \operatorname{tr} \rho(\gamma)\, L_k(s;x,\gamma x),$$

where $\mathcal{D}$ is a fundamental set in $\tilde{\mathbb{V}}$ for the π_1-action, and $L_k(s;x,y) = \exp(-s\ell(c))$ if there exists a path c in $(\tilde{\mathbb{V}},\tilde{\mathbb{E}})$ with $|c| = k$, $o(c) = x$, $t(c) = y$; otherwise we set $L_k(s;x,y) = 0$. It is straightforward to see that the right hand side equals

$$\sum_{\substack{c;\text{closed paths} \\ \text{in } (\mathbb{V},\mathbb{E}) \text{ with} \\ |c|=k}} \operatorname{tr} \rho(<c>) \exp(-s\ell(c)),$$

which also equals

$$\sum_{\substack{m,p \\ m|p|=k}} m^{-1} \operatorname{tr} \rho(<p>^m) \exp(-sm\ell(p)).$$

Hence $L(s,\rho) = \det(I - L_{s,\rho})^{-1}$.

<u>Proposition</u> D. $L(s,\rho)$ satisfies (L-1) (L-2) (L-3) and

(L'-4) if $n = \dim \rho \geq 2$ and ρ is irreducible, then $L(s,\rho)$ is holomorphic in $\operatorname{Re} s \geq h$,

(L'-5). for a character χ, $L(s,\chi)$ has a pole at $s = h + \sqrt{-1}t$ if and only if $\chi(<p>) = \exp\sqrt{-1}t\ell(p)$ for all p. In this case, $L(s,\chi) = L(s-\sqrt{-1}t, \mathbb{1})$, and every poles on $\operatorname{Re} s = h$ are simple.

In view of the above argument, the proof amounts to establishing a "twisted" Perron-Frobenius theorem for the operator $L_{s,\rho}$.

<u>Proposition</u> E. If $\operatorname{Re} s > 1$, then 1 is not an eigenvalue of $L_{s,\rho}$. If $n = \dim \rho \geq 2$ and ρ is irreducible, then 1 is not an eigenvalue of $L_{s,\rho}$ for any s with $\operatorname{Re} s = h$. For a character χ, 1 is an eigenvalue of $L_{h+\sqrt{-1}t,\chi}$ if and only if $\chi(<p>) = \exp\sqrt{-1}t\ell(p)$ for any prime cycle p.

III. We shall introduce L-functions associated to a class of topological graphs. A graph which we shall treat is a projective limit of finite graphs: $(\mathbb{V},\mathbb{E}) = \varprojlim (V_n, E_n)$ $(n = 1,2,\ldots)$ satisfying

a) each finite graph (V_n, E_n) is irreducible,

b) the morphisms $\pi_n : (V_n, E_n) \longrightarrow (V_{n-1}, E_{n-1})$ are surjective,

c) $\pi_n : o^{-1}(v) \longrightarrow o^{-1}(\pi_n(v))$ are bijective for any $v \in V_n$,

d) if $\pi_n(e) = \pi_n(e')$ $(e, e' \in E_n)$, then $t(e) = t(e')$.

The set $\mathbb{V}$ and $\mathbb{E}$ equip the metrics defined by

$$d(u,v) = \theta^{\sup\{n;\ \omega_n(u) = \omega_n(v)\}}$$

$$d(e,e') = \max\ \{d(o(e), o(e')),\ d(t(e), t(e'))\},$$

where $0 < \theta < 1$, and $\omega_n : (\mathbb{V}, \mathbb{E}) \longrightarrow (V_n, E_n)$ denotes the projection. We assume that the function ℓ on $\mathbb{E}$ is Lipschitz continuous with respect to this metric. The fundamental group of $(\mathbb{V}, \mathbb{E})$ is defined as the projective limit: $\pi_1(\mathbb{V}, \mathbb{E}) = \varprojlim \pi_1(V_n, E_n)$. A prime cycle p in $(\mathbb{V}, \mathbb{E})$ yields, in a natural manner, a conjugacy class in $\pi_1(\mathbb{V}, \mathbb{E})$, which we denote by $\langle p \rangle$. We finally let $\rho : \pi_1(\mathbb{V}, \mathbb{E}) \longrightarrow U(n)$ be a continuous homomorphism, and define the L-function with those data.

Theorem I ([2]). If $(\mathbb{V}, \mathbb{E})$ is not a circuit graph, then there exists a positive constant h such that $L(s,\rho)$ satisfies the same properties as in Proposition D.

The idea is to imitate the proof of finite graph case. Indeed, the operator $L_{s,\rho}$ can be also defined in this case, and satisfies much the same properties as in Proposition E. The Banach space on which $L_{s,\rho}$ acts is the space of Lipschitz continuous sections of a "flat line bundle" associated to the representation ρ. Then the proof can be accomplished by generalizing an approximation-technique developed by Ruelle [26].

An example of graphs satisfying the above conditions is a one-sided shift of finite type, that is, the space $\Sigma^+(V,E)$ of onesided infinite paths $c = (e_1, e_2, \cdots)$ in an irreducible finite graph (V,E). Edges in $\Sigma^+(V,E)$ are those pairs $(c, c') \in \Sigma^+(V,E) \times \Sigma^+(V,E)$ with $\sigma c = c'$, where σ is the shift operator.

IV. We now give another one-dimensional analogue of Selberg zeta function, which is much more arithmetic in its nature. As a special case, we obtain a simple interpretation of Ihara zeta functions associated to discrete subgroups of $SL_2(\mathbb{Q}_p)$ in terms of finite graphs (see Y. Ihara [14] and J. P. Serre [30]).

Let $(\mathbb{V},\mathbb{E})$ be a (non-oriented) finite graph, where $\mathbb{V}$ denotes the set of vertices and $\mathbb{E}$ denotes the set of edges. We assume that $(\mathbb{V},\mathbb{E})$ is connected as a one-dimensional CW-complex, and that the number of edges with a given origin v does not depend on v and is even, say $q+1$. A (two-sided infinite) path in $(\mathbb{V},\mathbb{E})$ is an element $c = (\cdots, v_{-1}, v_0, v_1, \cdots)$ such that $(v_i, v_{i+1}) \in \mathbb{E}$. A geodesic means a path without backtraking. We may define, in a natural way, a shift operator on the set of all geodesics. A periodic orbit p of the shift operator is called closed geodesics, and the least period is denoted by $|p|$. Let $\rho : \pi_1(\mathbb{V},\mathbb{E}) \longrightarrow U(n)$ be a unitary representation of the fundamental group of $(\mathbb{V},\mathbb{E})$. We define $\mathcal{P}$ to be the set of closed geodesic p in $(\mathbb{V},\mathbb{E})$, and put $N(p) = q^{|p|}$. We also define $\langle p\rangle$ to be the conjugacy class corresponding to the free homotopy class of a closed path in $(\mathbb{V},\mathbb{E})$ given by p in a usual manner. The L-function $L(s,\rho)$ is then defined by (1).

<u>Proposition</u> F (Ihara [14]). The function $Z(z,\rho) \equiv L(s,\rho)$, $z = q^{-s}$, is a rational function of z.

Proof. Let $(\tilde{\mathbb{V}},\tilde{\mathbb{E}})$ be the universal covering of $(\mathbb{V},\mathbb{E})$. The distance d on $\tilde{\mathbb{V}}$ is defined in an obvious way. Define the operators A_n, $n = 0, 1, 2, \cdots$, acting on the $\mathbb{C}^n$- valued functions on $\tilde{\mathbb{V}}$ by

$$A_0 = \mathrm{Id}, \qquad A_n\varphi(u) = \sum_{v;\, d(u,v)=n} \varphi(v).$$

Consider the vector space F_ρ defined in II and let $A_{n,\rho}$ denote the restriction of A_n to this space. What we shall prove is the following equality.

$$Z(z,\rho) = (1 - z^2)^{-g_\rho} \det(I - A_{1,\rho}z + q\, z^2)^{-1},$$

where $g_\rho = n(q-1)h/2$, $h = \#\,\mathbb{V}$. We first observe that

$$zZ'(z,\rho)/Z(z,\rho) + h = \sum_{m=0}^{\infty} N_{m,\rho} z^m,$$

where

$$N_{m,\rho} = \sum_{c;\,|c|=m} \operatorname{tr} \rho(<c>),$$

c running over all closed paths in $(\mathbb{V},\mathbb{E})$ without backtraking whose length are m. By an easy combinatrial argument, we find

$$\operatorname{tr} A_{m,\rho} = N_{m,\rho} + (q-1) \sum_{k=1}^{[(m-1)/2]} q^{k-1} N_{m-2k,\rho} \qquad (m > 0),$$

or

$$N_{m,\rho} = \operatorname{tr} A_{m,\rho} - (q-1) \sum_{k=1}^{[(m-1)/2]} \operatorname{tr} A_{m-2k,\rho}.$$

so that

$$\sum_m N_{m,\rho} z^m = \sum_m \operatorname{tr} A_{m,\rho} z^m - (q-1) \sum_m \sum_{k=1}^{[(m-1)/2]} \operatorname{tr} A_{m-2k,\rho}\, z^m.$$

We shall make use of the following universal identities (see [30]):

<u>Lemma</u> 1. Let Θ_m denote the correspondence which associates with each vertex $v \in \mathbb{V}$ the formal sum of the vertices v' such that $d(v,v') = m$. If we put

$$T_0 = \Theta_0 = \mathrm{Id}, \quad T_1 = \Theta_1$$

$$T_m = \sum_{k=0}^{[m/2]} \Theta_{m-2k},$$

then we have

$$\sum_{m=0}^{\infty} \Theta_m x^m = (1 - x^2) \sum_{m=0}^{\infty} T_m x^m = \frac{1 - x^2}{1 - T_1 x + q x^2},$$

where x is an indeterminate.

Since $\Theta_m \longrightarrow A_{m,\rho}$ gives a representation, we obtain

$$\sum N_{m,\rho} z^m = \mathrm{tr}\ \frac{1 - qz^2}{1 - A_{1,\rho} z + qz^2} + h(q-1)\ \frac{z^2}{1 - z^2}$$

which equals

$$h + z \frac{d}{dz} \log\{(1 - z^2)^{-g_\rho} \det(I - A_{1,\rho} z + qz^2)^{-1}\},$$

whence the proof is complete.

A graph satisfying the condition in the above proposition appears in the study of discrete groups in p-adic SL_2. Let K be a field with a discrete valuation ν, and let $\mathbb{O}$ denoete the valuation ring. Assume that the residue field k is finite, and denote $q = \# k$. The homogeneous space $\tilde{\mathbb{V}} = PSL_2(K)/PSL_2(\mathbb{O})$ can be identified with the set of homothetic equivalence classes of lattices in K^2, and one can define a distance function d on $\tilde{\mathbb{V}}$, by means of which one obtain a graph (acturally a tree) $(\tilde{\mathbb{V}}, \tilde{\mathbb{E}})$. Let Γ be a discrete subgroup of $PSL_2(K)$. If Γ is torsion free and co-compact, then Γ acts freely on $\tilde{\mathbb{V}}$ as automorphisms of graph, and the qutient graph $(\mathbb{V}, \mathbb{E})$, $\mathbb{V} = \Gamma\backslash\tilde{\mathbb{V}}$, is finite. The number of adjacent vertices to a given vertex is $q + 1$. We call an element $\gamma \neq 1$ of Γ or a conjugacy class $[\gamma]$ in Γ containing γ *primitive* if γ generates its centralizer in Γ. Put $\deg[\gamma] = |\nu(\lambda_\gamma \cdot \lambda'_\gamma)|$, where $\lambda_\gamma, \lambda'_\gamma \in K$ are the eigenvalues of a representative modulo K^* of γ. It is easy to see that there is a natural identification between the set $\mathcal{P}$ and the set of all primitive conjugacy classes. If $p \longleftrightarrow [\gamma]$, then $\deg[\gamma] = |p|$. This implies that the zeta function

$$\prod_{[\gamma];\mathrm{primitive}} \det(I - \rho(\gamma) z^{\deg[\gamma]})^{-1}$$

introduced by Ihara coincides with the above L-function.

It is interesting to note that the locations of poles of the L-function $L(s,\rho) = Z(q^{-s},\rho)$ are described by the eigenvalues of the hermitian matrix $A_1(\rho)$, which resembles the case of Selberg zeta functions. Let $\lambda_1(\rho) \geq \cdots \geq \lambda_m(\rho)$ be the eigenvalues of $A_1(\rho)$. In case $\rho = \mathbb{1}$, we find that the eigenvalue $\lambda_1(\mathbb{1}) = q + 1$ is simple and $L(s,\mathbb{1})$ has a simple pole at $s = 1$. Put $1 - \lambda_i(\rho) z + qz^2 =$

$(1 - \omega_i z)(1 - \omega_i^* z)$, and $\omega_i = q^{r_i + 1/2}$, $\omega_i^* = q^{r_i - 1/2}$, so that $s = r_i \pm 1/2$ are poles of $L(s,\rho)$. This implies that $L(s,\mathbb{1})$ satisfies "the Riemann hypothesis" if r_i are purely imaginary, or what is the same,

(R) $|\lambda_i(\mathbb{1})| \leq 2\, q^{1/2}$ for any $i > 1$.

Ihara gave some examples of discrete subgroups Γ which are constructed arithmetically and satisfy the (R). He, in fact, observed that his zeta functions are closely related to the congruence zeta functions of curves defined on finite number fields for which the Riemann hypothesis is valid. He also construct an example of Γ not satisfying (R). From our view point, this is easy to see, because $H_1(V,\mathbb{E})$ is a free abelian group with non-zero rank, so that by the same reason as in the existence of small eigenvalues of the Laplacian, we find, for any Γ, a subgroup $\Gamma_1 \subset \Gamma$ of finite index such that the eigenvalue $\lambda_1(\mathbb{1})$ of $A_{1,\mathbb{1}}$ defined on $\Gamma_1 \backslash \mathfrak{V}$ is close to $q + 1$, hence the zeta function $L(s,\mathbb{1})$ of $\Gamma_1 \backslash \mathfrak{V}$ does not satisfy the Riemann hypothesis.

V. Let (X,φ_t) be an Anosov flow on a compact smooth manifold X. We assume that the nonwandering set of φ_t is X. Put $\Gamma = \pi_1(X)$. As a set $\mathfrak{P}$, we take the set of all closed orbits of the flow, and put

$$N(p) = \exp(\ell(p)),$$

where $\ell(p)$ denotes the least period of p. We also denote by $\langle p \rangle$ the conjugacy class corresponding to the free homotopy class of p. Given a unitary representation $\rho : \Gamma \longrightarrow U(n)$, define $L(s,\rho)$ by (1). We should note that $L(s,\rho)$ associated to the geodesic flow on the unit sphere bundle on a compact Riemann surface with constant negative curvature is just the L-function given in I.

<u>Theorem</u> II. If (X,φ_t) is of Anosov type, then $L(s,\rho)$ satisfies the same properties as in Proposition D, where h should be rep-

laced by the topological entropy of the flow φ_t. If, in addition, (X,φ_t) is topological mixing and $\mathrm{Im}\,\rho$ is finite, then $L(s,\rho)$ is nice. In particular, this is the case for the L-function associated to the geodesic flow on the unit tangent bundle on a negatively curved manifold.

This was proven by Parry and Pollicott [23] for the case that the image $\rho(\Gamma)$ is finite, and by Adachi and Sunada [1] for general case. The key of the proof is to reduce Theorem II to Theorem I in the following way. First take a Markov family of sufficiently small size, which gives rise to an embedded finite oriented graph (V,E) in X. By the means of Bowen's symbolic dynamical system, (X,φ_t) is almost isomorphic to a suspension flow $(\Sigma(V,E,f),\sigma(f)_t)$, where

$$\Sigma(V,E,f) = \{ (\xi,\ t)\ ;\ \xi \in \Sigma(V,E),\ 0 \le t \le f(\xi) \}$$

$$\sigma(f)_t(\xi,\ s) = (\xi,\ s+t).$$

$\Sigma(V,E)$ denotes the set of paths $(\cdots, e_{-1}, e_0, e_1, \cdots)$ in (V,E). As is usual, one can associate a subshift $\Sigma^+(V,E)$ and a positive function f^+ on $\Sigma^+(V,E)$ which is cohomologous to f. Since the function ℓ defined by $\ell(e) = f^+(\ell(e))$ is Lipschitz continuous in the sense in IV for some $\theta \in (0,1)$, we can define L-functions $L(s,\overline{\rho})$ associated with the profinite graph $\Sigma^+(V,E)$, the length function ℓ and a representation $\overline{\rho}$. The relation between $L(s,\overline{\rho})$ and the dynamical L-function is:

<u>Lemma</u> 2. The topological entropy h of (X,φ_t) coincides with the critical exponent of $L(s,\overline{\rho})$. Moreover, if $\overline{\rho}$ be the representation of $\pi_1(\Sigma^+(V,E))$ given by the composition

$$\pi_1(\Sigma^+(V,E)) \longrightarrow \pi_1(V,E) \longrightarrow \pi_1(X) \xrightarrow{\ \rho\ } U(n),$$

then the ratio $L(s,\rho)/L(s,\overline{\rho})$ is a non-vanishing holomorphic function in a neighborhood of $\mathrm{Re}\, s \ge h$.

This can be proven, by applying an idea due to R. Bowen [4]. In the course of the proof of Theorem II, we make use of the following which refines a result on generations of homology groups given by Fried [7].

Proposition G (T. Adachi [3]). The fundamental group $\pi_1(X)$ is generated by homotopy classes of closed orbits of (X, φ_t).

One of consequences of Proposition B applied to the dynamical L-function is that any coset of a finite quotient group $H_1(X, \mathbb{Z})/H$ contains infinitely many homology classes represented by closed orbits (analogue of the Dirichlet theorem for arithmetic progressions). It is natural to ask if this is true for a general quotient group $H_1(X, \mathbb{Z})/H$. The similar question for a finite graph has a negative answer. The graph associated to the matrix $\begin{pmatrix} 1, 1 \\ 1 & 0 \end{pmatrix}$ provides an example that some $\alpha \in H_1(\mathbb{V}, \mathbb{E})$ does not contain any closed path. But we can prove the following.

Theorem III([1],[16]). If the geodesic flow on the unit tangent bundle of N is of Anosov type, or if N is non-positively curved and of rank one, then each homology class $\in H_1(N, \mathbb{Z})$ contains infinitely many prime closed geodesics.

We can say much more about the growth rate of number of closed geodesics with respect to the length. To explain this it is covenient to introduce the following notation. Given $\alpha \in H_1(N, \mathbb{Z})$, we set

$$\pi(x, \alpha) = \#\{\text{ prime geodesics } p \text{ ; } \ell(p) < x, \text{ homology class } [p] = \alpha \}.$$

If $H_1(N, \mathbb{Z})$ is of finite order, we have, as a special case of Proposition B,

$$\pi(x, \alpha) \sim (\# H_1(N, \mathbb{Z}))^{-1} e^{hx}/hx, \text{ as } x \nearrow \infty.$$

In the case that $H_1(N, \mathbb{Z})$ is infinite, we have a bit weak result:

(2) $\lim_{x \to \infty} \frac{1}{x} \log \pi(x, \alpha) = h$, if N has a geodesic flow of Anosov type.

The proof of (2) is rather combinatrial, and make use of a special feature of a geodesic flow φ_t that has a reversible property : $-\varphi_t(v) = \varphi_{-t}(-v)$ (note that the classical argument in number theory can not apply to the ideal class group of infinite order).

Remark 1. In our paper [1], we have proved

(3) $\liminf \frac{1}{x} \log \pi(x,\alpha) \geq \frac{h}{2}.$

A little effort is required to show (2). We thank M. Pollicott and A. Katsuda for communicating an idea of how to modify our proof to get (2). When N is non-positively curved and of rank one, (3) was shown by Katsuda [16]. In any case, we conjecture that there is a positive constant c not depending on α such that

$$\pi(x,\alpha) \sim c\ e^{hx}/\ x^{b_1(N)+1},$$

where $b_1(N)$ denotes the first betti number of N.

Remark 2. The statement of Theorem III is not always true for general Riemannian manifolds. The simplest counter example is the flat torus $\mathbb{R}^2/\mathbb{Z}^2$. In fact one may easily observe that the class $\alpha \in H_1(\mathbb{R}^2/\mathbb{Z}^2,\mathbb{Z}) \simeq \mathbb{Z}^2$ contains a prime geodesic if and only if α is primitive. Recently, P. Pansu [22] has taken up a problem on the growth rate of number of closed geodesics in a nil-manifolds. See also M. Gromov [11], who considers the following problem. Let $c(\alpha)$ be the shortest closed geodesics whose homology class is α (if there are several such geodesics, we choose one of them). Consider the number $N(x)$ of those geodesics $c = c(\alpha)$ for all $\alpha \in H_1(N,\mathbb{Z})$, for which

$$\ell(c) \leq x.$$

Then one has as $x \longrightarrow \infty$

$$N(x) \sim c'\ x^{b_1(N)},$$

for some computable constant c. This can be considered a complementary result to the above conjecture.

References

1. T. Adachi and T. Sunada: Homology of closed geodesics in a negatively curved manifold, preprint.
2. T. Adachi and T. Sunada: L-functions of dynamical systems and topological graphs, preprint.
3. T. Adachi: Closed orbits of an Anosov flow and the fundamental group, preprint.
4. R. Bowen: Symbolic dynamics for hyperbolic flows, Amer. J. Math. 95 (1973), 429-460.
5. R. Brooks: The first eigenvalue in a tower of coverings, preprint.
6. R. Brooks: Combinatorial problems in spectral geometry, preprint.
7. D. Fried: Flow equivalence, hyperbolic systems and a new zeta function for flows, Comment. Math. Helvetici 57 (1982), 237-259.
8. D. Fried: The zeta functions of Ruelle and Selberg I, preprint.
9. D. Fried: Analytic torsion and closed geodesics on hyperbolic manifolds, preprint.
10. R. Gangolli: Zeta functions of Selberg's type for compact space forms of symmetric spaces of rank one, Ill. J. Math. 21 (1977), 1-42.
11. M. Gromov: Filling Riemannian manifolds, J. Diff. Geom. 18 (1983), 1-147.
12. D. A. Hejhal: The Selberg trace formula and the Riemann zeta function, Duke Math. J. 43 (1976), 441-482.
13. D. A. Hejhal: *The Selberg Trace Formula for* $PSL(2,\mathbb{R})$, I. Springer Lecture Notes 548, 1976.
14. Y. Ihara: On discrete subgroups of the two by two projective linear group over p-adic fields, J. Math. Soc. Japan 18 (1966), 219-235.
15. S. Iyanaga and Y. Kawada (ed.): *Encyclopedic Dictionary of Mathematics*, MIT Press, Cambridge, 1977.

16 A. Katsuda: Homology of closed geodesics in a nonpositively curved manifold of rank one, preprint.

17. N. Kurokawa: On some Euler products. I, Proc. Japan Acad. 60 (1984) 335-338.

18. N. Kurokawa: On some Euler products. II, Proc. Japan Acad. 60 (1984), 365-368.

19. S. Lang: *Algebraic Number Theory*, Addison-Wesley, 1970.

20. A. Manning: Axiom A diffeomorphisms have rational zeta functions, Bull. London Math. Soc. 3 (1971), 215-220.

21. H. P. McKean: Selberg's trace formula as applied to a compact Riemann surface, Comm. Pure Appl. Math. 25 (1972), 225-246.

22. P. Pansu: *Croissance des boules et des géodésiques fermées dans les nilvariétés*, Ergod. Th. Dynam. Sys. 3 (1983), 415-445.

23. W. Parry and M. Pollicott: An analogue of the prime number theorem for closed orbits of Axiom A flows, Ann. of Math. 118 (1983), 573-591.

24. W. Parry and M. Pollicott: The Chebotarev theorem for Galois coverings of Axiom A flows, preprint.

25. M. Pollicott: Meromorphic extensions of generalized zeta functions, preprint.

26. D. Ruelle: *Thermodynamic Formalism*, Addison-Weasley, Reading, Mass., 1978.

27. D. Ruelle: Zeta functions for expanding maps and Anosov flows, Invent. Math. 34 (1976), 231-242.

28. P. Sarnak: Prime geodesic theorems, Ph. D. dissertation, Stanford University (1980).

29. A. Selberg: Harmonic analysis and discontinuous subgroups in weakly symmetric Riemannian spaces with applications to Dirichlet series, J. Indian Math. Soc. 20 (1956), 47-87.

30. J. P. Serre: *Tree*, Springer New York 1980.

31. S. Smale: Differentiable dynamical systems, Bull. AMS. 73 (1967), 747-817.

32. T. Sunada: Geodesic flows and geodesic random walks, Advanced Studies in Pure Math. 3 (Geometry of Geodesics and Related Topics)

(1984), 47-86.

33. T. Sunada: Riemannian coverings and isospectral manifolds, Ann. of Math. 121 (1985), 169-186.

34. T. Sunada: Number theoretic methods in spectral geometry, to appear in Proc. The 6th Symp. on Differential Geometry and Differential Equations held at Shanghai, China 1985.

STABILITY OF HARMONIC MAPS AND EIGENVALUES OF LAPLACIAN

Hajime Urakawa

Department of Mathematics
College of General Education
Tohoku University, Sendai, 980, Japan

§0. Introduction

The theory of harmonic maps has recently developped very much as we look excellent expository papers [E.L 1,2] of Eells and Lemaire. In this paper we focus on the eigenvalues of the second variation operator of harmonic maps.

A harmonic map ϕ from a compact domain Ω in a m-dimensional Riemannian manifold (M^m,g) into an n-dimensional Riemannian manifold (N^n,h) is a critical point of the energy

$$E(\Omega,\phi) = \int_\Omega e(\phi) *1 ,$$

where $e(\phi) = \frac{1}{2} h(d\phi,d\phi)$. That is, for every vector field V along ϕ with $V \equiv 0$ on $\partial\Omega$,

$$\frac{d}{dt}\Big|_{t=0} E(\Omega,\phi_t) = 0.$$

Here ϕ_t ; $\Omega \to N$ is a one-parameter family of smooth maps such that $\phi_0 = \phi$ and $\frac{d}{dt}\big|_{t=0}\phi_t(x) = V_x \in T_{\phi(x)}N$ for all x in Ω. In case of $\Omega = M$, we denote $E(\phi) = E(M,\phi)$ if defined.

Harmonic maps have a lot of examples (cf. [E.L 1,2]) :

<u>Example 1.</u> γ ; $[0,2\pi] \to (N^n,h)$, a geodesic.

Example 2. ϕ ; $(M^m,g) \to (N^n,h)$, an isometric minimal immersion.

Example 3. ϕ ; $(M^m,g) \to (N^n,h)$, a Riemannian submersion whose each fiber $\phi^{-1}(y)$, $y \in N$, is a minimal submanifold of (M,g). Here the Riemannian submersion ϕ ; $(M,g) \to (N,h)$ is, by definition, for each point x in M, the tangent space T_xM has the following orthogonal decomposition

$$T_xM = H_x \oplus V_x$$

with respect to g_x in such a way that (i) the subspace V_x is the kernel of the differential ϕ_* at x and (ii) the restriction of ϕ_* to the subspace H_x is an isometry of (H_x,g_x) onto $(T_{\phi(x)}N,h_{\phi(x)})$ (cf. [B.B]).

Example 4. A holomorphic map ϕ ; $(M,g) \to (N,h)$ between Kaehler manifolds (M,g), (N,h).

The second variation formula of the energy E for a harmonic map ϕ is given (cf. [Ma],[Sm]) as follows :

$$\frac{d^2}{dt^2}\Big|_{t=0} E(\Omega,\phi_t) = \int_\Omega h(V,J_\phi V) *1$$

where J_ϕ is a differential operator (called the *Jacobi operator*) acting on the space of all vector fields along ϕ which is identified with the space $\Gamma(E)$ of sections of the induced bundle $E = \phi^{-1}TN$ of the tangent bundle TN by ϕ. The Jacobi operator is written as

$$(0.1) \qquad J_\phi V = \tilde{\nabla}^*\tilde{\nabla}V - \sum_{i=1}^{m} {}^N R(\phi_* e_i,V)\phi_* e_i \ , \quad V \in \Gamma(E).$$

Here $\tilde{\nabla}$ is the connection of $E = \phi^{-1}TN$ which is defined by

$$\tilde{\nabla}_X V = {}^N\nabla_{\phi_* X} V \qquad , \quad V \in \Gamma(E),$$

for a tangent vector X on M, and ${}^N\nabla$, ${}^NR$ are the Levi-Civita connection, the curvature tensor of (N,h) , respectively :

$$^{N}R(Y,Z)W = {}^{N}\nabla_{[Y,Z]}W - [{}^{N}\nabla_Y , {}^{N}\nabla_Z]W ,$$

for tangent vectors Y,Z,W on N. The operator $\tilde{\nabla}^*\tilde{\nabla}$ is the rough Laplacian defined by

$$\tilde{\nabla}^*\tilde{\nabla}V = - \sum_{j=1}^{m}(\tilde{\nabla}_{e_j}\tilde{\nabla}_{e_j} - \tilde{\nabla}_{\nabla_{e_j}e_j})V , \quad V \in \Gamma(E),$$

where $\{e_j\}_{j=1}^{m}$ is a locally defined frame field on M and ∇ is the Levi-Civita connection of (M,g).

Now let us consider the following eigenvalue problems :

(I) In case of a compact manifold $\Omega = M$ without boundary,

$$J_\phi V = \lambda V , \qquad V \in \Gamma(E).$$

(II) In case of a relatively compact domain Ω in M,

$$\begin{cases} J_\phi V = \lambda V , & \text{on } \Omega , \\ V = 0 , & \text{on } \partial\Omega . \end{cases}$$

Since the Jacobi operator J_ϕ is a second order elliptic differential operator on $\Gamma(E)$, both the eigenvalue problems have the discrete spectra consisting of the eigenvalues with finite multiplicities. We denote for (I),

$$\mathrm{Spec}(J_\phi) = \{ \tilde{\lambda}_1 \leqq \tilde{\lambda}_2 \leqq \dots \} ,$$

and for (II),

$$\mathrm{Spec}_\Omega(J_\phi) = \{ \tilde{\lambda}_1(\Omega) \leqq \tilde{\lambda}_2(\Omega) \leqq \dots \} ,$$

respectively. Our main concern is *how the spectra* $\mathrm{Spec}(J_\phi)$ *or* $\mathrm{Spec}_\Omega(J_\phi)$ *reflect the geometry of harmonic maps* ϕ. Namely we will treat with the following two problems :

A) *How do the small , in paticular , non-positive eigenvalues of* J_ϕ *behave ?*

B) *How is the harmonic map* ϕ *characterized by the spectrum* $\mathrm{Spec}(J_\phi)$?

In connection with A), we will deal with the stability of harmonic maps. More precisely, let us define the index and nullity of a harmonic map ϕ following [E.L 1,2], which are the analogue of the Morse theory of geodesics :

$$\mathrm{Index}(\phi) = \text{sum of multiplicities of negative eigenvalues of the problem (I)},$$

$$\mathrm{Nullity}(\phi) = \dim \mathrm{Ker}(J_\phi).$$

We denote $\mathrm{Index}_\Omega(\phi)$, $\mathrm{Nullity}_\Omega(\phi)$, respectively in case of (II). A harmonic map ϕ form (M,g) (or Ω) is called *stable* (or *stable on* Ω) if $\mathrm{Index}(\phi) = 0$ (or $\mathrm{Index}_\Omega(\phi) = 0$), respectively. Then we will consider the following problems :

A1) *How can the index and nullity be estimated generally by the geometric quantities ?*

A2) *Can we expect something from a stable harmonic map ?* (application of stable harmonic maps)

A3) *What kind of harmonic maps are stable ?*

Concerning B), we will consider the following problem :

Characterize the typical harmonic maps appearing in Examples 1) ~ 4) *by the spectrum* $\mathrm{Spec}(J_\phi)$ *of the Jacobi operator* J_ϕ.

Partial answers will be obtained in §6.

This paper is mainly based on [U1,2,3,4] and [Oh].

Table of Contents

Part I. Stability of harmonic maps.

§1. Generic estimates of the index and nullity.

A classical Morse-Schoenberg theorem(cf.[G.K.M, p.177]) tells us the index and nullity of a geodesic γ ; $[0,2\pi] \to (N^n,h)$ can be estimated as follows :

Theorem (Morse-Schoenberg) *Assume that the sectional curvature* ${}^N K$ *of* (N^n,h) *satisfies* ${}^N K \leqq a$ *for some positive constant* a. *Then the nullity and index satisfy*

$$\mathrm{Index}_\Omega(\gamma) + \mathrm{Nullity}_\Omega(\gamma) \leqq n \left[L \frac{\sqrt{a}}{\pi} \right]$$

where L *is the length of the geodesic* γ *and* $[x]$ *is the integer part of* $x > 0$.

Remark 1.1. When (N,h) is the canonical sphere of constant curvature a, it is well-known that the equality holds. Then the above estimate is optimal.

Remark 1.2. If $L < \frac{\pi}{\sqrt{a}}$, the above inequality implies that $\mathrm{Index}_\Omega(\gamma) = \mathrm{Nullity}_\Omega(\gamma) = 0$ which says the stability of the geodesic γ.

Furthermore, if $L \leqq \frac{\pi}{\sqrt{a}}$, then $\mathrm{Index}_\Omega(\gamma) + \mathrm{Nullity}_\Omega(\gamma) \leqq n$.

Remark 1.3. Our definition of the index and nullity is slightly different from the one in [G.K.M] where the orthogonal vector fields to the tangent vector of the geodesic γ are adopted.

It would be natural to consider *whether or not analogous estimates of the index and nullity hold for a general harmonic map.* For estimations, we use the following geometric quantity :

$$D = {}^N R^\phi_\Omega \; C(M,g)^{-1} \, \mathrm{Vol}(\Omega)^{2/m} - 1$$

for a relatively compact domain Ω in M. Here ${}^N R^\phi_\Omega$ is defined by

$$\text{(i)} \qquad {}^N R^\phi_\Omega := \sup_{x \in \Omega} \; \sup_{0 \neq v \in T_{\phi(x)} N} \frac{\sum_{i=1}^m h({}^N R(\phi_* e_i, v)\phi_* e_i, v)}{h(v,v)}$$

In case of $\Omega = M$, we put ${}^N R^\phi = {}^N R^\phi_M$ when defined.

(ii) $\mathrm{Vol}(\Omega)$ is the volume of Ω in (M,g), and

(iii) $C(M,g)$ is the isoperimetric constant depending only on (M,g) which satisfies the following properties (cf. [Y, p.22], [B.G]) : The i-th eigenvalue $\lambda_i(\Omega)$ of the Dirichlet problem

$$\begin{cases} \Delta_M u = \lambda u & \text{on } \Omega , \\ u = 0 & \text{on } \partial\Omega \end{cases}$$

of the Laplace-Beltrami operator $\Delta_M = \delta d$ of (M,g) on $C^\infty(M)$ satisfies the inequalities

$$\lambda_i(\Omega) \geqq C(M,g) \, \mathrm{Vol}(\Omega)^{-2/m} \, i^{2/m} , \quad i=1,2,\dots .$$

For example, in case of the Euclidean space $(\mathbb{R}^m, \mathrm{can})$, we can take $C(\mathbb{R}^m, \mathrm{can}) = 4\pi^2 \omega_m^{-2/m}$, where $\omega_m = \pi^{m/2}/\Gamma(\frac{m}{2}+1)$ is the volume of the unit ball in $(\mathbb{R}^m, \mathrm{can})$.

Remark 1.4. If the sectional curvature ${}^{N}K$ of (N^n,h) satisfies ${}^{N}K \leqq a$ for some positive constant a, then

$$ {}^{N}R^{\phi}_{\Omega} \leqq 2\, a\, E^{\infty}(\Omega,\phi), $$

where $E^{\infty}(\Omega,\phi) = \sup_{x\in\Omega} e(\phi)$. Furthermore, if ϕ is an isometric minimal immersion of (M^m,g) into (N^n,h), then ${}^{N}R^{\phi}_{\Omega} \leqq ma$.

Then we have :

Theorem 1.1. (cf.[U1,Corollary 3.3]) *For a relatively compact domain* Ω *in* M, *and every harmonic map* $\phi;\ \Omega \to (N^n,h)$, *if* $D < 0$, *then*

$$ \mathrm{Index}_{\Omega}(\phi) = \mathrm{Nullity}_{\Omega}(\phi) = 0. $$

In particular, assume that the sectional curvature ${}^{N}K$ *of* (N,h) *satisfies* ${}^{N}K \leqq a$ *for some positive constant* a . *If* $C(M,g)\, \mathrm{Vol}(\Omega)^{-2/m} > 2\, a\, E^{\infty}(\Omega,\phi)$, *then*

$$ \mathrm{Index}_{\Omega}(\phi) = \mathrm{Nullity}_{\Omega}(\phi) = 0. $$

Since the constant ${}^{N}R^{\phi}_{\Omega}$ is bounded for a small domain Ω , this theorem implies that a harmonic map $\phi;\ \Omega \to (N^n,h)$ is stable if Ω shrinks so small as $D < 0$.

The next theorem tells us the index and nullity can be estimated by the quantity D if $D \geqq 0$:

Theorem 1.2. (cf.[U1,Theorem 3.4]) *Let* Ω *be a relatively compact domain in* (M^m,g) *and* $\phi\ ;\ \Omega \to (N^n,h)$ *any harmonic map. Assume that* $D \geqq 0$. *Then the index and nullity can be estimated as follows :*

(i) *In case of* $m = 1,2$,

$$ \mathrm{Index}_{\Omega}(\phi) + \mathrm{Nullity}_{\Omega}(\phi) \leqq n(1+\tfrac{1}{D})^{D}\{1+D\} . $$

(ii) In case of $m = 2(p+1)$, $p \geqq 1$,

$$\mathrm{Index}_\Omega(\phi) + \mathrm{Nullity}_\Omega(\phi) \leqq n(1+\tfrac{1}{D})^D\{1+P(D)\}$$

where $P(D) := (p+1)!\,\Sigma_{k=0}^{p}\frac{1}{k!}\{\frac{1}{\log(1+\frac{1}{D})}\}^{p+1-k}$.

(iii) In case of $m = 2p+1$, $p \geqq 1$,

$$\mathrm{Index}_\Omega(\phi) + \mathrm{Nullity}_\Omega(\phi) \leqq n(1+\tfrac{1}{D})^D\{1+Q(D)\}$$

where $Q(D) := \frac{m}{2}\,p!\,\Sigma_{k=0}^{p}\frac{1}{k!}\{\frac{1}{\log(1+\frac{1}{D})}\}^{p+1-k}$.

(iv) For every $m \geqq 1$,

$$\mathrm{Index}_\Omega(\phi) + \mathrm{Nullity}_\Omega(\phi) \leqq n\,\frac{\Gamma(\frac{m}{2}+1)\,e^{m/2}}{(m/2)^{m/2}}\{1+D\}^{m/2}.$$

<u>Remark 1.5.</u> $P(D)$, $Q(D)$ satisfy $\lim_{D\to 0}P(D) = \lim_{D\to 0}Q(D) = 0$, and $P(D) \sim (\frac{m}{2})!\,D^{m/2}$, $Q(D) \sim \frac{m}{2}(m-1)!\,D^{(m+1)/2}$ as $D \to \infty$. The function $(1+\frac{1}{D})^D$ satisfies $\lim_{D\to 0}(1+\frac{1}{D})^D = 1$, $(1+\frac{1}{D})^D < e$ and $\lim_{D\to\infty}(1+\frac{1}{D})^D = e$.

<u>Remark 1.6.</u> In case of a geodesic $\gamma\,;\,[0,2\pi] \to (N^n,h)$ whose sectional curvature ${}^N K$ satisfies ${}^N K \leqq a$ for some constant $a > 0$, $\sqrt{1+D} \leqq L\frac{\sqrt{a}}{\pi}$, and $\frac{\Gamma(\frac{m}{2}+1)e^{m/2}}{(m/2)^{m/2}} = (\frac{\pi e}{2})^{1/2} = 2.066\cdots$.

In case of a compact manifold $\Omega = M$ without boundary, we have:

<u>Theorem 1.3.</u> (cf.[U1,Theorem 2.5]) *Let* (M^m,g) *be a m-dimensional compact Riemannian manifold without boundary whose Ricci curvature* Ric_M *is bounded below by a positive constant :* $\mathrm{Ric}_M \geqq (m-1)\,\delta > 0$. *Let* $\phi\,;\,(M^m,g) \to (N^n,h)$ *be an arbitrary harmonic map. Then we have*

(i) In case of $m \geqq 3$,

$$\mathrm{Index}(\phi) + \mathrm{Nullity}(\phi) \leqq n(1+\tfrac{1}{A})^A\{1+(m-1)!\,m^{m-1}\,A(A+1)^{m-1}\}$$

where $A = {}^N\!R^\phi/m\delta$.

(ii) *In case of* $m = 2$,

$$\text{Index}(\phi) + \text{Nullity}(\phi) \leqq n(1+\tfrac{1}{B})^B\{1+4B^2\}$$

where $B = {}^N\!R^\phi/\delta$.

§2. Kaehler version of Lichnerowicz-Obata theorem about λ_1.

In the following we assume that M^m is compact without boundary. It is known (cf.[E.L2, Corollary 8.15]) that

(i) *each holomorphic map* ϕ; $(M^m,g) \to (N^n,h)$ *between Kaehler manifolds* $(M^m,g),(N^n,h)$ *is harmonic and an absolute minimum of the energy* E *in its homotopy. In particular,*

(ii) *such holomorphic map* ϕ *is stable, that is,* $\text{Index}(\phi) = 0$.

(iii) *Moreover if* ϕ_t ; $(M^m,g) \to (N^n,h)$ *is a smooth deformation of a* $\pm$*holomorphic map* ϕ_0 *through harmonic maps* ϕ_t , *then each* ϕ_t *is* $\pm$*holomorphic* (cf.[E.L2, Corollary 8.19]).

There are infinitesimal versions of these facts :

Proposition 2.1. *Let* ϕ; $(M^m,g) \to (N^n,h)$ *be a holomorphic map between Kaehler manifolds* $(M^m,g),(N^n,h)$. *Then*

$$\int_M h(J_\phi V,V) *1 = \frac{1}{2}\int_M h(DV,DV) *1 ,$$

where $DV(X) = {}^N\nabla_{\phi_* JX}V - J\,{}^N\nabla_{\phi_* X}V$, $V \in \Gamma(E)$, $X \in \Gamma(TM)$, *and* J *is a complex structure of* M^m *or* N^n. *In particular,*

(i) *such map* ϕ ; $(M^m,g) \to (N^n,h)$ *is stable, that is,*

$$\text{Index}(\phi) = 0.$$

(ii) $$\text{Ker}(J_\phi) = \{ V \in \Gamma(E);\ {}^N\nabla_{\phi_* JX}V = J\,{}^N\nabla_{\phi_* X}V \quad \textit{for all } X \in \Gamma(TM)\}.$$

Remark 2.1. In case of the identity map $\phi = id ; (M,g) \to (M,g)$, this is due to [Li,p.147]. In case of a holomorphic isometric immersion $\phi ; (M,g) \to (N,h)$, it is due to [Si,p.76]. T.Sunada obtained this proposition in [Su,p.164]. His proof is slightly different from ours which is similar as in [Si].

Proof. Let $m = 2p$, $p = \dim_{\mathbb{C}} M$. Let $\{E_i,F_i\}_{i=1}^{p}$ be a locally defined orthonormal frame field on M satisfying $JE_i = F_i$, $JF_i = - E_i$, $i=1,\ldots,p$. Then by (0.1), we have

$$h(J_\phi V,V) = \sum_{i=1}^{p} \{h(\tilde{\nabla}_{E_i} V,\tilde{\nabla}_{E_i} V) + h(\tilde{\nabla}_{F_i},\tilde{\nabla}_{F_i})$$
$$- h({}^N R(\phi_* E_i,V)\phi_* E_i,V) - h({}^N R(\phi_* F_i,V)\phi_* F_i,V)\}.$$

Lemma 2.2. ${}^N R(\phi_* E_i,V)\phi_* E_i + {}^N R(\phi_* F_i,V)\phi_* F_i = - J\, {}^N R(\phi_* E_i,\phi_* F_i)V.$

In fact, using that $\phi_* \circ J = J \circ \phi_*$, $J \circ {}^N\nabla = {}^N\nabla \circ J$, and $JE_i = F_i$, $JF_i = - E_i$, the left hand side of the above coincides with

$$- J\, {}^N R(\phi_* E_i,V)\phi_* F_i + J\, {}^N R(\phi_* F_i,V)\phi_* E_i$$
$$= J\, {}^N R(V,\phi_* E_i)\phi_* F_i + J\, {}^N R(\phi_* F_i,V)\phi_* E_i$$
$$= - J\, {}^N R(\phi_* E_i,\phi_* F_i)V.$$

We continue the proof of Proposition 2.1. Since

$$h(DV,DV) = 2 \sum_{i=1}^{p}\{h(\tilde{\nabla}_{E_i} V,\tilde{\nabla}_{E_i} V) - 2h(J\tilde{\nabla}_{E_i} V,\tilde{\nabla}_{F_i} V) + h(\tilde{\nabla}_{F_i} V,\tilde{\nabla}_{F_i} V)\},$$

we have , using Lemma 2.2,

$$(2.1)\quad h(J_\phi V,V) - \tfrac{1}{2}h(DV,DV) = \sum_{i=1}^{p}\{h(J\, {}^N R(\phi_* E_i,\phi_* F_i)V,V)$$
$$+ 2h(J\tilde{\nabla}_{E_i} V,\tilde{\nabla}_{F_i} V)\}.$$

So we have only to show that the integral over M of the right hand side of (2.1) vanishes. In fact, using the fact $X\,h(V,W) = h(\tilde{\nabla}_X V,W) + h(V,\tilde{\nabla}_X W)$ for $V,W \in \Gamma(E)$ and $X \in \Gamma(TM)$, the integral over M of the sum $\sum_{i=1}^{p} h(J\,{}^{N}R(\phi_* E_i,\phi_* F_i)V,V)$ coincides with

$$\int_M \Sigma_{i=1}^{p} \{h(\tilde{\nabla}_{E_i} V,\tilde{\nabla}_{F_i} JV) - h(\tilde{\nabla}_{F_i} V,\tilde{\nabla}_{E_i} JV)\} *1$$

$$+\int_M \Sigma_{i=1}^{p}\{E_i h(\tilde{\nabla}_{F_i} V,JV)-F_i h(\tilde{\nabla}_{E_i} V,JV)-h(\tilde{\nabla}_{\nabla_{E_i} F_i} V,JV)+h(\tilde{\nabla}_{\nabla_{F_i} E_i} V,JV)\}*1.$$

Here the second term is zero since the integrand coincides with the divergence of the vector field X on M defined by $g(X,Y) = h(\tilde{\nabla}_{JY}V,JV)$ for $Y \in \Gamma(TM)$. Hence the integral over M of (2.1) is

$$\int_M \Sigma_{i=1}^{p}\{h(\tilde{\nabla}_{E_i} V,\tilde{\nabla}_{F_i} JV)-h(\tilde{\nabla}_{F_i} V,\tilde{\nabla}_{E_i} JV)+2h(J\tilde{\nabla}_{E_i} V,\tilde{\nabla}_{F_i} V)\}*1 = 0.$$

Q.E.D.

Corollary 2.3. *Let $(M^m,g),(N^n,h)$ be Kaehler manifolds. For a relatively compact domain Ω in M^m, let ϕ be a holomorphic map from Ω into N^n. Then*

$$\mathrm{Index}_\Omega(\phi) = \mathrm{Nullity}_\Omega(\phi) = 0.$$

Proof. By the same way as the proof of Proposition 2.1, we have, for $V \in \Gamma(E)$ with $V|_{\partial\Omega} \equiv 0$,

$$\int_\Omega h(J_\phi V,V)\, *1 = \frac{1}{2}\int_\Omega h(DV,DV)\, *1 \geqq 0,$$

which impies $\mathrm{Index}_\Omega(\phi) = 0$. Moreover if we assume $V \in \mathrm{Ker}(J_\phi)$ with $V|_{\partial\Omega} \equiv 0$, then $DV = 0$ and $V|_{\partial\Omega} \equiv 0$. Therefore V vanishes identically on Ω. Q.E.D.

Remark 2.2. The similar theorem of Corollary 2.3 for a holomorphic isometric immersion was obtained in [Si, Theorem 3.5.1].

Let us consider the identity map id_M of a compact Kaehler manifold (M^m,g). In this special case, due to Proposition 2.1, we have (cf. [Li]) that

(i) the identity map id_M *is stable, i.e.,* $\mathrm{Index}(id_M) = 0$, *and*

(ii) $\mathrm{Ker}(J_{id_M}) \cong \mathfrak{a}(M)$,

where $\mathfrak{a}(M)$ is the space of all holomorphic vector fields on M^m. Together them and the equality

$$J_{id_M} = \Delta_H - 2\rho,$$

where Δ_H is the differential operator on $\Gamma(TM)$ corresponding to the Hodge Laplacian $d\delta+\delta d$ on 1-forms ,and ρ is the Ricci transform, we can estimate the first eigenvalue $\lambda_1(M,g)$ of the Laplace-Beltrami operator Δ_M of the compact Kaehler manifold (M^m,g) acting on $C^\infty(M)$:

<u>Theorem 2.4.</u> (cf.[U1,Theorem 4.2]) *Let* (M^m,g) *be a compact Kaehler manifold whose Ricci curvature* Ric_M *is bounded below by a positive constant* : $\mathrm{Ric}_M \geqq \alpha > 0$. *Then we have*

$$\lambda_1(M,g) \geqq 2\alpha .$$

If the equality holds, then $\mathfrak{a}(M) \neq \{0\}$.

<u>Remark 2.3.</u> In case of compact Kaehler *Einstein* manifolds (M^m,g) , Theorem 2.4 was obtained in [Ob]. In this case, the equality holds if and only if $\mathfrak{a}(M) \neq \{0\}$. A theorem of Lichnerowicz and Obata tells us (cf.[B.G.M]) that for a compact Riemannian manifold (M^m,g),

$$\mathrm{Ric}_M \geqq \alpha > 0 \quad \longrightarrow \quad \lambda_1(M,g) \geqq \frac{m}{m-1}\alpha ,$$

and the equality holds if and only if (M^m,g) is isometric to the standard unit sphere. Remark that $2 \geqq m/(m-1)$ and $2 = m/(m-1)$ if and only if $m = 2$.

§3. Instability theorem of Xin, Leung, and Ohnita.

What kind of Riemannian manifolds (M,g) admit stable harmonic maps ? It is easily shown that if the sectional curvature ${}^N K$ of the target manifold (N^n,h) is non-positive, then a harmonic map $\phi : (M^m,g) \to (N^n,h)$ is stable, i.e., $\mathrm{Index}(\phi) = 0$.

R.T.Smith [Sm] showed that

Proposition 3.1. *If (M,g) is Einstein, i.e., the Ricci tensor ρ satisfies $\rho = c\, g$, then the identity map id_M of (M,g) is stable if and only if the first eigenvalue $\lambda_1(M,g)$ of the Laplace-Beltrami operator Δ_M of (M,g) acting on $C^\infty(M)$ satisfies $\lambda_1(M,g) \geqq 2c$.*

Due to Proposition 3.1, we have : (cf. also [Sm])

Proposition 3.2. (cf. [U2], [Oh]) *(i) Let G be a compact simply connected simple Lie group, g the bi-invariant Riemannian metric on G induced from the Killing form of the Lie algebra of G. Then the identity map id_G of (G,g) is not stable, i.e., $\mathrm{Index}(id_G) > 0$ if and only if the type of G is one of the following :*

$$A_l \ (l \geqq 1),\ B_2,\ C_l \ (l \geqq 2),\ D_3.$$

(ii) Let $(G/K,h)$ be a simply connected irreducible Riemannian symmetric space of compact type. Then the identity map $id_{G/K}$ of $(G/K,h)$ is not stable, i.e., $\mathrm{Index}(id_{G/K}) > 0$ if and only if $(G/K,h)$ is one of the following :

$$S^n\ (n \geqq 3),\ Sp(l)/Sp(l-q)\times Sp(q)\ (l-q \geqq q \geqq 1),$$

$$E_6/F_4\ ,\ F_4/Spin(9).$$

Furthermore Y. Ohnita has recently obtained the following striking theorem :

Theorem 3.3. (Ohnita [Oh]) *Let* (M,g) *be the Riemannian product all of which factors are the one in the table of Proposition 3.2 (i),(ii). Then*

(i) each non-constant harmonic map ϕ *of* (M,g) *into arbitrary Riemannian manifold* (N,h) *is not stable, i.e.,*

Index$(\phi) > 0$.

(ii) Each non-constant harmonic map ϕ' *of arbitrary Riemannian manifold* (M',g') *into* (M,g) *is not stable, i.e.,* Index$(\phi') > 0$.

Remark 3.1. The statements *(i)* (resp. *(ii)*) for $(M,g) = S^n$, $n \geq 3$, the canonical unit sphere, were obtained by Y.L.Xin [X] (resp. P.F.Leung [Le]).

§4. Stability of Riemannian submersions.

Does there exist a deformation (M,g_t) *of a symmetric space* (M,g) *with* $g_0 = g$ *which have a stable harmonic map* ϕ; $(M,g_t) \to (N,h)$? In this section, we study stability of Riemannian submersions in order to sonsider this problem.

Let ϕ; $(M^m,g) \to (N^n,h)$ be the Riemannian submersion , that is , at each point x in M^m, the tangent space T_xM has the following orthogonal decomposition : $T_xM = H_x \oplus V_x$, where V_x is the kernel of ϕ_* and the restriction of ϕ_* to H_x is an isometry of (H_x,g_x) onto $(T_{\phi(x)}N, h_{\phi(x)})$. In this section, we furthermore assume each fiber $\phi^{-1}(\phi(x))$, $x \in M$, is a *totally geodesic* submanifold of (M^m,g).

Example 4.1. The Hopf fibering π_1 ; $(S^{2n+1},g) \to (\mathbb{C}P^n,h)$.

Example 4.2. The Hopf fibering π_2 ; $(S^{4n+3},g) \to (HP^n,h)$.

For each Riemannian submersion ϕ; $(M^m,g) \to (N^n,h)$, we consider the *vertical* (resp. *horizontal*) Jacobi operator

$$J^V_\phi := -\sum_{i=n+1}^{m} (\tilde\nabla_{e_i}\tilde\nabla_{e_i} - \tilde\nabla_{\nabla_{e_i}e_i}) , \text{ (resp. } J^H_\phi := J_\phi - J^V_\phi),$$

where the locally defined orthonormal frame field $\{e_i\}_{i=1}^{m}$ is chosen in such a way that $\{e_i\}_{i=n+1}^{m}$ is vertical, i.e., tangent to each fiber, and $\{e_i\}_{i=1}^{n}$ is basic associated to a locally defined orthonormal frame field $\{e'_i\}_{i=1}^{n}$ on N. Note that the definitions of J^V_ϕ and J^H_ϕ do not depend on the above choice of $\{e_i\}_{i=1}^{m}$.

Then we have :

Proposition 4.1. $[J^H_\phi, J^V_\phi] = 0$ *and* $J_\phi = J^V_\phi + J^H_\phi$. *Hence, the Hilbert space of all* L^2 *sections of* $E = \phi^{-1}TN$ *with respect to the inner product* $(V,W) = \int_M h(V,W)*1$, $V, W \in \Gamma(E)$, *has a complete orthonormal basis of simultanenous eigensections of* J^V_ϕ, J^H_ϕ *and* J_ϕ.

For the proof, see [U1, Theorem 6.5].

Now let us consider the following *canonical variation* g_t, $0 < t < \infty$, of g with $g_1 = g$ (cf.[B.B]) :

(i) $g_t(u,v) = g(u,v)$, $u,v \in H_x$, $x \in M$,

(ii) $g_t(u,v) = t^2\, g(u,v)$, $u,v \in V_x$, $x \in M$,

(iii) H_x *and* V_x *are orthogonal each other with respect to* g_t.

Then $\phi\,;\,(M,g_t) \to (N,h)$ is also a Riemannian submersion with totally geodesic fibers and the corresponding Jacobi operator, denoted by ${}^tJ_\phi$, satisfies

$${}^tJ_\phi = t^{-2}\, J^V_\phi + J^H_\phi .$$

Due to this equality and Proposition 4.1, we have :

Theorem 4.2. (cf. [U1, Theorem 7.3]) *Let* $\phi\,;\,(M,g) \to (N,h)$ *be the Riemannian submersion with totally geodesic fibers, and* g_t, $0 < t < \infty$, *the canonical variation of* g *with*

$g_1 = g$. *Then there exists a number* $\varepsilon > 0$ *such that the first eigenvalue* $\tilde{\lambda}_1({}^tJ_\phi)$ *of* ${}^tJ_\phi$ *satisfies*

$$\tilde{\lambda}_1({}^tJ_\phi) = \tilde{\lambda}_1(J_{id_N}),$$

for each $0 < t < \varepsilon$. *In particular, if the identity map* id_N *of* (N,h) *is stable, i.e.,* $\mathrm{Index}(id_N) = 0$, *or* $\tilde{\lambda}_1(J_{id_N}) \geqq 0$, *then the submersion* $\phi\ ;\ (M,g_t) \to (N,h)$ *is stable for each* $0 < t < \varepsilon$.

Theorem 4.3. (cf. [U1,Theorem 7.5]) *Let* $\phi\ ;\ (M,g) \to (N,h)$ *be the Riemannian submersion with totally geodesic fibers. Assume that the holonomy group of the Riemannian submersion does not act transitively on the fiber, and* $\mathrm{Index}(id_N) > 0$. *Then the index of the Riemannian submersion* $\phi;\ (M,g_t) \to (N,h)$ *goes to infinity when* $t \to \infty$.

Remark 4.1. In Example 4.1, the identity map of $(\mathbb{C}P^n,h)$ is stable (cf. Proposition 3.2), so the Riemannian submersion $\pi_1\ ;\ (S^{2n+1},g_t) \to (\mathbb{C}P^n,h)$ for the canonical variation g_t, is stable for each $0 < t < \varepsilon$. This gives an example which is contrast with the instability theorem (Theorem 3.3) of Xin, Leung and Ohnita.

Remark 4.2. Theorem 4.3 is a generalization of Corollary 3.3 in [Sm].

Part II. Spectral geometry of harmonic maps.

§5. Spectral invariants and the Jacobi operator.

Let us recall the spectral geometry of the Laplace-Beltrami operator. For a compact Riemannian manifold (M,g) without boundary, let $\mathrm{Spec}(M,g)$ be the spectrum of the Laplace-Beltrami operator Δ_M of (M,g). Then it is well-known (cf. [B.G.M], [Sa], [T]) that if the spectrum $\mathrm{Spec}(M,g)$ coincides with the one of the canonical unit sphere (S^n,can) $(n\leqq 6)$, then (M,g) is isometric to (S^n, can). Namely, *the spectrum characterize the canonical unit sphere* (S^n, can). In this part, we want to obtain the analogue of the spectral geometry for harmonic maps.

P.Gilkey [G1,2] calculated the first three terms of the asymptotic expansion of the heat kernel of a certain elliptic differential operator on a vector bundle. Using his results, H.Donnely [D], and T.Hasegawa [H] studied the spectral geometry of minimal submanifolds. In this part we study the spectral geometry of the Jacobi operator J_ϕ of a harmonic map ϕ.

Let $\phi ; (M^m, g) \to (N^n, h)$ be a harmonic map of a compact Riemannian manifold (M^m, g) into another Riemannian manifold (N^n, h). Let J_ϕ be the Jacobi operator acting on $\Gamma(E)$, $E = \phi^{-1}TN$. We denote the spectrum of the Jacobi operator J_ϕ by

$$\mathrm{Spec}(J_\phi) = \{ \tilde{\lambda}_1 \leqq \tilde{\lambda}_2 \leqq \dots \}$$

and let

$$\sum_{j=1}^{\infty} \exp(-t\tilde{\lambda}_j) \underset{t \to 0_+}{\sim} (4\pi t)^{-m/2} \{ a_0(J_\phi) + a_1(J_\phi) t + a_2(J_\phi) t^2 + \dots \}$$

be the asymptotic expansion. Using results of P.Gilkey [G1,2], we can determine the first three terms $a_0(J_\phi)$, $a_1(J_\phi)$, $a_2(J_\phi)$ of the asymptotic expansion :

<u>Theorem 5.1.</u> (cf. [U4, Theorem 2.1]) *For a harmonic map* $\phi ; (M^m, g) \to (N^n, h)$, *we have*

$$a_0(J_\phi) = n \,\mathrm{Vol}(M^m, g),$$

$$a_1(J_\phi) = \frac{n}{6} \int_M {}^M\tau \, *1 - \int_M \mathrm{Tr}_g(\phi^* \, {}^N\rho) \, *1,$$

$$a_2(J_\phi) = \frac{n}{360} \int_M \{ 5 \, {}^M\tau^2 - 2 \| {}^M\rho \|^2 + 2 \| {}^M R \|^2 \} *1$$

$$+ \frac{1}{360} \int_M \{ -30 \| \phi^* {}^N R \|^2 - 60 \, {}^M\tau \, \mathrm{Tr}_g(\phi^* {}^N\rho)$$

$$+ 180 \| L \|^2 \} *1 ,$$

where ${}^{M}R$, ${}^{M}\rho$, ${}^{M}\tau$; ${}^{N}R$, ${}^{N}\rho$, ${}^{N}\tau$ *are the curvature tensor, Ricci tensor, scalar curvature of* $(M^m,g),(N^n,h)$, *respectively. For tangent vectors* X,Y *at* x *in* M, $(\phi^*{}^{N}R)_{X,Y}$ *is the endomorphism of* $T_{\phi(x)}N$ *given by* $(\phi^*{}^{N}R)_{X,Y} = {}^{N}R_{\phi_*X,\phi_*Y}$, L *is the endomorphism of* $T_{\phi(x)}N$ *of the form*

$$L\, v = - \sum_{i=1}^{m} {}^{N}R(\phi_*e_i, v)\phi_*e_i \ , \quad v \in T_{\phi(x)}N,$$

and $\mathrm{Tr}_g(\phi^*\ {}^{N}\rho) = \sum_{i=1}^{m} {}^{N}\rho(\phi_*e_i,\phi_*e_i)$ *is the trace of the pull back of the Ricci tensor* ${}^{N}\rho$ *of* (N^n,h).

Then we have immediately :

<u>Corollary 5.2.</u> (cf.[U4, Corollary 2.2]) *Let* (M,g) *be a compact Riemannian manifold, and* (N,h) , *Einstein, that is* ${}^{N}\rho = c\,h$. *Let* ϕ , ϕ' *be two harmonic maps of* (M,g) *into* (N,h). *Assume that*

$$\mathrm{Spec}(J_\phi) = \mathrm{Spec}(J_{\phi'}).$$

Then we have

$$E(\phi) = E(\phi').$$

<u>Corollary 5.3.</u> (cf.[U4, Corollary 2.3]) *Let* $\phi;(M^m,g) \to (N^n,h)$ *be the Riemannian submersion with minimal fibers. Then the coefficients* $a_0(J_\phi),a_1(J_\phi),a_2(J_\phi)$ *of the asymptotic expansion for the Jacobi operator* J_ϕ *are given as follows:*

$$a_0(J_\phi) = n\,\mathrm{Vol}(M,g),$$

$$a_1(J_\phi) = \frac{n}{6}\int_M {}^{M}\tau \ {*1} - \int_M ({}^{N}\tau \circ \phi)\ {*1} \ ,$$

$$a_2(J_\phi) = \frac{n}{360}\int_M \{ 5\, {}^M\tau^2 - 2\|{}^M\rho\|^2 + 2\|{}^MR\|^2 \} *1$$

$$+ \frac{1}{360}\int_M \{-30\|{}^NR\|^2 \circ\phi - 60\, {}^M\tau({}^N\tau\circ\phi) + 180\|{}^N\rho\|^2\circ\phi\} *1.$$

§6. Spectral characterization of isometric minimal immersions.

6.1 Let $N^n(c)$ be an n-dimensional Riemannian manifold of constant curvature c. Then due to Theorem 5.1, we have :

Theorem 6.1. (cf.[U4, Theorem 3.1]) *Let* $\phi;(M^m,g) \to N^n(c)$ *be a harmonic map of a compact Riemannian manifold* (M^m,g) *into* $N^n(c)$. *Then the coefficients* $a_0(J_\phi)$, $a_1(J_\phi)$, *and* $a_2(J_\phi)$ *of the asymptotic expansion for the Jacobi operator* J_ϕ *are given as follows :*

$$a_0(J_\phi) = n\,\mathrm{Vol}(M^m,g),$$

$$a_1(J_\phi) = \frac{n}{6}\int_M {}^M\tau *1 - 2c(n-1)\,E(\phi),$$

$$a_2(J_\phi) = \frac{n}{360}\int_M \{5\,{}^M\tau^2 - 2\|{}^M\rho\|_g^2 + 2\|{}^MR\|_g^2\} *1$$

$$+ \frac{2c^2}{3}\int_M \{(3n-7)\,e(\phi)^2 + \|\phi^*h\|_g^2\} *1$$

$$- \frac{(n-1)}{3}\,c\int_M {}^M\tau\, e(\phi) *1 ,$$

where h *is the Riemannian metric of* $N^n(c)$, *and* $\|\ \|_g$ *is the norm of tensor fields with respect to* g.

Corollary 6.2. (cf. [U4, Corollary 3.2]) *Let* $c \neq 0$. *Let* ϕ, ϕ' *be two harmonic maps of a compact Riemannian manifold* (M^m,g) *with constant scalar curvature* ${}^M\tau$ *into the n-dimensional Riemannian manifold* $N^n(c)$ *of constant curvature* c. *Suppose that*

$$\mathrm{Spec}(J_\phi) = \mathrm{Spec}(J_{\phi'}).$$

Then we have

$$(6.1) \quad E(\phi) = E(\phi'),$$

$$(6.2) \quad \int_M \{(3n-7)e(\phi)^2 + \|\phi^*h\|_g^2\} *1$$

$$= \int_M \{(3n-7)e(\phi')^2 + \|\phi'^*h\|_g^2\} *1.$$

As an application of Corollary 6.2, we have :

<u>Theorem 6.3.</u> (cf.[U4, Theorem A]) *We assume the situations of Corollary 6.2 are preserved. Suppose that*

$$\mathrm{Spec}(J_\phi) = \mathrm{Spec}(J_{\phi'}).$$

If ϕ *is an isometric minimal immersion or an isometry, then so is* ϕ'.

<u>6.2.</u> In this subsection, we assume that (N^n,h) is an n-complex dimensional Kaehler manifold $P^n(c)$ of constant holomorphic sectional curvature c. Then we have :

<u>Theorem 6.4.</u> (cf.[U4, Theorem 4.1]) *Let* ϕ *be a harmonic map from a compact Riemannian manifold* (M^m,g) *into* $P^n(c)$. *Then the coefficients* $a_0(J_\phi)$, $a_1(J_\phi)$, $a_2(J_\phi)$ *of the asymptotic expansion for the Jacobi operator* J_ϕ *are given as follows :*

$$a_0(J_\phi) = 2n\,\mathrm{Vol}(M,g),$$

$$a_1(J_\phi) = \frac{n}{3}\int_M {}^M\tau \; *1 - (n+1)\, c\, E(\phi),$$

$$a_2(J_\phi) = \frac{n}{180}\int_M \{ 5\,{}^M\tau^2 - 2\|{}^M\rho\|_g^2 + 2\|{}^M R\|_g^2 \} *1$$
$$+ \frac{c^2}{24}\int_M \{(6n+10)e(\phi)^2 + 8\|\phi^*h\|_g^2 - (n+7)\|\phi^*\Phi\|_g^2\} *1$$
$$- \frac{(n+1)}{6}\, c \int_M {}^M\tau\, e(\phi) *1 ,$$

where Φ *is the Kaehler form of* $P^n(c)$, i.e., $\Phi(X,Y) = h(X,JY)$ *for tangent vectors* X,Y. *Here* J, h *are the complex structure, Kaehler metric on* $P^n(c)$, *respectively.*

Corollary 6.5. (cf.[U4, Corollary 4.2]) *Let* $c \neq 0$. *Let* ϕ, ϕ' *be two harmonic maps of a compact Riemannian manifold* (M^m,g) *with constant scalar curvature* ${}^M\tau$ *into the complex n-dimensional Kaehler manifold* $P^n(c)$ *of constant holomorphic sectional curvature* c. *Assume that*

$$\mathrm{Spec}(J_\phi) = \mathrm{Spec}(J_{\phi'}).$$

Then we have

$$(6.1) \qquad E(\phi) = E(\phi'),$$

$$(6.3) \qquad \int_M \{(6n+10)e(\phi)^2 + 8\|\phi^*h\|_g^2 - (n+7)\|\phi^*\Phi\|_g^2\} *1$$
$$= \int_M \{(6n+10)e(\phi')^2 + 8\|\phi'^*h\|_g^2 - (n+7)\|\phi'^*\Phi\|_g^2\} *1 .$$

As an application of Corollary 6.5, we have :

Theorem 6.6. (cf.[U4, Theorem B]) *Let* (M,g) *be a Kaehler manifold with constant scalar curvature* ${}^M\tau$. *Let* $P^n(c)$ *be a complex n-dimensional Kaehler manifold with constant holomorphic sectional curvature* $c \neq 0$. *Let* ϕ, ϕ' *be two* $\pm$ *holomorphic and weakly conformal maps from* M *into* $P^n(c)$. *Suppose that*

$$\mathrm{Spec}(J_\phi) = \mathrm{Spec}(J_{\phi'}).$$

If ϕ *is an isometric minimal immersion or an isometry, then so is* ϕ'.

Here a map ϕ of (M,g) into $P^n(c)$ is *weakly conformal* if the pull back ϕ^*h of the Kaehler metric h of $P^n(c)$ coincides with μg where μ is a C^∞ (not necessarily positive) function on M. In particular, we have :

Corollary 6.7. (cf.[U4, Corollary C]) *Let* ϕ, ϕ' *be two harmonic maps from the canonical 2 sphere* S^2 *into the complex n-dimensional Kaehler manifold* $P^n(c)$ *with constant holomorphic sectional curvature* $c \neq 0$. *Suppose that*

$$\mathrm{Spec}(J_\phi) = \mathrm{Spec}(J_{\phi'}).$$

If ϕ *is a* $\pm$ *holomorphic isometric immersion , then so is* ϕ' .

References

[B.B] L.Bérard Bergery & J.P.Bourguignon, Laplacians and Riemannian submersions with totally geodesic fibers, *Ill. J. Math., 26(1982), 181-200.*

[B.G] P.Bérard & S.Gallot, Inégalités isopérimetriques pour l'equation de la chaleur et application a l'estimation de quelques invariants, *Seminaire Goulaouic-Meyer-Schwartz, n° 15, 1983.*

[B.G.M] M.Berger, P.Gauduchon & E.Mazet, *Le spectre d'une variété riemannienne,* Lecture Notes in Math., n° 194, Springer, Berlin, 1971.

[D] H.Donnely, Spectral invariants of the second variation operator, *Ill. J. Math., 21(1977),185-189.*

[E.L1] J.Eells & L.Lemaire, A report on harmonic maps, *Bull. London Math. Soc.,10(1978),1-68.*

[E.L2] J.Eells & L.Lemaire, *Selected topics in harmonic maps,* Region.

Conf. series Math., Amer. Math. Soc., n° 50, 1982.

[G1] P.Gilkey, The spectral geometry of real and complex manifolds, *Proc. Sympos. Pure Math., 27(1975),265-280.*

[G2] P.Gilkey, The spectral geometry of a Riemannian manifold, *J. Diff. Geom., 10(1975),601-618.*

[G.K.M] D.Gromoll, W.Klingenberg & W.Meyer, *Riemansche Geometrie im Grossen,* Lecture Notes in Math., n° 55, Springer, Berlin,1968.

[H] T.Hasegawa, Spectral geometry of closed minimal submanifolds in a space form, real and complex, *Kodai Math. J.,3(1980),224-252.*

[Le] P.F.Leung, On the stability of harmonic maps, *Lecture Notes in Math., n° 949, Springer, Berlin,(1982),122-129.*

[Li] A.Lichnérowicz, *Géométrie des groupe de transformations,* Travaux Recherches Math., III, Dunod, Paris, 1958.

[Ma] E.Mazet, La formule de la variation seconde de l'energie au voisinage d'une application harmonique, *J. Diff. Geom.,8(1973),279-296.*

[Ob] M.Obata, Riemannian manifolds admitting a solution of a certain system of differential equations, *Proc. U.S.-Japan Semin. Diff. Geom., Kyoto, Japan (1965),101-114.*

[Oh] Y.Ohnita, Stability of harmonic maps and standard minimal immersions, *a preprint.*

[Sa] T.Sakai, On eigenvalues of Laplacian and curvature of Riemannian manifold, *Tohoku Math. J.,23(1971),589-603.*

[Si] J.Simons, Minimal varieties in Riemannian manifolds, *Ann. Math., 88(1968),62-105.*

[Sm] R.T.Smith, The second variation formula for harmonic mappings, *Proc. Amer. Math. Soc.,,47(1975),229-236.*

[Su] T.Sunada, Holomorphic mappings into compact quotient of symmetric bounded domains, *Nagoya Math. J., 64(1976),159-175.*

[T] S.Tanno, Eigenvalues of the Laplacian of Riemannian manifolds, *Tohoku Math. J., 25(1973),391-403.*

[U1] H.Urakawa, Stability of harmonic maps and eigenvalues of Laplacian, *a preprint.*

[U2] H.Urakawa, The first eigenvalue of the Laplacian for a positively curved homogeneous Riemannian manifold, *a preprint.*

[U3] H.Urakawa, Nullities and indicies of Yang-Mills fields over Einstein manifolds with positive Ricci tensor, *a preprint.*

[U4] H.Urakawa, Spectral geometry of the second variation operator of harmonic maps, *a prerint.*

[X] Y.L.Xin, Some results on stable harmonic maps, *Duke Math. J., 47(1980)609-613.*

[Y] S.T.Yau, Survey on partial differential equations in differetial geometry, *Seminar on Diff. Geom., Ann. Math. Studies, n° 102, Princeton, (1982),3-71.*

Uniformly locally convex filtrations on complete Riemannian manifolds

Takao Yamaguchi
Department of Mathematics
Saga University
Saga 840/ Japan

0. Introduction.

In the present paper, we define a concept of filtration by locally convex sets in a Riemannian manifold, and give a topological classification of manifolds admitting such filtrations. The result in this article extends that of author's previous work [10].

Let M be a complete Riemannian manifold without boundary. For $J = [\alpha, \beta)$ or (α, β) $(\alpha < \beta)$, let $\mathcal{F} = \{F^a\}_{a \in J}$ be a family of non-empty closed locally convex sets in M. For the terminology in this section, see section 1.

Definition. We say that $\mathcal{F}$ is a uniformly locally convex filtration (henceforth u.l.c.f.) on M if it satisfies the following conditions:

(i) $F^a \subset F^b$ $(a<b)$, $\cup_{a \in J} F^a = M$.

(ii) For each $p \in M$, there is a convex ball B_p around p such that $B_p \cap F^a$ is convex, or empty for every $a \in J$.

(iii) For every compact set K of M, $\varepsilon > 0$ and $a \in J - \{\alpha\}$, there are b and c in J such that

$F^c \subsetneq F^a \subsetneq F^b$, $F^b \cap K \subset U_\varepsilon(F^a)$, $F^a \cap K \subset U_\varepsilon(F^c)$.

(iv) $\cap_{a \in J} F^a = \emptyset$ if $J = (\alpha, \beta)$.

We denote by $U_\varepsilon(S)$ the ε-neighborhood of S.
A u.l.c.f. $\mathcal{F}$ is called condensed if $J = [\alpha, \beta)$, or expanding if $J = (\alpha, \beta)$. $\mathcal{F}$ is called singular if $\dim F^a < n$ for some $a \in J$, nonsingular if it is not singular, where n is the dimension of M and $\dim F^a$ means the topological dimension of F^a. In fact, it will be seen that F^a is connected, and hence has the topological manifold structure in the induced topology.

We shall prove the following topological structure theorems for M admitting a u.l.c.f. $\mathfrak{F}$.

Theorem 1. *M is homeomorphic to the normal bundle of F^a in M for all $a \in J$.*

In particular, if $\mathfrak{F}$ is singular, then the topology of M can be reduced to that of a lower dimensional locally convex set F^a. In the case when ∂F^a is empty, the assertion in Theorem 1 can be strengthened by "diffeomorphic".

Corollary. *If M admits a locally geodesically strictly quasiconvex function with minimum, then M is diffeomorphic to Euclidean space R^n.*

Theorem 2. *If $\mathfrak{F}$ is nonsingular and expanding, then there is a C^∞ manifold N such that M is diffeomorphic to $N \times R$.*

We note that the manifold N is not necessarily homeomorphic to the boundary of F^a in $\mathfrak{F}$ for any a. In fact, there is such an $\mathfrak{F}$ on the Poincare disk that ∂F^a consists of infinitely many connected components for every $a \in J$ (see [10], Example 9).

We now summarize historical background and motivation for the definition of u.l.c.f.. The concept of convexity provides a powerful tool in studying the global structure of complete manifolds. In fact, Cheeger and Gromoll [3] investigated the topological structure of complete manifolds of nonnegative sectional curvature by constructing a filtration by compact totally convex subsets, which is a special case of our u.l.c.f.. This method has greatly developed by Greene and Wu [5], Greene and Shiohama [4], Bangert [1] and others to study the structure of complete manifolds admitting convex functions, without any curvature assumption.

The sublevel sets of every convex function gives an example of u.l.c.f.. On the other hand, we have shown in [10] by exhibitting an example that the class of complete manifolds admitting u.l.c.f. includes that of complete manifolds admitting convex functions as a proper subset. As is seen in section 1, there is one to one correspondence between u.l.c.fs and locally geodesically quasiconvex functions of a certain type, which is a natural generalization of convex function.

1. Preliminary argument.

Let M be a complete Riemannian manifold without boundary of dimension n. A subset A of M is called convex if any two points in A are joined by a unique minimal geodesic, and if it lies in A. A subset A of M is called locally convex if any points in $\bar{A}$ has a neighborhood U in M such that $A \cap U$ is convex. For two subsets $A \subset B \subset M$, A is called totally convex in B if any two points in A can be joined by at least one geodesic lying in B and if each of these geodesics is necessarily included in A. If $B = M$, then A is simply called totally convex.

We summarize some local properties of a locally convex set. See [3] for detail. Let A be a closed connected locally convex set in M. Then in the induced topology, A is an imbedded submanifold of M with (possibly empty) boundary ∂A. The interior $\mathrm{Int}\, A = A - \partial A$ is a smooth totally geodesic submanifold of M. If $\gamma: [0, \infty) \longrightarrow M$ is a geodesic such that $\gamma([0,1)) \subset \mathrm{Int}\, A$ and $\gamma(1) \in \partial A$, then $\gamma(1+\varepsilon) \notin A$ for all sufficiently small $\varepsilon > 0$. The tangent cone $C_p(A)$ of A at a point $p \in A$ is by definition the set

$$\{v \in T_pM;\ \exp_p \varepsilon v \in \mathrm{Int}\, A \ \text{ for all sufficiently small } \varepsilon>0\},$$

where $\exp_p: T_pM \longrightarrow M$ is the exponential map on the tangent space T_pM. The tangent cone $C_p(A)$ is open and convex in the subspace of T_pM generated by $C_p(A)$. For $p \in A$, a tangent vector $v \in T_pM$ is called normal to A at p if $\langle v, w \rangle \leq 0$ for all $w \in C_p(A)$. The set $\nu_p(A)$ of all normal vectors to A at p is closed convex cone in T_pM. Consider the set $\nu(A) = \cup_{p \in A} \nu_p(A)$ with the induced topology from the tangent bundle TM of M. We call $\nu(A)$ the normal bundle of A in M. If A has no boundary, then $\nu(A)$ is the normal bundle in the usual sense.

For a closed locally convex set A in M, there is an open neighborhood U of A in M with the following properties (see [9]).

(i) For each $q \in U$, there is a unique minimal geodesic from q to A, and it is entirely included in U.

(ii) If $\pi(q)$ denotes the unique point of A with $dist(q, \pi(q)) = dist(q, A)$, then the mapping $\pi: U \longrightarrow A$ is locally Lipschitz continuous.

(iii) The function $f(q) = dist(q, A)$ is of class $C^{1,1}$ on $U - A$, that is, f is of class C^1 and its gradient ∇f is locally Lipschitz continuous. $(\nabla f)_q$ is given by the unit vector tangent to the minimal geodesic from q to $\pi(q)$.

(iv) U is homeomorphic to $\nu(A)$.

An open set with these properties is called a <u>tublar neighborhood</u> of A.

Let $\mathcal{F} = \{F^a\}_{a \in J}$ be a u.l.c.f. of M. We define the function $\varphi: M \longrightarrow R$ by $\varphi(p) = \inf\{ a;\ p \in F^a \}$. It is easily checked that $M^a(\varphi) = \varphi^{-1}((-\infty, a])$. The function φ satisfies the following conditions:

(I) φ is lower semicontinuous.

(II) φ is <u>locally geodesically quasiconvex</u> (henceforth l.g.q-convex), that is, every point p in M has a convex neighborhood B_p such that $\varphi\gamma(t) \leq \max\{ \varphi\gamma(a), \varphi\gamma(b)\}$ for every geodesic $\gamma: [a, b] \longrightarrow B_p$ and every $t \in [a, b]$. (We say that a function $\varphi: M \longrightarrow R$ is <u>locally geodesically strictly quasiconvex</u> (l.g.s.q-convex) if the above inequality is strict for nonconstant γ and $t \in (a,b)$).

(III) The restriction of φ to the set $M^a(\varphi) - \{$minimum set of $\varphi\}$ is <u>locally nonconstant</u> for all $a \in J - \{\alpha\}$, $\alpha = \inf J$.

A function is called locally nonconstant if it is non-constant on any open subset. (I) is clear. (II) and (III) follow from the definition (ii) and (iii) of u.l.c.f. in Introduction. Conversely, if φ is a l.g.q-convex function with (I) and (III), then the sublevel sets $M^a(\varphi)$ give a u.l.c.f. of M. Thus there is one to one correspondence $\mathcal{F} \longleftrightarrow \varphi$. We shall associate φ with $\mathcal{F}$ and consider F^a as a-sublevel set of φ implicitly. Clearly, every convex function has the properties (I), (II) and (III).

<u>Lemma 1.1.</u> $\varphi^{-1}(a) \subset \partial F^a$ for all $a \in J - \{\alpha\}$.

Proof. The l.g.q-convexity and local nonconstancy of φ imply that $\varphi^{-1}(-\infty, a)$ is locally convex and is dense in $M^a(\varphi)$. This yields $\mathrm{Int}\, M^a(\varphi) \subset \varphi^{-1}(-\infty, a)$ and $\varphi^{-1}(a) \subset \partial M^a(\varphi)$.

We set $\lambda_k = \inf\{a \in J;\ \dim F^a = n\}$. If $\lambda_k \in J - \{\alpha\}$, then the Baire Category Theorem (B.C.T.) implies $\dim F^{\lambda_k} < n$. Put $\lambda_{k-1} = \inf\{a \in J;\ \dim F^a = \dim F^{\lambda_k}\}$. If $\lambda_{k-1} \in J - \{\alpha\}$, then B.C.T. implies again $\dim F^{\lambda_{k-1}} < \dim F^{\lambda_k}$. Repeating this finitely many times and choosing k suitably, we obtain the sequence

$$\beta = \lambda_{k+1} > \lambda_k > \lambda_{k-1} > \cdots > \lambda_0$$

such that $\dim F^a = \dim F^{\lambda_i}$ for every $a \in (\lambda_{i-1}, \lambda_i]$ and $\dim F^{\lambda_i} > \dim F^{\lambda_{i-1}}$ for $i \geq 2$. The case $\dim F^{\lambda_1} = \dim F^{\lambda_0}$ and $\lambda_0 = \alpha$ may be occur.

Lemma 1.2. For every i, $1 \leq i \leq k+1$, and every $a \in (\lambda_{i-1}, \lambda_i]$, we have the following:

(1) F^a is connected.

(2) For every points $p \in \mathrm{Int}\, F^a$ and $q \in F^a$, $\mathrm{Int}\, F^a$ includes all geodesics from p to q lying in F^{λ_i} except the point q. In particular, $\mathrm{Int}\, F^a$ is totally convex in F^{λ_i}.

(3) For every two points in F^a, there is a minimal geodesic γ joining them such that $\gamma \subset F^a$.

Proof. We first consider the case $i = k+1$. Take a connected component U of F^a of dimension n. For $p \in U \cap \mathrm{Int}\, F^a$ and $q \in F^a$, suppose that there is a geodesic $\gamma : [0, 1] \longrightarrow M$ from p to q such that $\gamma([0, 1])$ is not included in F^a. Since φ is l.g.q-convex, it has a maximum b on γ. By Lemma 1.1, $\gamma(0,1)$ meets ∂F^b. Since $\gamma(0) \in \mathrm{Int}\, F^b$ and $\gamma([0, 1]) \subset F^b$, this is a contradiction. Hence F^a is connected. This argument also shows that $\gamma([0, 1)) \subset \mathrm{Int}\, F^a$ for all geodesics γ from p to q, and that $\mathrm{Int}\, F^a$ is totally convex. We now show (3). For p and q in F^a, choose sequences p_i and q_i in $\mathrm{Int}\, F^a$ with $p_i \longrightarrow p$ and $q_i \longrightarrow q$ as $i \longrightarrow \infty$. Take minimal geodesics γ_i joining p_i and q_i, which are included in $\mathrm{Int}\, F^a$. The limit geodesic of a sub- sequence of γ_i certainly lies in F^a. The assertion in the lemma follows by repeating this argument for $i = k, k-1, \ldots, 1$.

2. Construction of gradient-like vector fields

Let A be a closed connected locally convex set in M.

Lemma 2.1. There is a smooth submanifold $\tilde{A}$ of M without boundary suth that $\tilde{A} \supset A$ and $\dim \tilde{A} = \dim A$.

For the proof, see [2], 4.1.

From now on, all geodesics are assumed to have unit speed. For distinct points p and q in A, let $\mathcal{V}^A(q,p)$ denote the set of all initial tangent vectors to minimal geodesics in A from q to p. For a subset S of A, we set $\mathcal{V}^A(q,S) = \cup_{p \in S} \mathcal{V}^A(q,p)$. For a given u.l.c.f. $\mathcal{F}$ of M, we use the following notation for simplicity:

$M^i = F^{\lambda_i}$, $M = M^{k+1} \supset M^k \supset M^{k-1} \supset \ldots \supset M^1 \supset M^0$,
$\tilde{M}^i$ = a smooth extension of M^i of the same dimension as M^i.

Lemma 2.2. For every i, $a \in (\lambda_{i-1}, \lambda_i)$ and every compact set K in Int F^a, there is a C^∞ unit vector field X on a neighborhood of M^i in $\tilde{M}^i$ with the following properties:

(1) $X_q \in C_q(F^{\varphi(q)})$ for all q in $M^i - F^a$.
(2) $X_q \in C_q(F^a)$ for all q in ∂F^a.
(3) $\langle X_q, w \rangle > 0$ for all $w \in \mathcal{V}^{M^i}(q,K)$

As a result of Lemma 1.1, Lemma 1.2 (3) and Lemma 2.2, we see that for any integral curve $\phi(t)$ of X, $\varphi\phi(t)$ and dist(p, $\phi(t)$) are strictly decreasing as long as $\phi(t) \in M^i$ - Int F^a for all $p \in K$. In particular, $\phi(t)$ intersects ∂F^a in a single point.

Proof. For each $q \in M^i$ - Int F^a, we set $b = \varphi(q)$ if $q \in F^a$ and $b = a$ if $q \in \partial F^a$. From [10], Prop. 2.1, we can take a vector $v \in C_q(F^b)$ shch that $\langle v, w \rangle > 0$ for all $w \in C_q(F^b)$. By Lemma 1.2 (2), we have $\langle v, w \rangle > 0$ for all $w \in \mathcal{V}^{M^i}(q,K)$. Let V be a smooth extension of v on a neighbor- hood of q in $\tilde{M}^i$. From the argument using the lower semi- continuity of φ as in [10], Lemma 5.3, we see that the restriction of V to a small neighborhood U of q satisfies (1), (2) and (3) at each point of $U \cap (M^i$ - Int $F^a)$. We now obtain a locally finite open covering $\{U_\alpha\}$ of M^i - Int F^a in $\tilde{M}^i$ and C^∞ vector fields V_α on U_α which have the properties. Let $\{\rho_\alpha\}$ be a partition of unity dominated with $\{U_\alpha\}$. Then $X = \Sigma \rho_\alpha V_\alpha / |\Sigma \rho_\alpha V_\alpha|$ makes sense as a C^∞ unit vector field on a neighborhood of M^i - Int F^a. It is clear from the convexity of tangent cone that X has the properties (1), (2) and (3).

Using Lemma 2.2, we have the following proposition in the same way as in [10], Proof of Theorem A.

Proposition 2.3. For every i, there is a tublar neighborhood U_i of M^{i-1} in $\tilde{M}^i$ and a C^∞ vector field X_i over $\tilde{M}^i - M^{i-1}$ with the following properties:

(1) ∂U_i is of class C^1.

(2) The integral curve of X_i through each point of $M^i - U_i$ intersects ∂U_i transversally in a single point.

(3) For every $x \in U_i - M^{i-1}$, $(X_i)_x$ is the tangent to the minimal geodesic from x to M^{i-1}.

Theorem 1 is a direct consequence of the following

Theorem 2.4. For every $a \in J$, there is a tublar neighborhood U of F^a in M with C^1 boundary and a locally Lipschitz vector field X over $M - F^a$ such that the integral curve of X through each point of $M - U$ intersects ∂U trans- versally in a single point.

Proof. We modify the technique used in [2]. For the proof, it suffices to consider the case when $\mathcal{F}$ is condensed and $a = \lambda_0$. Let U_i and X_i be as in Proposition 2.3. We first extend X_1 to a vector field on $\tilde{M}^2$ using X_2. Set $V_1 = U_1$. Let V_2 be a tublar neighborhood of M^0 in $\tilde{M}^2$ such that $V_2 \subset U_2$ and $V_2 \cap \tilde{M}^1 \subset V_1$. Take a locally Lipschitz function $\eta : \tilde{M}^1 \longrightarrow R_+$ such that

$$\begin{aligned} &\exp_x \eta(x)(X_1)_x \in M^1 - M_0 && \text{if } x \in M^1 - V_2, \\ &\eta(x) = \tfrac{1}{2}\,\mathrm{dist}\,(x, M^0) && \text{if } x \in M^1 \cap V_2. \end{aligned}$$

For $x \in U_2$, we denote by x' the nearest point of M^1 from x, and by Z_x the initial tangent vector to the minimal geodesic from x to $\exp_{x'} \eta(x')(X_1)_{x'}$. Since $x \longrightarrow x'$ is locally Lipschitz, so is Z. If W is a sufficiently small tublar neighborhood of M^1 in $\tilde{M}^2$ with $\overline{W} \subset U_2$, we may assume by continuity that the integral curve of $Z/|Z|$ through each point of $W - V_2$ intersects ∂V_2 transversally. For a C^1 function $\delta : \partial W \longrightarrow R_+$, we denote by W^δ the tublar neighborhood of M^1 in $\tilde{M}^2$ defined as follows: For each $x \in \partial W$, let $\gamma_x : [0, r_x] \longrightarrow M$ the minimal geodesic from x to M^1 ($r_x = \mathrm{dist}(x, M^1)$). Then we put $W^\delta = \{\gamma_x(t);\ x \in \partial W,\ \delta(x) < t \leq r_x\}$. For a smooth function $\rho : \tilde{M}^2 \longrightarrow R$

with $0 \le \rho \le 1$ and $\rho = 1$ on W^{δ}, $\rho = 0$ on $\tilde{M}^2 - W$, we set $Y_2 = \rho Z/|Z| + (1 - \rho)X_2$. Let T be the unit field on $U_2 - M^1$ tangent to the minimal geodesics to M^1. Note that $\langle T, Y_2\rangle = \rho\langle T, Z/|Z|\rangle + (1 - \rho)$, and $\langle T, Z/|Z|\rangle > 0$ on $W - M^1$. Therefore by taking δ sufficiently small, we may assume that the integral curve of Y_2 through each point of $M^2 - W$ reaches W^{δ} and hence intersects ∂V_2. Repeating this, we have, for each i, a tublar neighborhood V_i of M^0 in $\tilde{M}^i$ and a locally Lipschitz continuous vector field Y_i such that the integral curve of Y_i through each point of $M^i - V_i$ intersects ∂V_i transversally. To complete the proof, it suffices to set $U = V_{k+1}$ and $X = Y_{k+1}$.

3. Proof of Corollary and Theorem 2.

For a function $f: M \longrightarrow R$, let $f^*: M \longrightarrow \underline{R \cup \{-\infty\}}$ be the lower limit function of f: $f^*(x) = \inf\{a: x \in \overline{f^{-1}((-\infty, a])}\}$. If φ is a l.g.q-convex function on M, then so is φ^*. Moreover, if φ is locally nonconstant, then so is φ^* (see [10], section 4).

Proof of Corollary. Let φ be a l.g.s.q-convex function with minimum β. We first note that $M^{\beta}(\varphi)$ is discrete. Setting

$$\lambda_k = \inf\{a;\ \dim M^a(\varphi^*) = n\},\ M^k = M^{\lambda_k}(\varphi^*),\ n_k = \dim M^k,$$

we have $n_k < n$. If $n_k > 0$, we put

$$\varphi_k = \varphi|M^k - \alpha_k,\quad \alpha_k = \sup_{M^k}\varphi - \lambda_k,$$
$$\lambda_{k-1} = \inf\{a;\ \dim (M^k)^a(\varphi_k^*) = n_k\},\ M^{k-1} = (M^k)^{\lambda_{k-1}}(\varphi_k^*).$$

Repeating this procedure, we have the following data:

$$\lambda_k > \lambda_{k-1} > \dots > \lambda_0,$$
$$M \supset M^k \supset M^{k-1} \supset \dots \supset M^0,\quad \dim M^0 = 0,\quad M^0 \supset M^{\beta}(\varphi),$$
$$(\varphi_k, \varphi_{k-1}, \dots, \varphi_1).$$

We set $J = [\lambda_0, \sup_M \varphi^*)$ and $F^a = (M^i)^a(\varphi_i^*)$ for $a \in [\lambda_{i-1}, \lambda_i)$. Then it is verified that $\mathfrak{F} = \{F^a\}_{a \in J}$ satisfies the conditions of u.l.c.f.. By Lemma 1.2 (1), $F^{\lambda_0} = M^0$ is connected. Thus $M^0 = M^{\beta}(\varphi)$ and F^{λ_0} consists of a single point. Therefore from Theorem 1, M is diffeomorphic to R^n.

Proof of Theorem 2. Let c_i ($i = 0, 1, 2, \dots$) be a decreasing sequence in J with $\lim c_i = \alpha$. Using the argument in section 2 (Cf. [10], Proof of Theorem C), we can construct a C^{∞} vector field X over M and C^1 hyper- surfaces L_i of M

near ∂F^{c_i} with the following properties:

(1) For every $p \in M$, the integral curve $\phi_p(t)$ of X with $\phi_p(0) = p$ intersects ∂F^{c_i} transversally for all i with $\varphi(p) > c_i$.

(2) The integral curves of X give rise to a homeomorphism of L_i onto ∂F^{c_i}.

We denote by Φ_i the projection onto L_i by means of integral curves of X. The restriction $\Phi_i \mid \partial F^{c_i}$ is a homemorphism of ∂F^{c_i} onto L_i. For the proof of the theorem, it suffices to construct a C^1 hypersurface N such that M is diffeomorphic to $N \times R$.

Let $\{L_{i,\mu}\}_{\mu \in A_i}$ be the family of connected components of L_i. For each $\mu \in A_i$, there is a unique $\mu' \in A_{i+1}$ such that $\Phi_{i+1}(L_{i,\mu}) \subset L_{i+1,\mu'}$. We set $A_i' = \{\mu \in A_i ;\ \Phi_{i+1}(L_{i,\mu}) \neq L_{i+1,\mu'}\}$. Let $\{K_{i,\mu}^{\alpha}\}_{\alpha=0,1,2,\ldots}$ be an exaustion of $L_{i,\mu}$ by compact subsets with $\operatorname{Int} K_{i,\mu}^{\alpha} \subset K_{i,\mu}^{\alpha+1}$ and $\Phi_{i+1}(K_{i,\mu}^{\alpha}) \subset K_{i+1,\mu'}^{\alpha}$. We first construct C^1 hypersurfaces N_i $(i = 1, 2, \ldots)$ diffeomorphic to L_i by means of integral curves of X inductively so that the sequence N_i gives rise to a connected C^1 hypersurface.

<u>First step.</u> <u>Case i)</u> $A_0' = \emptyset$.

Then we set $N_1 = L_0 \cup (L_1 - \cup_{\mu \in A_0} L_{1,\mu'})$

<u>Case ii)</u> $A_0' \neq \emptyset$.

Take a C^1 function ρ_1 on L_1 with $0 \leq \rho_1 \leq 1$ and

$$\rho_1 = \begin{cases} 1 & \text{on} \quad \cup_{\mu \in A_0'} \Phi_1(K_{0,\mu}^0) \\ 0 & \text{on} \quad L_1 - \cup_{\mu \in A_0'} K_{1,\mu}^1 . \end{cases}$$

For the C^1 function $s_1 : \Phi_1(L_0) \longrightarrow R$ with $\phi_x(s_1(x)) \in L_0$ $(x \in \Phi_1(L_0))$, we set

$$N_1 = \{ \phi_x(\rho_1(x) s_1(x)) ;\ x \in \Phi_1(L_0) \} \cup (L_1 - \Phi_1(L_0)).$$

<u>Second step.</u> We assume that N_i has been constructed.

<u>Case i)</u> $A_i' = \emptyset$.

Then we set $N_{i+1} = N_i \cup (L_{i+1} - \cup_{\mu \in A_i} L_{i+1,\mu'})$.

<u>Case ii)</u> $A_i' \neq \emptyset$.

Take a C^1 function ρ_{i+1} on L_{i+1} with $0 \leq \rho_{i+1} \leq 1$ and

$$\rho_{i+1} = \begin{cases} 1 & \text{on} \quad \cup_{\mu\in A_i{}'} \Phi_{i+1}(K^{\,i}_{i,\mu}) \\ 0 & \text{on} \quad L'_{i+1} - (\cup_{\mu\in A_i{}'} K^{\,i+1}_{i+1,\mu}). \end{cases}$$

For the C^1 function $s_{i+1}:\Phi_{i+1}(N_i) \longrightarrow R$ with $\phi_x(s_{i+1}(x)) \in N_i$ $(x\in\Phi_{i+1}(N_i))$, we set

$$N_{i+1} = \{\phi_x(\rho_{i+1}(x)s_{i+1}(x));\ x\in\Phi_{i+1}(N_i) \cup (L_{i+1} - \Phi_{i+1}(N_i))\}.$$

We note that each integral curve of X intersects N_i just constructed transversally in at most one point. We denote by N the subset of M consisting of all points belonging to N_i for infinitely many i. It is verified that N is a well-defined C^1 hypersurface. Every integral curve ϕ of X meets L_i and hence $K^{\alpha}_{i,\mu}$ for some i, μ and α. From the construction of N_i, ϕ meets N_j for all $j \geq \max\{i, \alpha\}$. Thus ϕ meets N and M is diffeo- morphic to $N \times R$.

Added in proof. Without assuming the uniformity (ii) in the definition of our filtration, we have obtained a result ([11]) on the structure of such filtrations. This filtration can exist on some compact manifolds.

References

[1] V. Bangert, Riemannsche Mannigfaltigkeiten mit nicht-konstanten konvexer Funktionen, Arch. Math., 31(1978), 163-170.

[2] Yu. D. Burago and V. A. Zalgaller, Convex sets in Riemannian spaces of non-negative curvature, Russian Math. Surveys, 32:3(1977), 1-57.

[3] J. Cheeger and D. Gromoll, On the structure of complete manifolds of nonnegative curvature, Ann. of Math., 96(1972), 413-443.

[4] R. E. Greene and K. Shiohama, Convex functions on complete noncompact manifolds: Topological structure, Invent. Math., 63(1981), 129-157.

[5] R. E. Greene and H. Wu, C^∞ convex functions and manifolds of positive curvature, Acta Math., 137(1976), 209-245.

[6] D. Gromoll and W. Meyer, On complete open manifolds of positive curvature, Ann. of Math., 90(1969), 75-90.

[7] M. Gromov, Curvature, diameter and Betti numbers, Comment. Math. Helv., 56(1981), 179-195.

[8] V. A. Sharafutdinov, Complete open manifolds of non-negative curvature, Siberian Math. J. 15(1974), 126-136.

[9] R. Walter, On the metric projection onto convex sets in Riemannian spaces, Arch. Math., 25(1974), 91-98.

[10] T. Yamaguchi, Locally geodesically quasiconvex functions on complete Riemannian manifolds, to appear in Trans. Amer. Trans. Amer. Math. Soc.

[11] T. Yamaguchi, On the structure of locally convex filtration on complete manifolds, in preparation.

EINSTEIN METRICS WITH POSITIVE SCALAR CURVATURE

M. WANG and W. ZILLER

Institut des Hautes Etudes Scientifiques
35, Rte de Chartres
91440 Bures-sur-Yvette (France)

December 1985

IHES/M/85/

In this paper, we review some recent examples of Einstein metrics. A Riemannian metric g on M is called Einstein if $Ric(g) = E.g$ for some constant E. The discussion of Einstein metrics depends very much on the sign of E and we restrict ourselves here to the case $E > 0$. For the case $E < 0$ and $E = 0$ and a thorough general discussion of Einstein metrics we refer to the recent book by A. Besse, [Be, 1985].

If the dimension of M is ≤ 3, an Einstein metric has constant sectional curvature, and hence we will assume that $\dim M \geq 4$. There are a few obstructions known to the existence of an Einstein metric with $E > 0$. $E > 0$ implies that M is compact and that the fundamental group is finite by a theorem of Bonnet-Meyers. If $\dim M = 4$, the existence of an Einstein metric (independent of the sign of E) implies that $|\tau| \leq 2/3\ \chi$ by a theorem of Thorpe [Th, 1969] and Hitchin [Hi 1, 1974], where τ is the signature and χ the Euler characteristic of M. The only other known obstruction to an Einstein metric with $E > 0$ is already an obstruction to positive scalar curvature. Lichnérowicz [Li 2, 1963] showed that positive scalar curvature implies the vanishing of the $\hat{A}$ genus for a 4n dimensional manifold, which was generalized by Hitchin [Hi 2, 1974] to the vanishing of the gene-

The first author is partially supported by a University Research Fellowship from the Natural and Engineering Research Council of Canada.

The second author is partially supported by a grant from the National Science Foundation and would like to thank the I.H.E.S. for its hospitality.

ralized $\hat{A}$ genus, which is a $\mathbb{Z}_2$ invariant for manifolds of dimension $8n+1$ and $8n+2$.

It is therefore of interest to obtain many examples of manifolds which admit Einstein metrics with $E > 0$. In the first part, we discuss the known examples which are homogeneous and in the second part the more recent examples of inhomogeneous Einstein metrics.

HOMOGENEOUS EINSTEIN METRICS.

In this section M^n is an n-dimensional manifold on which the compact Lie group G acts transitively. H is the isotropy group of a point p in M and hence M is diffeomorphic to G/H . For the Lie algebras we let $\mathfrak{g} = \mathfrak{h} + \mathfrak{m}$ where $\mathfrak{m}$ is an $Ad(H)$ invariant complement to $\mathfrak{h}$. Then T_pM can be naturally identified with $\mathfrak{m}$ and the isotropy action of H on T_pM gets identified with the action of $Ad(H)$ on $\mathfrak{m}$.

The oldest examples of Einstein metrics are given by the isotropy irreducible homogeneous spaces, where we assume that $Ad(H)$ acts irreducibly on $\mathfrak{m}$. M is then automatically Einstein since every eigenspace of Ric is invariant under the isometries of H . If $Ad(H_o)$ also acts irreducibly on $\mathfrak{m}$, where H_o is the connected component of H , we call M strongly isotropy irreducible. Irreducible symmetric spaces are strongly isotropy irreducible. The non-symmetric strongly isotropy irreducible homogeneous spaces were classified by Manturov [Ma 1, 1961], [Ma 2, 1961], [Ma3, 1966] and independently by Wolf [Wo1, 1969]. Actually, both Manturov's and Wolf's paper do not contain a complete list of examples, but the union of both lists is complete, see also [Wo 2, 1984] . It is not hard to see that if G/H is strongly isotropy irreducible and not symmetric, then G is a compact simple Lie group. The classification then breaks up into the different types of simple Lie groups. If G is an exceptional Lie group, the classification already follows immediately from Dynkin's tables [Dy, 1957]. If G is a classical Lie group a uniform description of these strongly isotropy irreducible spaces was given by C.T.C. Wall [Wo 1, 1969], p. 147, who noticed that they can be described in terms of hermitian symmetric spaces if $G = SU(n)$, quaternionic symmetric spaces if $G = Sp(n)$, and in terms of the other symmetric spaces if $G = SO(n)$. In [W-Z 2, 1984] a conceptual proof of these relationships was given.

Another general class of homogeneous Einstein metrics is given by the homogeneous Kähler Einstein metrics. A well known theorem [Borel, 1954] states that a compact, simply connected, homogeneous Kähler manifold is of the form $G/C(T)$, where G is compact, semisimple, and connected and $C(T)$ is the centralizer of a torus $T \subset G$. Equivalently, they are the orbits under the adjoint representation of G on $\mathfrak{g}$. $G/C(T)$ has a canonical G invariant complex structure and a G invariant Kähler Einstein metric compatible with this complex structure which has positive scalar curvature and is unique up to scaling. The existence of the Kähler Einstein metric seems to have been first observed by Matsushima [Mat2, 1972], but essentially goes back to Koszul [Kos, 1955]. In [Mat2, 1972] it was also shown that any Kähler Einstein metric, compatible with the canonical complex structure, is isometric to a G invariant Kähler Einstein metric by some automorphism of the complex structure. These manifolds include the hermitian symmetric spaces, which is in fact the only intersection with the previous examples, i.e. every homogeneous Kähler Einstein metric which is isotropy irreducible is hermitian symmetric [Lichnérowicz, 1952]. Some other examples are the flag manifolds G/T where T is a maximal torus in G.

To these homogeneous Kähler Einstein manifolds one can apply a theorem of Kobayashi [Ko 2, 1963] to obtain a further class of homogeneous Einstein metrics. He showed that if M is a Kähler Einstein metric with positive scalar curvature (not necessarily homogeneous) and $P \to M$ the principal circle bundle with Euler class equal to $c_1(M)$ (or any rational multiple of $c_1(M)$), then P admits an S^1 invariant Einstein metric. If M is homogeneous, then the metric on P will also be homogeneous. Applied to $P^n\mathbb{C}$, this gives the constant curvature metric on S^{2n+1}, but otherwise these metrics are not covered by previous examples. E.g. if M is the complex quadric $SO(n+2)/SO(n)\cdot SO(2)$, then P is the unit tangent bundle of S^{n+1}. The Einstein metric becomes unique, if one requires that $P \to M$ is a Riemannian submersion with totally geodesic fibres and that the metric on the base is the given Kähler Einstein metric on M. We will later come back to Kobayashi's theorem in the case where M is not homogeneous.

A number of homogeneous Einstein metrics were obtained by the method of Riemannian submersions. Let $M \to B$ be a Riemannian submersion. If we assume that the fibres F are totally geodesic, we can speak of a metric on F, which is then well determined up to isometry. The Einstein condition on M can be expressed in terms of the O'Neill tensor, the Ricci tensor on M and F and the horizontal connection (it must be a Yang-Mills connection). In general, these equations are complicated (the metric on M and F does not have to be Einstein), but in many special cases one can obtain Einstein metrics by this method. If $H \subset K \subset G$, then the fibration $K/H \to G/H \to G/K$ can be made into a Riemannian submersion with totally geodesic fibres. In [Jensen, 1973] and [D'Atri-Ziller, 1979] it was shown that if $G/H\cdot H'$ is strongly isotropy irreducible, then one obtains an Einstein metric on G/H by using the submersion $H' \to G/H \to G/H.H'$. In this way, Jensen constructed a non-standard Einstein metric on $S^{4n+3} = Sp(n+1)/Sp(n)$ and in [D-Z] it was shown that $SO(n)$ carries at least n distinct Einstein metrics if $n \geq 12$.

The following general theorem, which was observed by Bérard-Bergery, see [Be, 1985], and independently by Matsuzawa [M, 1983], can then be applied to obtain further Einstein metrics on M : Let $F \to M \to B$ be a Riemannian submersion with totally geodesic fibres and assume that the metric on F, M and B are Einstein with Einstein constant E_F, E_M, and E_B. Then if $E_F > 0$, $E_F \neq 1/2\, E_B$, and if the submersion is not locally a Riemannian product, one obtains another Einstein metric on M by scaling the metric on M in the direction of the vertical subspaces by the factor $t = E_F/(E_B - E_F)$. One can apply this theorem e.g. to the Hopf fibrations $S^3 \to S^{4n+3} \to P^n H$, $S^7 \to S^{15} \to S^8$, and $S^2 \to P^{2n+1}\mathbb{C} \to P^n H$ to obtain non-standard Einstein metrics on S^{4n+3}, S^{15}, and $P^{2n+1}\mathbb{C}$ which were obtained earlier by using different methods in [Jensen, 1973], [Bourguignon-Karcher, 1978], and [Ziller 1, 1982]. For further examples using Riemannian submersions see [Je, 1973], [D-Z, 1979], [W-Z1, 1985], [W-Z 3, 1985], and [Zi 2, 1984].

We mention here that the same method of using Riemannian submersions also gives rise to compact, simply connected homogeneous spaces G/H with no G-invariant Einstein metrics [W-Z 3, 1985]. Since such spaces always carry a G-invariant metric with positive Ricci curvature, it was believed for a while that they might always carry a G-invariant Einstein metric. A general class is obtained as follows : Let $SO(n)/H$ be a strongly isotropy irreducible homogeneous space which is not symmetric and $H \neq G_2$. Using the inclusions $H \subset SO(n) \subset SO(n+1)$ we obtain a homogeneous space $SO(n+1)/H$ and one can show that there exists no $SO(n+1)$ invariant Einstein metric on it. In fact, $SO(n+1)/H \to SO(n+1)/SO(n) = S^n$ is a submersion with fibres $SO(n)/H$ and one easily shows that every $SO(n+1)$ invariant metric is a Riemannian submersion metric for appropriate invariant metrics on base and fibre. Since these metrics are uniquely determined up to a multiple, the Einstein condition becomes a quadratic equation and a computation shows that it has no real solutions. But it is not known if these manifolds carry a non-homogeneous Einstein metric or not.

We mention two other types of homogeneous Einstein metrics that have been studied. In [Wang-Ziller 1, 1985] the Einstein metrics were determined that are induced on G/H by the Killing form of G in the case where G is compact and simple. Theyinclude the strongly isotropy irreducible homogeneous spaces. In [Wang, 1982] it was shown that $SU(3)/S^1$ has an $SU(3)$ invariant Einstein metric for every subgroup S^1 . For two different embeddings of S^1 one obtains in general different cohomology rings and hence there are infinitely many homotopy types of compact simply connected Einstein manifolds in dimension seven, which were the first examples of this type.

The Einstein metrics are also critical points of the total scalar curvature functional on the space of Riemannian metrics of volume one. If we let M_G be the set of G-invariant metrics of volume one on G/H , then the critical points of the scalar curvature functional S on M_G are again the G-invariant Einstein metrics on G/H . Hence one should examine the global behaviour of S on M_G in order to deduce the existence of critical points. In [Wang-Ziller 3, 1985] the homogeneous spaces were determined where S is bounded from above or from below. S is bounded from below only in the trivial cases where G/H is at least locally a product of strongly isotropy irreducible homogeneous spaces and possibly a euclidean space. If S is bounded from above, one has to distinguish two cases. S is bounded from above and not proper only in the special case where $H \cdot S^1$ is a subgroup of G and $G/H \cdot S^1$ is an irreducible hermitian symmetric space not equal to $SO(n+2)/SO(n) \cdot SO(2)$. But it happens quite frequently that S is bounded from above and proper. It is shown that this is the case iff $\mathfrak{h}$ is maximal in $\mathfrak{g}$. Hence the maximum of S on M_G is an Einstein metric on G/H.

It is easy to see that if $\mathfrak{h}$ is not maximal in $\mathfrak{g}$, then S is either not bounded from above or not proper. Indeed if $H \subset K \subset G$, we can use the Riemannian submersion $F = K/H \to G/H \to G/K = B$ and scale the metric on G/H by multiplying it by a factor t in the vertical direction. Then one has for the scalar curvature of this metric g_t:

$$S(g_t) = 1/t\ S(F) + S(B) - t\,||A||^2$$

where A is the O'Neill tensor. If we normalize the metric g_t to have volume one, we get

$$S(\tilde{g}_t) = t^{\dim F/\dim B}(1/t\ S(F) + S(B) - t\,||A||^2) .$$

Hence if $S(F) > 0$, then $S(\tilde{g}_t) \to +\infty$ as $t \to 0$ and if $S(F) = 0$, then S is not proper. This proves our claim since F always has a metric with $S \geq 0$.

To see that S is bounded from above and proper if $\mathfrak{h}$ is maximal in $\mathfrak{g}$, requires a careful examination of the scalar curvature functional on a general homogeneous space and is quite delicate.

It would be interesting to obtain the existence of further critical points from the global behaviour of S. In general, the structure of the subgroups K with $H \subset K \subset G$ will come into play. But it is not even known if the function S on $\mathfrak{M}_G$ satisfies the Palais-Smale condition or not.

NON HOMOGENEOUS EINSTEIN METRICS.

The first non-homogeneous Einstein metric with positive scalar curvature was discovered by Page [Pa, 1979] on $P^2\mathbb{C} \# - P^2\mathbb{C}$. This was generalized by Bérard-Bergery [BB, 1982] as follows : Let (M,g) be a Kähler Einstein metric with positive scalar curvature. These are simply connected [Kobayshi, 1961] and hence $H^2(M,\mathbb{Z})$ has no torsion. Write $c_1(M) = q.\alpha$ where α is an indivisible integer cohomology class. Let P_k be the principle circle bundle with Euler class $k.\alpha$ and N_k the 2-sphere bundle over M associated with P_k for the usual action of $S^1 = SO(2)$ on S^2. Then, if $1 \leq k < q$, N_k admits an Einstein metric with positive scalar curvature. If $M = S^2$, we have $q = 2$ and N_1 is the unique non-trivial 2sphere bundle over S^2, which is diffeomorphic to $P^2\mathbb{C} \# - P^2\mathbb{C}$. The metric in this case is the same as the Page metric. More generally for $M = P^n\mathbb{C}$ we have $q = n+1$ and for $k = 1$ N_k is diffeomorphic to $P^{n+1}\mathbb{C} \# - P^{n+1}\mathbb{C}$.

To describe the metric, one regards N_k as $[0,1] \times P_k$ with identification on the boundary given by the projection $\pi : P_k \to M$. On P_k one constructs a 2-parameter family of metrics as follows : Let θ be the principal connection on P_k with $d\theta = 2\pi k.\pi^*\omega$ where

ω is the harmonic two form with $[\omega] = \alpha$, θ defines a horizontal distribution on P_k and we define the metric $g(a,b)$ on P_k by declaring the vertical and the horizontal directions as perpendicular, the metric on the horizontal space as the pullback under π of $b.g$ on M, and the metric on the vertical space such that the fibres S^1 have length $2\pi a$. Then, if $f,h : [0,1] \to \mathbb{R}$ are two positive functions, we can define the metric $\tilde{g} = dt^2 + g(f(t), h(t))$ on $[0,1] \times P_k$. The Einstein condition for $\tilde{g}$ then reduces to three ordinary differential equations in f and h with certain boundary conditions at 0 and 1 to guarantee smoothness of the metric on N_k. In [BB] it is then shown that these equations have a solution iff $k < q$. See also [P-P, 1985] for a slightly different description of these metrics.

The manifolds N_k are always complex manifolds and the Einstein metric is hermitian, but never Kähler, although it is conformal to a Kähler metric (which is not Einstein).

If the metric on M is homogeneous, i.e. M is of the form $G/C(T)$ as in the previous chapter, then P_k is also homogeneous, in fact G acts transitively on P_k and by isometries in the metric $g(a,b)$ for every a,b. Hence G acts on N_k preserving $\tilde{g}$, and the codimension of the principal orbits is equal to one. On the other hand $\tilde{g}$ can never be homogeneous and hence the metric is of cohomogeneity one. Presumably, in most cases the manifolds N_k do not have the homotopy type of a homogeneous space either.

Recently, some non-homogeneous Kähler-Einstein metrics with positive scalar curvature were constructed by Sakane [Sa, 1985] and generalized by Koiso-Sakane [K-S, 1985]. They again start with a Kähler Einstein metric (M,g) and write $c_1(M) = q.\alpha$ with α indivisible. Let L be the complex line bundle over M with $c_1(L) = k.\alpha$ and let $L \times L$ be the $\mathbb{C}^2$ bundle over $M \times M$. Define N_k to be the projectivised bundle $P(L \times L)$ which is hence a $P^1\mathbb{C} = S^2$ bundle over $M \times M$. $c_1(N_k) > 0$ iff $k < q$ and they show that in this case N_k admits a Kähler Einstein metric with positive scalar curvature. If M is homogeneous, N_k will be of cohomogeneity one. In principle, one can now repeat the construction, the only difficulty being that one needs $c_1(N_k)$ to be divisible. In [K-S] an example of cohomogeneity two is constructed by taking $M = P^n\mathbb{C}$, $q = n+1$. Then $c_1(N_k)$ is divisible by two if $k \leq n$ and $k-n$ is odd and hence the construction can be repeated once. See [K-S] for some more general existence and non-existence theorems for Kähler Einstein metrics with positive scalar curvature. Unlike in the case $c_1 < 0$ or $c_1 = 0$, there are obstructions to the existence of Kähler Einstein metrics if $c_1 > 0$, see [Mat 1, 1957] and [Fu, 1983].

To the examples of Koiso and Sakane, one can apply the construction of Bérard-Bergery (if $c_1(M)$ is divisible) and the Kobayashi theorem mentioned in the previous chapter and obtain new examples of (non-Kähler) Einstein metrics.

We finally describe a generalization of Kobayashi's theorem [Wang-Ziller 4, 1985] which gives rise to many new homogeneous and non-homogeneous Einstein metrics in odd dimensions.

Let (M_i, g_i), $i=1,\ldots,m$, be Kähler Einstein metrics with positive scalar curvature and let $c_1(M_i) = q_i \cdot \alpha_i$ where α_i is an indivisible integer class. Set $M = M_1 \times \ldots \times M_m$ and denote by π_i the projection of M onto M_i. Let $\pi : P_{k_1 \ldots k_m} \to M$ be the principle circle bundle over M with Euler class $\Sigma k_i \pi_i^* \alpha_i$. Then for any choice of non-zero integers k_i, $P_{k_1 \ldots k_m}$ admits an Einstein metric.

If $m=1$ this is Kobayashi's theorem. But in this case the manifolds, one gets for different values of k are only covering of each other. Indeed, since M is always simply connected, P_k is simply connected iff $k=1$ and hence $P_k = P_1/\mathbb{Z}_k$ where $\mathbb{Z}_k \subset S^1$. If $m > 1$, $P_{k_1 \ldots k_m}$ is simply connected iff $k_1, \ldots, k_m$ are relatively prime.

We construct these metrics as follows. Let ω_i be the harmonic two form on M_i with $[\omega_i] = \alpha_i$ and let θ be the principle connection on $P_{k_1 \ldots k_m}$ with $d\theta = 2\pi(\Sigma k_i \pi_i^* \omega_i)$. Then for each set of positive real numbers $x_1, \ldots, x_m$ we define a metric on $P_{k_1 \ldots k_m}$ by declaring the vertical space and horizontal space as perpendicular, letting the metric on the vertical space be such that the fibres have length 2π and the metric on the horizontal space given by the pullback under π of the metric $x_1 g_1 \perp \ldots \perp x_m g_m$ on M. The Einstein condition for this metric on $P_{k_1 \ldots k_m}$ then turns out to be m coupled quadratic equations in the x_i and one easily shows that there is at least one positive solution and hence an Einstein metric with positive scalar curvature on $P_{k_1 \ldots k_m}$.

If all the M_i are homogeneous, then the metric on $P_{k_1 \ldots k_m}$ is also homogeneous. In general, the isometries of M may not lift to isometries of P and not all isometries of P may be lifts of isometries on M .Note also that the circle action on P consists of isometries, in fact, they are the only isometries on P which are the identity on M . We can show that for the above Einstein metrics, at least if $\Sigma|k_i|$ is large, the only isometries of the metric on P are induced by isometries on M . Thus if the metric on M_i has cohomogeneity n_i , the metric on P has cohomogeneity at least Σn_i . Hence, by using the examples of Koiso and Sakane, we obtain Einstein metrics of arbitrarily large cohomogeneity. Of course, it is not a priori, clear, that these manifolds do not have smaller cohomogeneity for some other group action, although this should happen only in special cases.

In general, for given manifolds M_i , the manifolds $P_{k_1 \ldots k_m}$ will have different homotopy type for different values of $k_1, \ldots, k_m$ and hence give rise to infinitely many homotopy types of Einstein metrics in odd dimensions (≥ 7) which can be chosen to be all homogeneous or all nonhomogeneous. The only previous examples of this type where the seven dimensional homogeneous Einstein metrics [Wa] mentioned earlier. But, as we will see, there are also some interesting exceptions to this rule.

The lowest dimensional examples one can obtain by this method are in dimension 5. Let $M = S^2 \times S^2$ with standard Kähler Einstein metric on each S^2 and $P_{k\ell}$ the principle circle bundle with Euler class $k\alpha_1 + \ell\alpha_2$ where α_i are the two generators in dimension 2. As mentioned earlier, we may as well assume that k and ℓ are relatively prime, in which case $P_{k\ell}$ is simply connected. The Einstein metrics on $P_{k\ell}$ are all homogeneous, in fact, as a homogeneous space $P_{k\ell} = S^3 \times S^3/S^1$ where the circle is given by

$(e^{i\ell\theta}, e^{-ik\theta}) \subset S^3 \times S^3$. By using the classification of compact simply connected 5-manifolds [Ba, 1965], one can show that $P_{k\ell}$ is always diffeomorphic to $S^3 \times S^2$. Furthermore, if we normalize the volume to be one, the Einstein constant goes to 0 as $k^2+\ell^2$ goes to ∞ . Hence there are ∞ many non-isometric Einstein metrics on $S^3 \times S^2$ with positive scalar curvature, which are the first examples of this type. This also implies that the moduli space of Einstein metrics on $S^3 \times S^2$ must have infinitely many components, since the Einstein constants are distinct. Furthermore, it shows how the Palais-Smale condition fails for the total scalar curvature functional on $S^3 \times S^2$.

One should note, that, although these Einstein metrics are homogeneous, they are homogeneous for different transitive group actions of $S^3 \times S^3$ on $S^3 \times S^2$. It would be interesting to see these group actions more explicitly. Equivalently, one can say that there are infinitely many distinct free circle actions on $S^3 \times S^2$, all with quotient $S^2 \times S^2$. Again we do not know an explicit description of these circle actions.

More generally, we can examine the circle bundles $P_{k\ell}$ over $M = P^n\mathbb{C} \times P^m\mathbb{C}$ which can also be described as $S^{2n+1} \times S^{2m+1}/S^1$ where S^1 acts by $e^{i\theta}(p,q) = (e^{i\ell\theta}p, e^{-ik\theta}q)$ using the standard circle actions on S^{2n+1} and S^{2m+1} .

If $n = m > 1$, it turns out that there are infinitely many diffeomorphism types among the $P_{k\ell}$. Indeed $H^*(P_{k\ell}, \mathbb{Z}) = H^*(P^n\mathbb{C} \times S^{2n+1}, \mathbb{Z})$ independent of k and ℓ , but the first Pontrayagin class is given by $p_1 = (n+1)(k^2+\ell^2)x^2$ where x is a generator in H^2 .

If $n < m$, then $H^{2n+2}(P_{k\ell}, \mathbb{Z}) = \mathbb{Z}_{\ell^{n+1}}$ and $p_1 = [(n+1)k^2 + (m+1)\ell^2]x^2$. If $n > 1$, x^2 has infinite order since $H^4 = \mathbb{Z}$, and hence, we again obtain infinitely many diffeomorphism types. But if $1 = n < m$, it follows from [SU, 1977] that there are only finitely many diffeomorphism types among the $P_{k\ell}$ for each fixed value of ℓ. If $\ell = 1$ it is easy to see directly that $P_{k\ell}$ is diffeomorphic to $S^2 \times S^{2m+1}$ if $k(m+1)$ is even, or diffeomorphic to the unique non-trivial S^{2m+1} bundle over S^2 if $k(m+1)$ is odd. Hence we obtain infinitely many non-isometric Einstein metric on $S^2 \times S^{2m+1}$ and on the non-trivial S^{4m+1} bundle over S^2 .

We finally remark that the above construction will in general fail, if we replace the principle circle bundle by a principle G bundle for some other group G . For example, the homogeneous manifolds $S^3 \times S^3 \times S^3/S^1$, which are principle T^2 bundles over $S^2 \times S^2 \times S^2$ carry no $S^3 \times S^3 \times S^3$ invariant Einstein metric for every embedding of S^1 , as long as the projection onto every S^3 factor is non-trivial.

A particularly interesting open problem for Einstein metrics with positive scalar curvature is the question if there can be families of non-isometric Einstein metric. For negative and zero Einstein constants, one obtains such families from Aubin's and Yau's solution of the Calabi conjecture, but no examples are known for positive scalar curvature. One does not even know if this is possible or not among the G-invariant metrics for a given transitive action of a compact group G . In all cases examined so far, there were always only finitely many G-invariant Einstein metrics.

Of course, a second interesting problem is, if there are any compact simply connected manifolds in dimension ≥ 5 which cannot carry any Einstein metric, or equivalently if there are any topological obstructions to the existence of Einstein metrics (no prescribed sign for the Einstein constant) in dimension ≥ 5, similar to the ones in dimension 4.

BIBLIOGRAPHY

[Ba] D. Barden, Simply connected five-manifolds, Ann. of Math. 82 (1965), 365-385.

[BB] L. Bérard Bergery, Sur des nouvelles variétés riemanniennes d'Einstein, Publications de l'Institut E. Cartan 4, Nancy, (1982), 1-60.

[Be] A. Besse, Einstein manifolds, to appear in Springer-Verlag.

[Bo] A. Borel, Kählerian coset spaces of semisimple Lie groups, Proc. Nat. Acad. Sciences 40 (1954), 1147-1151.

[B-K] J.P. Bourguignon-H. Karcher, Curvature operators : pinching estimates and geometric examples, Ann. Scient. Ecole Normale Supérieure 11 (1978), 71-92.

[D-Z] J. D'Atri-W. Ziller, Naturally reductive metrics and Einstein metrics on compact Lie groups, Memoir Amer. Math. Soc. 215 (1979).

[Dy] E.B. Dynkin, Semisimple subalgebras of semisimple Lie algebras, Transl. Amer. Math. Soc. Series 2, Vol. 6 (1957), 111-244.

[Fu] A. Futaki : An obstruction to the existence of Einstein Kähler metrics, Inv. Math. 73 (1983), 437-443.

[Hi1] N. Hitchin, On compact four dimensional Einstein manifolds, J. Diff. Geom. 9 (1974), 435-442.

[Hi2] N. Hitchin, Harmonic spinors, Adv. in Math. 14 (1974), 1-55.

[Je] G. Jensen, Einstein metrics on principal fibre bundles, J. Diff. Geom. 8 (1973), 599-614.

[Ko1] S. Kobayashi, On compact Kähler manifolds with positive definite Ricci tensor, Ann. of Math. 74(1961), 570-774.

[Ko2] S. Kobayashi, Topology of positively pinched Kähler manifolds, Tohoku Math. J. 15 (1963), 121-139.

[K-S] N. Koiso-Y. Sakane, Non-homogeneous Kähler Einstein metrics on compact complex manifolds, Preprint 1985.

[Kos] J.L. Koszul, Sur la forme hermitienne canonique des espaces homogènes complexes, Can. J. Math. 7 (1955), 562-576.

[Li1] A. Lichnérowicz, Variétés pseudokählerian à courbure de Ricci non nulle; applications aux domaines bornés homogènes de $\mathbb{C}^n$, C.R. Acad. Sci. Paris 235 (1952), 12-14.

[Li2] A. Lichnérowicz, Spineur harmonique, C.R. Acad. Sci. Paris 257 (1963), 7-9.

[Ma1] O.V. Manturov, Homogeneous asymmetric Riemannian spaces with an irreducible group of rotations, Dokl. Akad. Nauk. SSSR 141 (1961), 792-795.

[Ma2] O.V. Manturov, Riemannian spaces with orthogonal and symplectic groups of motions and an irreducible group of rotations, Dokl. Akad. Nauk. SSSR 141 (1961), 1034-1037.

[Ma3] O.V. Manturov, Homogeneous Riemannian manifolds with irreducible isotropy group, Trudy Sem. Vector and Tensor Analysis 13 (1966), 68-145.

[Mac1] Y. Matsushima, Sur la structure du groupe d'homéomorphismes analytiques d'une certaine variété Kählérienne, Nagoya Math. J. 11 (1957), 145-150.

[Mat2] Y. Matsushima, Remarks on Kähler Einstein manifolds, Nagoya Math. J. 46 (1972), 161-173.

[M] T. Matsuzawa, Einstein metrics on fibred Riemannian structures, Kodai Math. J. 6 (1983), 340-345.

[Pa] D. Page, A compact rotating gravitational instanton, Phys. Letters 79 B (1979), 235-238.

[P-P] D. Page- C.N. Pope, Inhomogeneous Einstein metrics on complex line bundles, Preprint 1985.

[Sa] Y. Sakane, Examples of compact Kähler Einstein manifolds with positive Ricci tensor, Preprint 1985.

[Su] D. Sullivan, Infinitesimal computations in topology, Publ. I.H.E.S. 47 (1977), 269-332.

[Th] J. Thorpe, Some remarks on the Gauss-Bonnet integral, J. Math. Mech. 18 (1969), 779-786.

[Wa] M. Wang, Some examples of homogeneous Einstein manifolds in dimension seven, Duke Math. J. 49 (1982), 23-28.

[W-Z1] M. Wang-W. Ziller, On normal homogeneous Einstein manifolds, to appear in Ann. Scient. Ec. Norm. Sup.

[W-Z2] M. Wang-W. Ziller, On the isotropy representation of a symmetric space, to appear in Rend. Sem. Mat. Univers. Politecn. Torino.

[W-Z3] M. Wang-W. Ziller, Existence and non-existence of homogeneous Einstein metrics, to appear in Inv. Math.

[W-Z4] M. Wang-W. Ziller,New examples of Einstein metrics on principal circle bundle, in preparation.

[Wo1] J. Wolf, The geometry and structure of isotropy irreducible homogeneous spaces,Acta. Math. 120 (1968), 59-148.

[Wo2] J. Wolf, Correction to "The geometry and structure of isotropy irreducible homogeneous spaces", Acta Math. 152 (1984), 141-142.

[Zi1] W. Ziller, Homogeneous Einstein metrics on spheres and projective spaces, Math. Ann. 259 (1982), 351-358.

[Zi2] W. Ziller, Homogeneous Einstein metrics, Global Riemannian geometry, ed. Willmore and Hitchin, Ellis Horwood Limited, Chichester 1984, p. 126-135.

Mc Kenzie WANG
Mc Master University
Hamilton, Ontario
Canada, L85 4K1

Wolfgang ZILLER
University of Pennsylvania
Philadelphia, PA 19104,
U.S.A.